Lecture Notes in Mathematics

Edited by A. Dold, B. Eckmann and F. Takens

1411

Boju Jiang (Ed.)

Topological Fixed Point Theory and Applications

Proceedings of a Conference
held at the Nankai Institute of Mathematics
Tianjin, PR China, April 5–8, 1988

Springer-Verlag
Berlin Heidelberg New York London Paris Tokyo Hong Kong

Editor

Boju JIANG
Department of Mathematics, Peking University
Beijing 100871, P. R. China

Mathematics Subject Classification (1980): 55M, 57R, 57S, 58F, 58G, 58B, 35J

ISBN 3-540-51932-7 Springer-Verlag Berlin Heidelberg New York
ISBN 0-387-51932-7 Springer-Verlag New York Berlin Heidelberg

Printing and binding: Druckhaus Beltz, Hemsbach/Bergstr.
2146/3140-543210 – Printed on acid-free paper

Foreword

The Conference on Topological Fixed Point Theory and Applications was held at Nankai Institute of Mathematics, Tianjin, China, during April 5–8, 1988. Its aim was to bring together topologists working in various areas of fixed point theory as well as analysts interested in the applications, to discuss recent progress and current trends in research.

This conference was sponsored by the Nankai Institute of Mathematics and supported by a grant from the Chinese Ministry of Education.

We would like to thank all the participants for their enthusiasm. We gratefully acknowledge the assistance of many people who helped make the conference a success. In particular, Prof. Zixin Hou who was one of the organizers of this meeting and Mr. Xiuhua Dong who looked after the logistical details.

Boju Jiang

TABLE OF CONTENTS

BIFURCATION THEORY FOR METRIC PARAMETER SPACES

Thomas Bartsch
Mathematisches Institut der Universität Heidelberg
Im Neuenheimer Feld 288, 6900 Heidelberg, W. Germany

Introduction

In [6] we introduced an index BI for bifurcation of fixed points. $BI(f)$ is defined for maps $f : \mathcal{O} \to X$ where

- X is a Banach space, $\mathcal{O} \subset \mathbb{R}^k \times X$ is open,
- f is completely continuous,
- $f(p,0) = 0$ for all $p \in \mathcal{T} = \mathcal{T}(f) := \mathcal{O} \cap (\mathbb{R}^k \times \{0\})$,
- the set $\mathcal{B} = \mathcal{B}(f) \subset \mathcal{T}$ of bifurcation points is compact. Here $\mathcal{B} = \mathcal{T} \cap clos(\mathcal{F})$, $\mathcal{F} = \mathcal{F}(f) := \{(p,x) \in \mathcal{O}: f(p,x) = x, x \neq 0\}$.

$BI(f)$ is an element of π_{k-1}^s , the stable (k-1)-stem. It has properties analogous to the fixed point index. In this note we shall first show how to define BI if the parameter space is not $\mathbb{R}^k$ but any metric space P. For simplicity we consider only the finite-dimensional case $X = \mathbb{R}^n$. The bifurcation index will then lie in $\tilde{\omega}^1(P)$ where $\tilde{\omega}^*$ is reduced stable cohomotopy theory. If $P = S^k = \mathbb{R}^k \cup \{\infty\}$ then $\tilde{\omega}^1(S^k) \cong \pi_{k-1}^s$ and the new definition reduces to the old one. In addition to the properties of BI proved in [6] we show that BI is natural in P and is commutative. This last property allows to replace X by an ENR or even $P \times X$ by an ENR_P $E \to P$ where the trivial fixed points are given by a section $P \to E$.

Obviously BI has a lot in common with the fixed point index. This similarity is not purely formal as will be shown by a formula relating BI and Dold's fixed point index for fibre-preserving maps (cf. [9]). This formula gives a new way of calculating BI.

1. Construction and properties of BI

We first define $BI(f)$ in the following situation. P is a metric space, $E := P \times \mathbb{R}^n$. We consider a continuous map $f : \mathcal{O} \to E$ over P, i. e. $f(p,x) = (p, f_2(p,x))$; $\mathcal{O} \subset E$ open. Assume $f(p,0) = (p,0)$ for all $p \in \mathcal{T} := \mathcal{O} \cap P$. (We identify P and $P \times \{0\}$.)

This paper is in final form. No version of it will be submitted for publication elsewhere.

We are interested in the set

$$\mathcal{F} = \mathcal{F}(f) := \{(p,x) \in \mathcal{O} : f(p,x) = (p,x), x \neq 0\}$$

of nontrivial fixed points and the set

$$\mathcal{B} = \mathcal{B}(f) := \mathcal{T} \cap clos(\mathcal{F})$$

of bifurcation points. Of course, $\mathcal{B}$ is a closed subset of $\mathcal{T}$.

If $\mathcal{B}$ is a closed subset of P we define $BI(f)$ as follows. Let d denote the metric of E given by the sum of the metrics of P and $\mathbb{R}^n$. Then

$$\begin{aligned} &\varphi : \mathcal{T} \to \mathbb{R}_0^+,\ \varphi(p) := min\Big\{1, \frac{1}{2} \cdot d(\{p\}, clos(\mathcal{F}) \cup \partial\mathcal{O})\Big\}, \\ (1.1)\quad &A_0 := \big\{(p,x) \in \mathcal{O}\colon p \in \mathcal{T}, \|x\| = \varphi(p)\big\} \subset \mathcal{O} \setminus \mathcal{F}, \\ &A_1 := \big\{(p,x) \in A_0\colon p \in \mathcal{T} \setminus \mathcal{B}\big\} \subset \mathcal{O} \setminus (\mathcal{T} \cup \mathcal{F}), \\ &\rho : (\mathcal{T}, \mathcal{T} \setminus \mathcal{B}) \times S^{n-1} \to (A_0, A_1),\ (p,x) \mapsto (p, \varphi(p) \cdot x). \end{aligned}$$

(1.2) Definition. $BI(f) \in \tilde{\omega}^1(P)$ is the image of $1 \in \omega^n(\mathbb{R}^n, \mathbb{R}^n \setminus \{0\}) \cong \mathbb{Z}$ under the following sequence of homomorphisms ($\iota : P \times \mathbb{R}^n \to \mathbb{R}^n$ the projection).

$$\omega^n(\mathbb{R}^n, \mathbb{R}^n \setminus \{0\}) \xrightarrow{(\iota - f_2)^*} \omega^n(A_0, A_1) \xrightarrow{\rho^*} \omega^n((\mathcal{T}, \mathcal{T} \setminus \mathcal{B}) \times S^{n-1})$$

$$\xrightarrow{/\Sigma_{n-1}} \omega^1(\mathcal{T}, \mathcal{T} \setminus \mathcal{B}) \xleftarrow{i^*} \omega^1(P, P \setminus \mathcal{B}) \longrightarrow \tilde{\omega}^1(P)$$

Here $\Sigma_{n-1} \in \omega_{n-1}(S^{n-1})$ is induced by the identity on S^{n-1}.

$$/ : \omega^n((\mathcal{T}, \mathcal{T} \setminus \mathcal{B}) \times S^{n-1}) \otimes \omega_{n-1}(S^{n-1}) \longrightarrow \omega^1(\mathcal{T}, \mathcal{T} \setminus \mathcal{B})$$

is the slant product (cf. Switzer [15], Chapter 13). i^* is an excision isomorphism since $\mathcal{B}$ is closed and $\mathcal{T}$ is open in P.

$BI(f)$ can be thought of as a stable map $P \wedge S^{n-1} \longrightarrow S^n$. If $P = S^k$ this map can be considered both as an element of $\tilde{\omega}^1(S^k)$ and of π^s_{k-1}. In [6] we chose the latter version. Also observe that in the case $\mathcal{O} \subset \mathbb{R}^k \times \mathbb{R}^n$ $\mathcal{B}$ is closed in $S^k = \mathbb{R}^k \cup \{\infty\}$ iff it is a compact subset of $\mathbb{R}^k$.

(1.3) BI enjoys a number of properties.
Let E^+ denote the space $P \times (\mathbb{R}^n \cup \{\infty\})/P \times \{\infty\}$ and $\mathcal{C}^+\mathcal{O} := E^+ \setminus \mathcal{O}$.

Existence. *If $BI(f) \neq 0$ then $\mathcal{B}$ and $\mathcal{C}^+\mathcal{O}$ cannot be separated in $\mathcal{B} \cup \mathcal{F} \cup \mathcal{C}^+\mathcal{O}$. This means that $\mathcal{B} \cup \mathcal{F} \cup \mathcal{C}^+\mathcal{O}$ cannot be written as the disjoint union $U \sqcup V$ of open subsets U and V with $\mathcal{B} \subset U$ and $\mathcal{C}^+\mathcal{O} \subset V$. If P is compact this implies the existence of a connected set $\mathcal{S} \subset \mathcal{F}$ with $\mathcal{B} \cap clos(\mathcal{S}) \neq \emptyset \neq \mathcal{C}^+\mathcal{O} \cap clos(\mathcal{S})$ (cf. [3], Proposition 5). $\mathcal{C}^+\mathcal{O} \cap clos(\mathcal{S}) \neq \emptyset$ is equivalent to the statement that the projection $\mathcal{B} \cap \mathcal{S} \subset E \longrightarrow P$ is not proper. In particular, f is not compactly fixed (cf. [9]). Thus $BI(f) \neq 0$ implies that the fixed point index $I(f)$ of f in the sense of [9] is not defined. Even more, there does not exist an open neighbourhood V of $\mathcal{B} \in \mathcal{O}$ such that $I(f|V)$ is defined.*

Localisation. *If $\mathcal{O}' \subset \mathcal{O}$ is open and $\mathcal{B} \subset \mathcal{O}'$ then $BI(f) = BI(f|\mathcal{O}')$.*

Additivity. *If $\mathcal{O} = \mathcal{O}_1 \cup \mathcal{O}_2$, $\mathcal{O}_1, \mathcal{O}_2$ open and $\mathcal{B} \cap \partial(\mathcal{O}_1 \cap \mathcal{O}_2) = \emptyset$ then $BI(f|\mathcal{O}_1)$, $BI(f|\mathcal{O}_2)$ and $BI(f|\mathcal{O}_1 \cap \mathcal{O}_2)$ are defined and*

$$BI(f) = BI(f|\mathcal{O}_1) + BI(f|\mathcal{O}_2) - BI(f|\mathcal{O}_1 \cap \mathcal{O}_2).$$

Homotopy invariance. *If $\mathcal{P} \subset F = [0,1] \times P \times \mathbb{R}^n$ is open and $h: \mathcal{P} \to F$ is a map over $[0,1] \times P$ with $h(t,p,0) = (t,p,0)$ and $\mathcal{B}(h)$ closed in $[0,1] \times P$ then $BI(h_t)$ is independent of $t \in [0,1]$. Here $h_t(p,x) = h(t,p,x)$ is the part of h over t.*

Stability. *If $0: \mathbb{R} \to \mathbb{R}$ is the constant map then $BI(f \times 0) = BI(f)$. Of course, $(f \times 0)(p,x,y) = (f(p,x),0)$.*

Naturality. *Let Q be another metric space and $\psi: Q \to P$ continuous. Let $\psi^* f: \psi^*\mathcal{O} \to \psi^* E$ denote the pullback of $f: \mathcal{O} \to E$ (cf. Dold [9], (2.8)). Then*

$$BI(\psi^* f) = \psi^*(BI(f)) \in \tilde{\omega}^1(Q).$$

Commutativity. *Let $\mathcal{O} \subset E = P \times \mathbb{R}^n$, $\mathcal{P} \subset F = P \times \mathbb{R}^m$ be open and $f: \mathcal{O} \to F$, $g: \mathcal{P} \to E$ be continuous maps over P such that $f(p,0) = (p,0)$, $g(p,0) = (p,0)$. Then $\mathcal{B}(f \circ g) = \mathcal{B}(g \circ f) =: \mathcal{B}$ and if this set is closed in P $BI(f \circ g) = BI(g \circ f)$.*

Only the last two properties have not been proved in [6]. The proofs in [6], though only for the special case $\mathcal{O} \subset \mathbb{R}^k \times \mathbb{R}^n \subset S^k \times \mathbb{R}^n$ and working with stable homotopy ω_* instead of ω^*, can easily be adopted to the situation here. (The statement of the homotopy invariance in [6] is not correct. There it was only assumed that $BI(h_t)$ is defined for all $t \in [0,1]$. This does not imply $BI(h_0) = BI(h_1)$ as can be seen from the example

$$h_t: \mathbb{R} \times \mathbb{R} \to \mathbb{R} \times \mathbb{R},\ h_t(p,x) = (p, tpx).$$

$BI(h_0) = 0$, $BI(h_t) = 1$ for $t \neq 0$.)

To prove **Naturality** observe that $\mathcal{B}(\psi^* f) = \psi^{-1}(\mathcal{B}(f))$, hence $BI(\psi^* f)$ is defined if $BI(f)$ is defined. Furthermore ψ induces homomorphisms connecting the corresponding terms in the defining sequences of $BI(f)$ and $BI(\psi^* f)$ making everything commute.

The proof of the **Commutativity** property is as in the case of the fixed point index (cf. [11], VII.5.9). For the convenience of the reader we give the necessary deformations and check the bifurcation sets.

$$\alpha: I \times \mathcal{O} \times \mathbb{R}^m \to I \times E \times \mathbb{R}^m = I \times E \times_P F,$$
$$(t,p,x,y) \mapsto (t,p,g_2 \circ f(p,x), tf_2(p,x)),$$

is a homotopy between $\alpha_0 = (g \circ f) \times 0$ and α_1. The fixed point set of α is

$$Fix(\alpha) = \{(t,p,x,y): g \circ f(p,x) = (p,x),\ y = tf_2(p,x)\}.$$

Thus $\mathcal{B}(\alpha) = I \times \mathcal{B}(g \circ f) \subset I \times P$ is closed.

$$\beta: I \times \mathcal{O} \times_P \mathcal{P} \to I \times E \times_P F,$$
$$(t,p,x,y) \mapsto (t,p,(1-t)g_2 \circ f(p,x) + tg_2(p,y), f_2(p,x)),$$

is a homotopy between $\beta_0 = \alpha_1 \mid I \times \mathcal{O} \times_P \mathcal{O}'$ and β_1. The fixed point set of β is

$$Fix(\beta) = \{(t,p,x,y): f(p,x) = (p,y),\ g(p,y) = (p,x)\}$$
$$= \{(t,p,x,y): (p,x) \in Fix\ g \circ f,\ y = f_2(p,x)\}.$$

Again $\mathcal{B}(\beta) = I \times \mathcal{B}(g \circ f) \subset I \times P$ is closed. Applying **Stability, Homotopy invariance** and **Localisation** yields $BI(g \circ f) = BI(\beta_1)$. Symmetric deformations yield $BI(f \circ g) = BI(\beta_1)$.

We use **Commutativity** to define BI in more general situations. We remind the reader that an ENR_P is a space over P, $\pi: E \to P$, such that there exists a euclidean space $\mathbb{R}^n$, an open subset U of $P \times \mathbb{R}^n$ and (continuous) maps $E \xrightarrow{i} U \xrightarrow{r} E$ over P with $r \circ i = id_E$.

Let E be an ENR_P, $\mathcal{O} \subset E$ open and $f: \mathcal{O} \to E$ a map over P. Suppose there exists a section $\sigma: P \to E$ such that $f \circ \sigma = \sigma$ on $\mathcal{T} = \sigma^{-1}(\mathcal{O})$. $\mathcal{T}$ is the set of trivial fixed points of f (we identify P and $\sigma(P)$). As usual we set $\mathcal{F} = Fix(f) \setminus \mathcal{T}$ and $\mathcal{B} = \mathcal{T} \cap clos(\mathcal{F})$. If $\mathcal{B}$ is a closed subset of P we define $BI(f)$ as in the case of the fixed point index.

(1.4) Definition. Choose $E \xrightarrow{i} U \xrightarrow{r} E$ as in the definition of an ENR_P. We may assume $i \circ \sigma = i_0: P \to P \times \{0\} \subset U$. This can always be achieved by a translation in each fibre $\{p\} \times \mathbb{R}^n$. Then

$$BI(f) := BI(i \circ f \circ r | r^{-1}(\mathcal{O}))$$

is the bifurcation index of f. It is independent of the choice of i and r (**Commutativity**).

The properties of $BI(f)$ continue to hold in the general situation. The space E^+ needed to formulate the **Existence** property is $E^+ := E \cup \{\infty\}$. A neighbourhood basis of ∞ consists of the complements $E^+ \setminus A$ of subsets $A \subset E$ such that $\pi|A$ is proper. As a consequence of the **Existence** property $BI(f)$ has to be 0 if $\mathcal{O} = E$ and $\pi: E \to P$ is proper.

2. Relation to the fixed point index

As in Part 1 we consider an ENR_P $\pi: E \to P$ with a section $\sigma: P \to E$ and a map $f: \mathcal{O} \to E$ over P, $\mathcal{O} \subset E$ open, with $f \circ \sigma = \sigma$ on $\mathcal{T} = \sigma^{-1}(\mathcal{O})$.

If the parameter space is one-dimensional, $P = \mathbb{R} \cup \{\infty\}$, and if $\mathcal{B}$ is contained in an interval, $\mathcal{B} \subset]p_0, p_1[\subset [p_0, p_1] \subset \mathcal{T}$, then $BI(f)$ is simply the difference of the local fixed point indices $I(f_{p_1}, 0) - I(f_{p_0}, 0)$. Here $f_p: \mathcal{O}_p \to E_p$ is the part of f over p. For $p \notin \mathcal{B}$ 0 is an isolated fixed point, hence $I(f_p, 0)$ is well defined. We shall generalize this formula to the multiparameter situation considered here.

(2.1) Definition. If there is no bifurcation from $\mathcal{T}$, i. e. $\mathcal{B} = \emptyset$, then the local fixed point index of f at $\mathcal{T}$ is defined as

$$I(f, \mathcal{T}) := I(f|W) \in \omega^0(P)$$

where $W \subset \mathcal{O}$ is an open neighbourhood of $\mathcal{T}$ in $\mathcal{O}$ with $W \cap Fix(f) = \mathcal{T}$ and $I(f|W)$ is the fixed point index of $f|W$ in the sense of Dold [9] or [10].

(2.2) Theorem. *If $\mathcal{B} \subset \mathcal{T}$ is closed in P consider open neighbourhoods U, V of $\mathcal{B}$ in P such that $clos(V) \subset U \subset \mathcal{T}$. (Such neighbourhoods exist since P is normal as a metric space.) Then*

$$BI(f) = -i^* \circ (j^*)^{-1} \circ \delta(\mathrm{I}(f|\pi^{-1}(U \setminus V), U \setminus V))$$

where

$$\omega^0(U \setminus V) \xrightarrow{\delta} \omega^1(U, U \setminus V) \xleftarrow{j^*} \omega^1(P, P \setminus V) \xrightarrow{i^*} \tilde{\omega}^1(P).$$

j^* is an excision isomorphism.

PROOF: Since both I and BI are commutative we may assume that $E = P \times \mathbb{R}^n$ and $\mathcal{T} = \mathcal{O} \cap P$.

Claim: $BI(f)$ is the image of $1 \in \omega^n(\mathbb{R}^n, \mathbb{R}^n \setminus \{0\})$ under the following sequence of homomorphisms.

$$\omega^n(\mathbb{R}^n, \mathbb{R}^n \setminus \{0\}) \xrightarrow{(\iota - f_2)^*} \omega^n(B_0, B_1) \xrightarrow{\rho^*} \omega^n((U, U \setminus V) \times (D^n \setminus \{0\}))$$

$$\xrightarrow{/\Sigma_{n-1}} \omega^1(U, U \setminus V) \longrightarrow \tilde{\omega}^1(P)$$

Here

$$D^n := \{x \in \mathbb{R}^n : \|x\| \le 1\},$$
$$B_0 := \{(p, x) \in \mathcal{O} : p \in U,\ 0 < \|x\| \le \varphi(p)\} \subset \mathcal{O} \setminus \mathcal{F},$$
$$B_1 := \{(p, x) \in B_0 : p \in U \setminus V\} \subset \mathcal{O} \setminus (\mathcal{T} \cup \mathcal{F}).$$

For later use we set

$$B_2 := \{(p, x) \in \mathcal{O} : p \in U \setminus V,\ \|x\| \le \varphi(p)\}.$$

$\varphi : \mathcal{T} \to \mathbb{R}_0^+$ and $\rho : (U, U \setminus V) \times (D^n \setminus \{0\}) \longrightarrow (B_0, B_1)$ are defined as in **(1.1)**. For notational convenience we do not distinguish $\omega_{n-1}(D^n \setminus \{0\})$ and $\omega_{n-1}(S^{n-1})$. Thus $\Sigma_{n-1} \in \omega_{n-1}(D^n \setminus \{0\})$. The last map in the above sequence is $i^* \circ (j^*)^{-1}$.

The proof of the above claim is an easy excision argument. The theorem is now a consequence of the commutativity of the following diagram. (We write $\dot{D}^n$ for $D^n \setminus \{0\}$.)

$$\begin{array}{ccccc}
\omega^n(\mathbb{R}^n, \mathbb{R}^n \setminus \{0\}) & \xleftarrow{\delta} & \omega^{n-1}(\mathbb{R}^n \setminus \{0\}) & \xrightarrow{\delta} & \omega^n(\mathbb{R}^n, \mathbb{R}^n \setminus \{0\}) \\
\downarrow{\scriptstyle (\iota - f_2)^*} & & \downarrow{\scriptstyle (\iota - f_2)^*} & & \downarrow{\scriptstyle (\iota - f_2)^*} \\
\omega^n(B_2, B_1) & \xleftarrow{\delta} & \omega^{n-1}(B_1) & \xrightarrow{\delta} & \omega^n(B_0, B_1) \\
\downarrow{\scriptstyle \rho^*} & & \downarrow{\scriptstyle \rho^*} & & \downarrow{\scriptstyle \rho^*} \\
\omega^n((U \setminus V) \times (D^n, \dot{D}^n)) & \xleftarrow{\delta} & \omega^{n-1}((U \setminus V) \times \dot{D}^n) & \xrightarrow{\delta} & \omega^n((U, U \setminus V) \times \dot{D}^n)
\end{array}$$

$$\begin{array}{ccccc} \Big\downarrow /\Delta_n & & \Big\downarrow /\Sigma_{n-1} & & \Big\downarrow /\Sigma_{n-1} \\ \omega^0(U \setminus V) & \xrightarrow{-1} & \omega^0(U \setminus V) & \xrightarrow{\delta} & \omega^1(U, U \setminus V) \end{array}$$

Here $\Delta_n \in \omega_n(D^n, D^n \setminus \{0\})$ corresponds to Σ_{n-1}, i. e. $\Sigma_{n-1} = \partial\Delta_n$. The last two squares commute according to Switzer [15], 13.56(v), (vi). Now

$$\mathrm{I}(f|\pi^{-1}(U \setminus V), U \setminus V) = (\rho^* \circ (\iota - f_2)^*(1))/\Delta_n$$

by definition (cf. [9], (2.3), (2.14)). Together with the first claim the theorem follows.

(2.3) Corollary. *Let $f: \mathbb{R} \times E \longrightarrow \mathbb{R} \times E \subset S^1 \times E$ be a map over $S^1 \times P$ and assume that the set $\mathcal{B}$ of bifurcation points is contained in $]-R, R[\times P$ and is closed in $S^1 \times P$. Let $f_t: E \to E$ be the part of f over $t \in \mathbb{R}$. Then*

$$I_+ := I(f_t, P) \in \omega^0(P),\ t \geq R,$$

and

$$I_- := I(f_t, P),\ t \leq -R,$$

are well defined. If $I_+ \neq I_-$ then $BI(f) \neq 0 \in \tilde{\omega}^1(S^1 \times P)$.

PROOF: Apply **Theorem (2.2)** to $U = \mathbb{R} \times P$, $V =]-R, R[\times P$ and observe that the kernel of

$$\omega^0(U \setminus V) \xrightarrow{\delta} \omega^1(U, U \setminus V) \xleftarrow{\cong} \omega^1(S^1 \times P, (S^1 \times P) \setminus V)$$

is the diagonal in $\omega^0(U \setminus V) \cong \omega^0(P) \oplus \omega^0(P)$. Thus $\delta(\mathrm{I}_+, \mathrm{I}_-) \neq 0$. Furthermore, the restriction homomorphism $\omega^1(S^1 \times P, S^1 \times P \setminus V) \longrightarrow \tilde{\omega}^1(S^1 \times P)$ is injective since $\omega^1(S^1 \times P, S^1 \times P \setminus V) \cong \omega^1(S^1 \times P, \{\infty\} \times P)$ and $\{\infty\} \times P$ is a retract of $S^1 \times P$ which implies that $\omega^0(S^1 \times P) \longrightarrow \omega^0(\{\infty\} \times P)$ is surjective.

(2.4) Example. If $P = S^1$ consider

$$f_+: S^1 \times \mathbb{C} \longrightarrow S^1 \times \mathbb{C},\ (p, z) \mapsto (p, p - p \cdot z),$$

and

$$f_-: S^1 \times \mathbb{C} \longrightarrow S^1 \times \mathbb{C},\ (p, z) \mapsto (p, 0).$$

Then $I(f_+) = 1$ and $I(f_-) = 0$ in $\tilde{\omega}^0(S^1) \cong \mathbb{Z}/2\mathbb{Z}$. Only the first assertion requires some work since we cannot use the Lefschetz-Hopf formula (6.18) of [10]. Instead we observe that the map

$$\begin{aligned} &S^1 \to SO(2), \\ &p \mapsto (z \mapsto p \cdot z = (\iota - f_{2+})(p, z)) \end{aligned}$$

induces a generator of $\pi_1(SO(2)) \cong \mathbb{Z}$ and hence the nonzero element

$$\gamma_{f_+} = 1 \in \pi_1(SO(n)) \cong \mathbb{Z}/2\mathbb{Z},\ n \geq 3.$$

Now the J-homomorphism

$$J: \pi_1(SO(n)) \longrightarrow \pi_{n+1}(S^n) \cong \tilde{\omega}^0(S^1)$$

is an isomorphism (cf. Adams [1]) and $J(\gamma_{f_+})$ is the fixed point index of f_+. This can be seen by using the Hopf construction.

Now consider a map $f: \mathbb{R} \times S^1 \times \mathbb{C} \longrightarrow \mathbb{R} \times S^1 \times \mathbb{C}$ over $\mathbb{R} \times S^1$ such that $f(t,p,0) = (t,p,0)$. If $f_{\pm t} \simeq f_{\pm}$ for $|t|$ big (controlling the fixed points during the deformation) then **Corollary (2.3)** gives $BI(f) \neq 0 \in \tilde{\omega}^1(S^1 \times S^1)$. Thus there must exist a global branch of fixed points bifurcating from $\mathbb{R} \times S^1$.

In all previous computations of the bifurcation index it was either assumed that there is just one bifurcation point in a disc in the parameter space (compare [2], [8], [12] and the references therein) or a more general situation has been reduced to that case using the properties of the bifurcation index (cf. [6]). **Theorem (2.2)** provides a different way of calculating BI.

3. Problems, Remarks

(3.1) If the domain O of $f: O \to E$ contains the whole section P then the defining sequence of homomorphisms in **Definition (1.2)** factors through $\tilde{\omega}^n(\mathbb{R}^n)$. Consequently, $BI(f) = 0$. In that case one can use a refined version as follows. Suppose one knows that the set of bifurcation points is contained in a subset Q of P. Then one can define $BI(f,Q) \in \omega^1(P, P \setminus Q)$. For example, if P is a compact manifold with boundary ∂P the bifurcation points should be restricted to $P \setminus \partial P$, thus $\mathrm{BI}(f) \in \omega^1(P, \partial P)$. Another variation for locally compact P is to replace P by its one point compactification.

(3.2) In [4] and [12] (possibly infinite-dimensional) Banach spaces P occur as parameter spaces. There one assumes the existence of a disc D^l in a finite-dimensional subspace of P with $\mathcal{B} \cap D^l = \{0\}$. Then one considers the bifurcation index of f restricted to the part over D^l. If it is different from 0 the bifurcating branch of solutions has dimension at least $dim\ P - l$. This type of result has also been proved in [7] where P is a finite-dimensional manifold. In addition, $S^{l-1} = \partial D^l$ and $\mathcal{B}$ are linked, i. e. the inclusion $S^{l-1} \hookrightarrow P \setminus \mathcal{B}$ is topologically nontrivial. It should be easy to replace D^l by more general subspaces Λ of P and look at $\partial\Lambda \hookrightarrow P \setminus \mathcal{B}$. Nonlinear parameter spaces have also been considered in [14] (finite CW-complexes, in particular compact differentiable manifolds) and — in a continuation setting — in [5] (metric spaces).

(3.3) For applications it is necessary to replace $\mathbb{R}^n$ by a normed linear space and ENR_P by ANR_P. One has to approximate f, so one needs some compactness conditions. In [13] Nussbaum uses the fixed point index to prove bifurcation into an ANR X. He considers maps $f: J \times X \to X$ which are strict set contractions on a neighbourhood of the fixed point set. J is an open interval of reals. Of particular importance is the case where X is a cone in a Banach space.

(3.4) Another problem (Dold) is to define $BI(f)$ for maps $f: M \to M$ where M is a manifold and $f|N = id|N$ on a submanifold N. This situation is different from the one considered in this paper since M is not a space over N and f not a map over N.

Bibliography

[1] J. F. Adams, On the groups J(X), *Topology* **5** (1966), 21–71.

[2] J. C. Alexander, Bifurcation of zeroes of parametrized functions, *J. Func. Anal.* **29** (1978), 37–53.

[3] J. C. Alexander, A primer on connectivity, *Proc. Conf. Fixed Point Theory* 1980, E. Fadell and G. Fournier eds., 455–483.

[4] J. C. Alexander and S. Antman, Global behavior of solutions of nonlinear equations depending on infinite-dimensional parameters, *Indiana Univ. Math. J.* **32** (1983), 39–62.

[5] J. C. Alexander and M. Reeken,On the topological structure of the set of generalized solutions of the catenary problem, *Proc. Royal Soc. Edinburgh* **98A** (1984), 339–348.

[6] T. Bartsch, A global index for bifurcation of fixed points, *J. reine angew. Math.* **391** (1988),181–197.

[7] T. Bartsch, Global bifurcation from a manifold of trivial solutions, *Forschungsschwerpunkt Geometrie, Universität Heidelberg, Heft Nr.* **13** (1987).

[8] T. Bartsch, The role of the J-homomorphism in multiparameter bifurcation theory, *Bull. Sciences Math., 2e série* **112** (1988), 177–184.

[9] A. Dold, The fixed point index of fibre-preserving maps, *Inventiones math.* **25** (1974), 281–297.

[10] A. Dold, The fixed point transfer of fibre-preserving maps, *Math. Z.* **148** (1976), 215–244.

[11] A. Dold, Lectures on Algebraic Topology, Berlin-Heidelberg-New York 1980.

[12] J. Ize, I. Massabò, J. Pejsachowicz and A. Vignoli, Structure and dimension of global branches of solutions to multiparameter nonlinear equations, *Trans. AMS* **291** (1985), 383–435.

[13] R. Nussbaum,The fixed point index and some applications, *Sém. de Math. Sup., Univ. Montréal* 1985.

[14] J. Pejsachowicz, K-theoretic methods in bifurcation theory, *Proc. Conf. Fixed Point Theory and its Applications* 1986, R. F. Brown ed.

[15] R. M. Switzer, Algebraic Topology—Homotopy and Homology, Berlin-Heidelberg-New York 1975.

A Fixed Point Index Approach to some Differential Equations

Dedicated to Professor Andrzej Granas
on the Occasion of his 60th Birthday

R.Bielawski and L.Górniewicz

(Institute of Mathematics, University of Nicholas Copernicus,
Chopina 12/18, 87–100 Toruń, Poland)

The aim of this paper is to show that, using the fixed point index method for compact maps as a tool, many types of differential equations with the right side depending on the derivative can be reduced very easily to differential inclusions with right sides not depending on derivative. It is shown in fact, that the fixed point theory for condensing-type maps, which was used to obtain existence results or topological characterizations of the set of solutions in such situations (cf. [14]) is not needed. We apply our method to the following types of differential equations only, but some other applications are also possible:

(i) ordinary differential equations of first or higher order (e.g., the satellite equation, see [15]),

(ii) hyperbolic differential equations,

(iii) elliptic differential equations.

We shall mention these in section 3. General statements needed for applications will be given in section 2. This paper is a continuation of [2] and gives a generalization of respective results in [1,3,13 and 14]. Finally, note that the class of maps for which we are able to obtain an existence result is quite rich (see Theorem (2.7)).

1. Preliminaries. In this paper all topological spaces are assumed to be metric. Let X, Y be two spaces and assume that for every $x \in X$ a nonempty subset $\varphi(x)$ of Y is given; in this case we say that $\varphi : X \to Y$ is a multivalued map. In what follows the symbols φ, ψ, χ are reserved for multivalued maps; singlevalued maps we shall denote by $f, g, h, \cdots$.

A multivalued map $\varphi : X \to Y$ is called *upper semi continuous* (*lower semi continuous*) if for each open $U \subset Y$ the set $\{x \in X; \varphi(x) \subset U\}$ ($\{x \in X; \varphi(x) \cap U \neq \emptyset\}$) is an open subset of X. An upper semi continuous (lower semi continuous map) $\varphi : X \to Y$ we shall write shortly as u.s.c. (l.s.c.). A multivalued map $\varphi : X \to Y$ is called *continuous* if it is both u.s.c. and l.s.c. It is evident that for $\varphi = f$ a singlevalued map the above three notions coincide. An u.s.c. map $\varphi : X \to Y$ is called *compact* provided the closure $\mathrm{cl}(\varphi(X))$ of $\varphi(X) = \bigcup\{\varphi(x); x \in X\}$ in Y is a compact set.

Let $\varphi, \psi : X \to Y$ be two maps. We shall say that ψ is a selector of φ (written $\psi \subset \varphi$), if for each $x \in X$ we have $\psi(x) \subset \varphi(x)$. For a map $\varphi : X \to X$ by $\mathrm{Fix}(\varphi)$ we shall denote the set of all fixed points of φ, i.e.,

$$\mathrm{Fix}(\varphi) = \{x \in X; x \in \varphi(x)\}.$$

We recommend [1,7,10] for details concerning multivalued maps. Let A be a subset of X, then by δA we shall denote the boundary of A in X, by cl$A$ the closure of A in X and by dim$A$ the covering dimension of A (cf. [8]). It is well known (see [9]) that for a compact set A we have: $\dim A = 0$ if and only if A has an open-closed basis.

It immediately implies the following:

(1.1) Proposition. Let A be a compact subset of X such that $\dim A = 0$. Then for every $x \in A$ and for every open neighbourhood U of x in X there exists an open neighbourhood $V \subseteq U$ of x in X such that $\delta V \cap A = \emptyset$.

2. General statements. In this section we shall present the topological material needed in our applications to the theory of differential equations. Let X be a metric absolute neighbourhood retract (written $X \in$ ANR) and let $g : X \to X$ be a compact map. Assume further that U is an open subset of X such that $\delta U \cap \mathrm{Fix}(g) = \emptyset$, then (following [11]) we shall denote by $i(g, U)$ the fixed point index of g with respect to U.

We shall start with the following:

(2.1) Proposition. Let $X \in$ ANR and let $g : X \to X$ be a compact map. Assume further that the following two conditions are satisfied:

(2.1.1) $\dim \mathrm{Fix}(g) = 0$,

(2.1.2) there exists an open subset $U \subset X$ such that $\delta U \cap \mathrm{Fix}(g) = \emptyset$ and $i(g, U) \neq 0$.

Then there exists a point $z \in \mathrm{Fix}(g)$ for which we have:

(2.1.3) for every open neighbourhood U_z of z in X there exists an open neighbourhood V_z of z in X such that: $V_z \subset U_z$, $\delta V_z \cap \mathrm{Fix}(g) = \emptyset$ and $i(g, V_z) \neq 0$.

Proof. Let $\Gamma = \{A \subset \mathrm{Fix}(g) \cap U; A$ is compact nonempty and for every open neighbourhood W of A in X there is an open neighbourhood V of A in X which satisfies the following three conditions: $V \subset W$, $\delta V \cap \mathrm{Fix}(g) = \emptyset$ and $i(g, V) \neq 0\}$.

It follows from (2.1.2) that Γ is a nonempty family. We consider in Γ the partial order given by the inclusion between subsets of X. We are going to apply the famous Kuratowski-Zorn Lemma (cf. [6]). To do it let us assume that $\{A_i\}_{i \in I}$ is a chain in Γ. We put $A_0 = \bigcap\{A_i; i \in I\}$. To prove that $A_0 \in \Gamma$ assume that W is an open neighbourhood of A_0 in X. We claim that there is $i \in I$ such that $A_i \subset W$. Indeed, if we assume on the contrary, then we get a family $B_i = (X \setminus W) \cap A_i$, $i \in I$, of compact nonempty sets which has nonempty compact intersection B_0. Then $B_0 \subset X \setminus W$ and $B_0 \subset A_0$ so we obtain a contradiction and hence $A_0 \in \Gamma$. Consequently, in view or Kuratowski-Zorn Lemma, we get a minimal element A_* in Γ. We claim that A_* is a singleton. Let $z \in A_*$. It is sufficient to prove that $\{z\} \in \Gamma$. Since $A_* \in \Gamma$ we obtain an open neighbourhood U_* of A_* in X with the following properties: $U_* \subset U$, $\delta U_* \cap \mathrm{Fix}(g) = \emptyset$ and $i(g, U_*) \neq 0$. Let W be an arbitrary open neighbourhood of z in X. Using (1.1) we can choose an open neighbourhood U_z of z in $U_* \cap W$ such that $\mathrm{Fix}(g) \cap \delta U_z = \emptyset$. Since A_* is a minimal element of Γ the compact set $A_* \setminus U_z$ is not in Γ and hence there exists an open set $V \subset U_*$ such that $(A_* \setminus U_z) \subset V \subset U_*$, $\mathrm{Fix}(g) \cap \delta V = \emptyset$, $V \cap U_z = \emptyset$, $i(g, V) = 0$ and $i(g, U_*) = i(g, V \cup U_z)$. Now from the additivity property of the fixed point index we have

$$i(g, U_*) = i(g, U_z) + i(g, V) \neq 0$$

and consequently $i(g, U_z) \neq 0$. It implies that $\{z\} \in \Gamma$ and the proof is complete.

Now, we are going to consider a more general situation. Namely, let Y be a locally arcwise connected space, $X \in \mathrm{ANR}$ and let $f : Y \times X \to X$ be a compact map. In what follows we shall assume that f satisfies the following condition:

(2.2) $\forall y \in Y \exists U_y : U_y$ is open in X and $i(f_y, U_y) \neq 0$,

where $f_y : X \to X$ is given by the formula $f_y(x) = f(y, x)$ for every $x \in X$. Observe that in particular, if X is an absolute retract, then (2.2) holds automatically. We associate with a map $f : Y \times X \to X$ satisfying the above conditions the following multivalued map:

$$\varphi_f : Y \to X, \quad \varphi_f(y) = \mathrm{Fix}(f_y).$$

Then from (2.2) follows that φ_f is well defined. Moreover, we get:

(2.3) Proposition. Under all of the above assumptions the map $\varphi_f : Y \to X$ is u.s.c.

Let us remark that, in general, φ_f is not a l.s.c. map. Below we would like to formulate a sufficient condition which guarantees that φ_f has a l.s.c. selector. To get it we shall add one more assumption. Namely, we assume that f satisfies the following condition:

(2.4) $\forall y \in Y : \dim \mathrm{Fix}(f_y) = 0$.

Now, in view of (2.2) and (2.4), we are able to define the map $\psi_f : Y \to X$ by putting $\psi_f(y) = \mathrm{cl}\{z \in \mathrm{Fix}(f_y)$; for z condition (2.1.3) is satisfied$\}$, for every $y \in Y$.

We prove the following:

(2.5) Theorem. Under all of the above assumptions we have:

(2.5.1) ψ_f is a selector of φ_f,

(2.5.2) ψ_f is a l.s.c. map.

Proof. Since (2.5.1) follows immediately from the definition we shall prove (2.5.2). To do it we let:

$$\eta_f : Y \to X, \quad \eta_f(y) = \{z \in \mathrm{Fix}(f_y); \quad z \text{ satisfies (2.1.3)}\}.$$

For the proof it is sufficient to show that η_f is l.s.c. Let U be an open subset of X and let $y_0 \in Y$ be a point such that $\eta_f(y_0) \cap U \neq \emptyset$. Assume further that $x_0 \in \eta_f(y_0) \cap U$. Then there exists an open neighbourhood V of x_0 in X such that $V \subset U$ and $i(f_{y_0}, V) \neq 0$. Since φ_f is an u.s.c. map and Y is locally arcwise connected we can find an open arcwise connected W in Y such that $y_0 \in W$ and for every $y \in W$ we have:

$$(*) \qquad \mathrm{Fix}(f_y) \cap \delta V = \emptyset.$$

Let $y \in W$ and let $\sigma : [0,1] \to W$ be an arc joining y_0 with y, i.e., $\sigma(0) = y_0$ and $\sigma(1) = y$. We define a homotopy $h : [0,1] \times V \to X$ by putting: $h(t,x) = f(\sigma(t), x)$. Then it follows form (*) that h is a well defined homotopy joining f_{y_0} with f_y and hence we get: $i(f_{y_0}, V) = i(f_y, V) \neq 0$; so $\mathrm{Fix}(f_y) \cap V \neq \emptyset$ and our assertion follows from (2.1).

(2.6) Remark. Let us remark that the above results remain true for admissible multivalued maps (cf. [7] and [10]); proofs are completely analogous.

Observe that condition (2.4) is quite restrictive. Therefore it is interesting to characterize the topological structure of all mappings satisfying (2.4). We shall do it in the case of Euclidean spaces (which is sufficient from the point of view of our applications), but in fact it is possible (cf. [12]) for arbitrary smooth manifolds. Let A be a closed subset of the Euclidean space R^m. By $C(A \times R^n, R^n)$ we shall denote the Banach space of all compact (singlevalued) maps from $A \times R^n$ into R^n with the usual supremum norm. Let

$$Q = \{f \in C(A \times R^n, R^n);\ f \text{ satisfies (2.4) for } Y = A \text{ and } X = R^n\}.$$

As an easy application of Theorem 4.1 in [12] we obtain:

(2.7) Theorem. The set Q is dense in $C(A \times R^n, R^n)$.

(2.8) Remark. Let us observe that using (2.7) one can easily prove more, namely the set Q is residual in $C(A \times R^n, R^n)$.

3. Applications. In this section we shall show how to apply results of section 2 to ordinary and partial differential equations with right hand sides depending on derivatives (cf. [2,3,5,6,14]). Our results provide some extensions and modification of methods presented in ([2,3,13,14]).

(3.1) (Ordinary differential equations of first order). According to section 2 we let $Y = [0,1] \times R^n$, $X = R^n$ and let $f : Y \times X \to X$ be a compact map. Then f satisfies condition (2.2) automatically so we shall assume only (2.4). Let us consider the following equation:

$$x'(t) = f(t, x(t), x'(t)), \tag{3.1.1}$$

where the solution is understood in the sense of almost everywhere $t \in [0,1]$ (a.e., $t \in [0,1]$).

We shall associate with (3.1.1) the following two differential inclusions:

$$x'(t) \in \varphi_f(t, x(t)) \tag{3.1.2}$$

and

$$x'(t) \in \psi_f(t, x(t)) \tag{3.1.3}$$

where φ_f and ψ_f are defined in the last section and by solution of (3.1.2) or (3.1.3) we mean an absolutely continuous function which satisfies (3.1.2) (resp. (3.1.3)) in the sense of a.e., $t \in [0,1]$.

Denote by $S(f), S(\varphi_f)$ and $S(\psi_f)$ the set of all solutions of (3.1.1), (3.1.2) and (3.1.3) respectively. Then we get:

$$S(\psi_f) \subset S(f) = S(\varphi_f). \tag{3.1.4}$$

But in view of (2.5) the map ψ_f is l.s.c., so (see [4] or [9]) we obtain:

$$S(\psi_f) \neq \emptyset. \tag{3.1.5}$$

Thus we have proved:

$$\emptyset \neq S(\psi_f) \subset S(\varphi_f) = S(f). \tag{3.1.6}$$

Observe that in (3.1.2) and (3.1.3) the right side doesn't depend on derivative.

(3.2) (Ordinary differential equations of higher order). According to section 2 we let $Y = [0,1] \times R^{kn}$, $X = R^n$ and let $f : Y \times X \to X$ be a compact map. Then, similarily as in (3.1) f satisfies (2.2) so we shall assume only (2.4). To study the existence problem for the following equation:

$$x^{(k)}(t) = f(t, x(t), x'(t), \cdots, x^{(k)}(t)) \tag{3.2.1}$$

we consider the following two differential inclusions (cf. [1] or [13]):

$$x^{(k)}(t) \in \varphi_f(t, x(t), x'(t), \cdots, x^{(k-1)}(t)) \tag{3.2.2}$$

and

$$x^{(k)}(t) \in \psi_f(t, x(t), x'(t), \cdots, x^{(k-1)}(t)). \tag{3.2.3}$$

Then the existence problem for (3.2.1) can be reduced very easily to (3.2.2) or (3.2.3).

(3.3) (Hyperbolic equations). Now let $Y = [0,1] \times [0,1] \times R^{3n}$, $X = R^n$ and let $f : Y \times X \to X$ be a compact map. Again it is easy to see that f satisfies (2.2) so we shall assume (2.4). Now let us consider the following hyperbolic equation:

$$u_{ts}(t,s) = f(t, s, u(t,s), u_t(t,s), u_s(t,s), u_{ts}(t,s)), \tag{3.3.1}$$

where the solution $u : [0,1] \times [0,1] \to R^n$ is understood in the sense of a.e., $(t,s) \in [0,1] \times [0,1]$.

As above we associate with (3.3.1) the following two differential inclusions (cf. [13]):

$$u_{ts}(t,s) \in \varphi_f(t, s, u(t,s), u_t(t,s), u_s(t,s)) \tag{3.3.2}$$

and

$$u_{ts}(t,s) \in \psi_f(t, s, u(t,s), u_t(t,s), u_s(t,s)). \tag{3.3.3}$$

Then it is evident that the set of all solutions of (3.3.1) is equal to the set of all solutions of (3.3.2) and every solution of (3.3.3) is a solution of (3.3.2). So inclusions (3.3.2) and (3.3.3) give us full information about (3.3.1).

(3.4) (Elliptic differential equations). Let $K(0,r)$ denotes the closed ball in R^n with center at 0 and radius r. Now, according to section 2 we put $Y = K(0,r) \times R^{2n}$, $X = R^n$ and let $f : Y \times X \to X$ be a compact map. Since (2.2) is satisfied we assume only (2.4). We consider the following elliptic equation:

$$\Delta(u)(z) = f(z, u(z), D(u)(z), \Delta(u)(z)), \quad \text{a.e., } z \in K(0,r), \tag{3.4.1}$$

where Δ denotes the Laplace operator and $D(u)(z) = u_{z_1}(z) + \cdots + u_{z_n}(z)$; $z = (z_1, \cdots, z_n)$. Then we can consider the following two differential inclusions (cf. [13]):

(3.4.2) $$\Delta(u)(z) \in \varphi_f(z, u(z), D(u)(z)).$$

(3.4.3) $$\Delta(u)(z) \in \psi_f(z, u(z), D(u)(z)).$$

and we have exactly the same situation as in (3.3) or (3.2).

We shall end our applications by making the following three remarks.

(3.5) Remark. Observe that all results of this section, except (3.4), remain true if we replace the Euclidean space R^n by an arbitrary Banach space.

(3.6) Remark. Let us observe (cf. (2.6)) that if we replace (3.1.1), (3.2.1), (3.3.1) and (3.4.1) by the respective differential inclusions then we get all results of this section without any change.

(3.7) Remark. We recommend [2] and [3] for some other applications of our approach. In particular, for the case when f is not a compact map.

References

[1] J.P.Aubin and A.Cellina, Differential inclusions, Springer-Verlag 1984.

[2] R.Bielawski, An application of the fixed point index to the differential equation $x'(t) = f(t, x(t), x'(t))$, Torun 1987 (in Polish).

[3] R.Bielawski and L.Górniewicz, Some applications of the Leray-Schauder alternative to differential equations, S.P.Singh (ed.), Reidel, 1986, 187-194.

[4] A. Bressan, Solutions of lower semicontinuous differential inclusions on closed sets, Rend. Sem. Univ. Padova, 55 (1976), 99-107.

[5] K. Deimling, Ordinary differential equations on Banach Spaces, Lecture Notes in Math., 596, Springer-Verlag 1977.

[6] J.Dugundji and A.Granas, Fixed point theory 1, PWN, Warszawa 1981.

[7] Z.Dzedzej, Fixed point index theory for a class of nonacyclic multivalued maps, Dissertationes Mathematicae, 253 (1985), 1-5.

[8] R.Engelking, Dimension theory, PWN, Warszawa, 1979.

[9] A.Fryszkowski, Continuous selections for a class of non-convex multivalued maps, Studia Math., 76 (1983), 163-174.

[10] L.Górniewicz, Homological methods in fixed point theory of multivalued mappings, Dissertationes Mathematicae, 129 (1976), 1-71.

[11] A.Granas, The Leray-Schauder index and the fixed point theory for arbitrary ANR-s, Bull. Soc. Math. France, 100 (1972), 209–228.

[12] H.Kurland and J.Robbin, Infinite codimension and transversality, Lecture Notes in Math., 468, Springer-Verlag 1974, 135-150.

[13] T.Pruszko, Some applications of the topological degree theory to multivalued boundary value problem, Dissertationes Mathematicae, 229 (1984), 1-48.

[14] B.N.Sadovskij, Limit-compact and condensing operators, Uspehi Mat. Nauk, 27 (1971), 81-146.

[15] A.P.Torzewskij, Periodic solutions of the equation of plane perturbations of a satellite on an elliptic orbit, Kosmiceskije issliedovanija, T.2, 5 (1964).

Topological Characteristics of Infinite - Dimensional Mappings and the Theory of Fixed Points and Coincidences

Yu.G.Borisovich

Department of Mathematics
Voronezh State University
394693, Voronezh, USSR

In global analysis the classical topological problem of the existence of a fixed point of a mapping $f : X \to X$ or a coincidence for a pair of mappings $f, g : X \to Y$

(1) $$x = f(x), \quad f(x) = g(x),$$

often arising from a nonlinear boundary problem, requires the consideration of infinite dimensional manifolds X, Y; in case f or g are multivalued we have, instead of equations (1), "inclusions"

(2) $$x \in f(x), \quad f(x) \in g(x).$$

It is important to obtain existence theorems for the solutions and to study the topological structure of the set of all solutions. In this direction various generalizations of the theory of J.Leray, J.Schauder and M.A.Krasnosel'skii are being developed which include "monotone" operators (F.Browder, I.V.Skrypnik), Fredholm operators (R.Caccioppoli, A.S.Shvarts, S.Smale, K.D.Elworthy, A.J.Tromba), multivalued operators both with convex images (S.Kakutani, J.von Neumann) and with images having a more complex topological structure (S.Eilenberg, D.Montgomery) (these references relate to the beginning of the theory; for the subsequent development see [1–6]). A review of the results obtained by the author and his colleagues in this subject is given below.

1. Fredholm operators and their perturbations.

A present-day interest in Fredholm mappings is explained by the fact that, firstly, nonlinear Fredholm equations naturally appear in different problems of mechanics and physics, geometry and analysis and, secondly, Fredholm mapping is a mathematical object, rich in contents and worth the attention of both topologists and analysts.

Denote by $\Phi_n C^r$, $n \geq 0$, $r > n + 1$, the class of Fredholm mappings $f : X \to Y$ (index $n \geq 0$, smoothness C^r, $r > n + 1$) of a Banach manifold X into a Banach manifold Y; consider a mapping f coordinated with Fredholm structures X_ϕ, Y_ϕ and proper. K.D.Elworthy, A.J.Tromba introduced a generalized degree $\deg(f, X_\phi, Y_\phi)$ with values in GL_c-equipped bordism classes $F_n(X)$. However, boundary problems showed the necessity to reduce the smoothness $r > n + 1$ and to consider Fredholm operators with compact (non-smooth) perturbations $k : F = f - k$ (in this case $Y = E_1$ is a Banach space) which was done in the works by the author, V.G.Zvyagin, Yu.I.Sapronov,

N.M.Ratiner ([1], [6], [7]). In order to consider Fredholm mappings of a negative index $n < 0$ the author and Yu.I.Sapronov introduced intersection indices $\gamma(f, s)$ where $s : Z^m \to Y$ is a finite dimensional mapping and $n + m \geq 0$ ([1], [8]). It should be noted that negative index operators often occur in applied problems: boundary problems on domains with cone points at the boundary, boundary problems of Riemann and Poincaré, boundary problems with an oblique derivative and also problems with non-local boundary conditions which have been recently investigated ([9]) may have a negative index. In this case an intersection index is a correctly defined topological characteristic for a corresponding boundary problem operator.

Next, noncompact and non-single-valued perturbations of a Fredholm operator f (see, e.g. [10]–[16] and also [17]) were considered. On the one hand, these investigations were inspired by the task of extending the method of Leray and Schauder of studying nonlinear boundary problems to the case when an elliptic quasilinear operator is perturbed by members of the same differential order, on the other hand by the theory of differential inclusions and variational inequalities. In [15], [16] the author formulated a general method based on the principle of compact contraction for common multivalued mappings [18] and considered different variants studied before. Let us formulate the general principle.

Let $X = \overline{\Omega}$ be the closure of an open set in an E-manifold M of class C^0, where E is a separable locally convex space (LCS); E_1 is a complete separable LCS and let $f : X \to E_1$ be a continuous single-valued mapping and $g : X \to E_1$ be an upper semicontinuous mapping with compact images, g should fulfil some condition (α) which is preserved at approximations $\tilde{g}$ and linear homotopies of g to $\tilde{g}$ (and which is specified depending on the choice of the class of mappings g, e.g. by fixing a topological structure of images). The pair (f, g) will be called admissible if the condition of "convex 0-separability" on the boundary $\partial X : (f - \overline{co} g)(\partial X) \cap \{0\} = \emptyset$ is satisfied, if $f : X \to E_1$ is a proper mapping and $g : X \to E_1$ is f-condensing with respect to a semi-additive measure of non-compactness ψ on E_1 (i.e., $\psi(g \cdot f^{-1}(M)) \ngeq \psi(M)$ for any $M \subset E_1$, $\psi(M) \neq 0$), and if the image $g \cdot f^{-1}(M)$ is precompact for any compact $M \subset E_1$. Fix the mapping f and consider homotopies of the form $F_t = f - g_t$, $t \in I$. Admissible pairs (f, g) generate the collection $D^\alpha_\psi[X, \partial X]$ of vector fields $F = f - g$ and their homotopic classes $\{[f - g]^\alpha_\psi\}$; by choosing compact $g = k$ we shall obtain the collection $D^\alpha_c[X, \partial X]$ of vector fields $G = f - k$ and the homotopic classes $\{[f - k]^\alpha_c\}$. The following "principle of bijectivity" holds:

Theorem I. The natural inclusion $i : D^\alpha_c[X, \partial X] \to D^\alpha_\psi[X, \partial X]$ induces a bijective correspondence of homotopic classes

$$i_* : \{[f - k]^\alpha_c\} \longrightarrow \{[f - g]^\alpha_\psi\}.$$

Theorem II. If E_1 is a metrizable complete LCS then theorem I holds true without the conditions of properness and continuity of f.

The "principle of bijectivity" generalizes the analogous principle of Yu.I. Sapronov, V.G.Zvyagin, V.T.Dmitrienko established in partial situations. It allows one to determine topological characteristics $\deg(F, X, 0)$ of "f-condensing" vector fields $F = f - g$ by topological characteristics $\deg(G, X, 0)$ if the latter are defined (see [10], [15]). This will take place, for example, when $M \in C^1$, E and E_1 are Banach spaces and if one of the following conditions is satisfied: 1) $f \in \Phi_n C^1, n \geq 0$ is a proper operator, g is a convex-valued operator; 2) $f \in \Phi_0 C^1$ is a proper mapping and a homeomorphism onto its own image, g is a generalized acyclic mapping; 3) condition 1) but with $n < 0$, then the intersection index $\gamma(F, S)$ is determined where $S : Z^m \to E_1$ is continuous, $n+m \geq 0$ and $F(\partial X) \cap S(Z^m) = \emptyset$ (condition of "separability"); 4) $M = E$, $E_1 = E^*$, f is a demi-continuous operator satisfying the condition of Browder-Skrypnik, and g has convex images.

2. Fredholm operators and boundary problems.

Recently Fredholm operators have found interesting applications in the study of branching of solutions to boundary problems depending on a parameter $\lambda \in R^k$ and in the investigation of solvability and finite multiplicity of their solutions "for almost all" right-hand sides and boundary conditions.

In the study of branching and bifurcation classical methods of Lyapunov-Schmidt are connected with the construction of a normal form and with the action of a symmetry group. In [19]–[21] Yu.I.Sapronov considerably uses the fact that the operators determining the equations of the problem are potential operators. In [19] he studies a potential Fredholm operator dependent on the parameters $(\delta, \varepsilon) \in R^l \times R^p$ — "load-perfection". It is established that the symmetry (at $\varepsilon = 0$) and breaking of the symmetry ($\varepsilon \neq 0$) may result in a "cascade" of bifurcations. The number of cascade steps is estimated by the adjointness of the singularity to the singularity of the type A_{2K+1}. In [20] Euler-Lagrange equations for the potential $V(w, \lambda, a)$ of a rectangular lengthwise contracted plate $\overline{\Omega}_a = (0 \leq x \leq a, 0 \leq y \leq 1)$ with joint fixing at the edge are investigated. At critical values λ_m of the upper load satisfying $a_m = \sqrt{m(m+1)}$, $m = 1, 2, \cdots$, Karman's equation has a two - dimensional degeneration at the point $w = 0$. Yu.I.Sapronov proved that at $\lambda = \lambda_m, a = a_m$ the functional V has a strict local minimum. He constructed a bifurcation diagram in the neighbourhood of the points (λ_m, a_m) in the plane of the parameters (λ, a) and calculated quantities, Morse indices and first asymptotics of solutions bifurcating from $w = 0$ using the methods of [19]. The results obtained are more complete than those of Knightly, Holder E.F., Schaeffer who did not use potentiality. In [21] methods used in [19] are applied to the study of the stability of stationary rotations of a multidimensional solid in the gravitational field; for a non-symmetrical top sufficient conditions of stability generalizing the classical ones are obtained.

Let us consider another complicated problem — the problem of bifurcation of a minimal surface in R^3 bounded by a given contour. In [22] – [24] a functional-operational method for the study of this problem based on the theory of nonlinear Fredholm operators is developed, sufficient conditions of bifurcation on critical contours are obtained and the existence of bifurcations for a number of classical minimal surfaces and the type of the arising peculiarities are established.

In boundary problems without uniqueness the question on finite multiplicity of solutions is of interest. For a stationary equation of Navier-Stokes the results of R.Teman, J.C.Saut are known; for Plateau's problem final results are not yet obtained but intensive investigations are being carried out (M.Beeson, F.Tomi, A.J.Tromba [25]). V.G.Zvyagin in [26] considered the nonlinear Dirichlet boundary problem:

$$F(x, u, \cdots, D^{2m}u) = h(x), \; x \in \Omega; \; D^\alpha u(x) = 0, \; x \in \partial\Omega, \; \alpha = 0, \cdots, m-1,$$

where F is a smooth function, and posed a question on the finiteness of the number of solutions depending on $h(x)$. In [26] by methods of smooth topology of Fredholm operators it was shown that if there is an a priori estimate of solutions

$$\|u\|_X \leq C(\|h\|_Y),$$

where $C : R_+ \to R_+$ is a bounded function on bounded subsets, then there exists an open dense set $O \subset Y$ such that for each $h \in O$ the set of solutions of the boundary problem in X is finite. Here $X = W_2^{2m+1}(\Omega) \cap W_2^m(\Omega)$, $Y = W_2^1(\Omega)$, $1 > \frac{n}{2} + 1$.

3. Topology of multivalued mappings.

It was shown above how the perturbations of Fredholm operators are connected with multivalued mappings. Here we shall directly consider the degree theory of multivalued mappings $F : X \to Y$ of topological spaces X, Y. It is convenient to consider a multivalued mapping as a single-valued mapping $F : X \to K(Y)$ into the space of compact subsets in Y provided with the topology $\mathcal{N}$ or Exp ([4]). Normally the continuity of F is assumed and limitations on the images are imposed. Let us consider, in a more general sense, the mapping $F : X \to \mathcal{N}$ where $\mathcal{N} \subset K(Y)$ is a closed subspace in $K(Y)$. In order to construct a topological characteristic let us choose X to be an open set with boundary ∂X, a distinguished point $m_* \in \mathcal{N}$ and let us demand $F : \partial X \to \mathcal{N} \backslash m_*$. There arises the problem of homotopic classification of the pairs $(X, \partial X) \to (\mathcal{N}, \mathcal{N} \backslash m_*)$ and the construction of topological invariants. To date this problem has been studied in partial situations: 1) Exp $K(Y)$ is linearly connected and homotopically weakly equivalent to a point if Y is a separable linearly connected and locally contractible space; 2) let Y be a subset of a Banach space E and let $\mathcal{N} = K_v(Y)$ be the subspace of convex subsets in $K(Y)$; then $\mathcal{N}K_v(Y)$ is homotopically weakly equivalent to Y if Y is an open and linearly connected subset of some closed convex set $T \subset E$; 3) for any subset $Y \subset E$ Exp$K_v(Y)$ is homotopically equivalent to Y. (We formulated the results obtained by B.D.Gel'man and N.Benkafader, see [4].) On properties 2), 3) one may base a degree theory of vector fields $F = I - k$ with convex images in Euclidean and Banach spaces, LCS, where k is a compact operator (see, for example, [27], [3]). For more complex images $F(x)$ (acyclic, generalized acyclic, almost acyclic and others), starting from Eilenberg and Montgomery, homological methods were developed using exact sequence for the graph (or its generalization) of a multivalued mapping (Granas A., Jaworovski J., Sklyarenko E.G., Skordev G., Górniewicz L., Bryszewski J., Obukhovskii V.V., Gel'man B.D. e.a., see [3–5]); the most general construction belongs to Gel'man B.D. ([27], [4]).

B.D.Gel'man on the basis of the topological characteristic of a multivalued mapping constructed by him has recently obtained new fixed point theorems which generalize the well-known theorems of Kakutani and Eilenberg-Montgomery. Let $F : B \to R^{n+1}$, where $B \subset R^{n+1}$ is a closed ball with boundary S; let $M_k(F) = \{x \in B | \overline{H}^k(t^{-1}(x), G) \neq 0\}$, where $\overline{H}$ are reduced cohomologies of Alexander-Čech over the group G, $t : \Gamma_B(F) \to B$ is the canonical projection from the graph. Denote by $d_k(M_k)$ the relative dimension of $M_k(F)$ with respect to B, $k \geq 0$. Let us formulate the results obtained by Gel'man B.D. [28].

<u>Theorem III</u>. Let $F : B \to R^{n+1}$, $F(S) \subset B$, $n \geq 1$ be an upper semicontinuous multivalued mapping with compact images satisfying the conditions $H^n(\Gamma_s(F)) \neq 0$

and $d_k(M_k) \le n-k-1$, $k \ge 1$. Here, $\Gamma_s(F) := \Gamma(F|S)$. Then there exists a fixed point $x_* \in F(x_*)$ in the ball B.

In particular, if $d_k(M_k \cap S) \le n-k-2$, $k \ge 0$, then the condition $H^n(\Gamma_s(F)) \ne 0$ is satisfied.

Theorem IV. Let $F : B \to K(R^{n+1})$ be an upper semicontinuous multivalued mapping satisfying the conditions: a) $F(S) \subset B$, b) if $x \in \text{Int } B$, then the set $F(x)$ is acyclic. Then F has a fixed point in the ball B.

Theorem V. Instead of condition b) let the following condition be satisfied: b') for any ball $B_0 \subset \text{Int } B$ and any $\varepsilon > 0$ the restriction of F to B_0 admits a continuous single-valued ε-approximation. Then F has a fixed point in the ball B.

As it was shown, the condition of acyclicity may be violated on the boundary of the ball without violating the condition to have a fixed point.

V.V.Obukhovskii developed the theory of "essential" multivalued mappings. Let us consider the class $D_U^{k,\alpha}$- of upper semicontinuous (k,α) bounded multivalued mappings $H : U \to K_v(E_1)$, i.e.

$$\alpha(H(A)) \le k \cdot \alpha(A)$$

for any bounded $A \subset U$, where $k \in R_+$, α is the Kuratovskii measure of non-compactness, $U \subset E$; E, E_1 are Banach spaces.

Definition. An upper semicontinuous multivalued mapping $\Phi : \overline{U} \to K(E_1)$ is called $D_U^{k,\alpha}$ - essential if 1) $\Phi^{-1}(0) \subset U$; 2) any $H \subset D_U^{k,\alpha}$ has at least one point of coincidence with Φ in $U : \Phi(X_0) \cap H(x_0) \ne \emptyset$.

The following fixed point theorem holds.

Theorem VI. Let $\Phi : \overline{U} \to K(E_1)$ be $D_U^{k,\alpha}$-essential, $0 \le k < 1$ and let $Q \subset E_1$ be a starshaped subset with respect to zero such that $Q \cap \Phi(\partial U) = \emptyset$. Then for any (ℓ,α)-bounded mapping $F : \overline{U} \to K_v(E_1)$ where $0 \le \ell \le k$ such that $F(\partial U) \subset Q$ there exists a point of coincidence $x_0 \in U : \Phi(x_0) \cap F(x_0) \ne \emptyset$. In particular, $Q \subset \Phi(U)$.

Note that the theory of $D_U^{k,\alpha}$-essential m-mappings is a development of a topological degree theory of (k,α)-bounded m-mappings.

4. Calculation of topological characteristics.

When one is doing practical calculations of topological characteristics it is sometimes useful to restrict the operator to a finite-dimensional mapping. For example, the degree of a Fredholm mapping is reduced, by Sapronov, to the mapping degree of finite-dimensional manifolds. In comparatively simple cases the degree (rotation) of multivalued mappings is also reduced to a finite-dimensional one.

In this respect the task of the effective calculation of a mapping degree in a finite-dimensional situation becomes important. The existing methods did not make it possible to calculate the degree effectively (the integral formula of L.Kronecker in R^n, algorithm of P.P. Zabreiko in R^2 and others) except for the simplest situations (for example in the case of linear non-degenerate fields). N.M.Bliznyakov obtained effective algebraic algorithms and formulas for calculating the index of a singular point of a vector field by its Taylor coefficients [29]. In particular, for the case R^2 he pointed out an algorithm allowing to calculated the index of a singular point of almost any vector field (except for a set of infinite- co-dimension in the space of germs of vector fields) by its Taylor coefficients as the result of a finite number of arithmetic and logical operations. The algorithm reduces the calculation of the index of a singular point to the calculation of classical Cauchy indices $I_a^b R(x)$ of rational functions defined as the difference between the number of discontinuities of the rational function $R(x)$ with the transition from $-\infty$ to $+\infty$ and the number of discontinuities with the transition from $+\infty$ to $-\infty$ and the change of x from a to b. The calculation techniques for Cauchy indices have been thoroughly developed by the classics (Hermite, Hurwitz, Frobenius).

In [30] some generalizations and refinements of the well-known theorem on the addition of rotations are established. Let us give the following result (theorems 1, 3, [30]).

Theorem VII. Let the continuous vector field $\Phi : \partial\Omega^n \to R^n \setminus 0$ where $\partial\Omega^n$ is the connected boundary of a compact polyhedral connected domain be presented in the form $\Phi(x) = A(x) \cdot F(x)$, where $F : \partial\Omega^n \to R^n \setminus 0$ is a continuous vector field, $A : \partial\Omega^n \to C(R^n \setminus 0, R^n \setminus 0)$ is a continuous mapping. Then for the rotation $\gamma(\Phi)$ of vector field Φ on $\partial\Omega^n$ the following relation takes place $\gamma(\Phi) = \deg A(\overline{x}) \cdot \gamma(F) + \gamma(A_e)$, where $\deg A(\overline{x})$ is the degree of the mapping $A(\overline{x})$ of the homological $(n-1)$ sphere $R^n \setminus 0$ into itself, $\gamma(F)$ is the rotation of the vector field F on $\partial\Omega^n$, $\gamma(A_e)$ is the rotation on $\partial\Omega^n$ of the vector field A_e given by the formula $A_e(x) = A(x) \cdot e$ (here $\overline{x}$ is an arbitrarily given point on $\partial\Omega^n$, e is an arbitrarily given vector from $R^n \setminus 0$).

Note that in the case when the polyhedron Ω^n is homeomorphic to a ball in R^n and $|\deg A(x)| = 1$ the quantity $\gamma(A_0)$ can assume only the following values: a) $\gamma(A_e) = 0$ at odd $n > 1$; b) at even n not equal to 2, 4, 8 the quantity $\gamma(A_e)$ may be equal to any integer.

The analogous statement is established in [30] for completely continuous fields in a Hilbert space; in this case the component of the type $\gamma(A_e)$ equals zero.

Let us consider an important (for applications) problem of the investigation of singular points in the iterative processes of the solution of nonlinear equations in finite- and infinite-dimensional spaces. The results obtained in [30] are applied to the investigation of singular points of iterative processes of Newtonian type $\dot{x} = -\Gamma(x) \cdot f(x)$ ($\dot{x} = -(D_x f)^{-1} \cdot f(x)$ and equations close to it) for the solutions to equations of the form $f(x) = 0$ in Euclidean and Hilbert spaces. Statements are obtained (theorems 5, 6, 7 [30]) on the sum of indices of singular points (zeroes of the right-side parts of the field $T(x) \cdot f(x)$ which are solutions to the equation $f(x) = 0$) and on the existence of "branches" of singular points traversing the isolated multiple solution to the equation.

Another important task is the calculation of the degree (and also Lefschetz number) for equivariant mappings with respect to the action of group G.

Let actions T_1, T_2 of a finite group G on spaces X_1, X_2 be given and let the continuous mapping $f : X_1 \to X_2$ be equivariant with respect to these actions, i.e. $fT_1 = T_2 f$. It is necessary to obtain the most complete information on the degree of the mapping f in terms of (co)homological characteristics of the actions T_1, T_2 and information on the restriction of the mapping f to the appropriate invariant subsets of the space X.

The most interesting results in this respect are obtained for equivariant mappings of (co)homological spheres.

In the case of $G = Z_k$, the author, Izrailevich Ya.A. and Fomenko T.N. obtained a number of theorems in this direction of different degrees of generality and in line with generalizations of the classical results of Smith and Krasnosel'skii M.A. In [31] the main stages of these investigations are described and the most general results (for a finite group G) obtained by Fomenko T.N. are formulated. Let us summarize the latter.

Note first of all that with the help of a spectral sequence of a Borel fibration with the fibre (Y, X) ($Y = C_f$ is a cylinder of the equivariant mapping $f\colon X \to Y$, $G = Z_k$) the equivariant index $I(f)$ is determined in the case when X, Y are Z_k- cohomological spheres of arbitrary dimensions [32]; this index generalizes spectral indices for a mapping f which is equivariant with respect to Z_k introduced earlier by the author and Ya.A.Izrailevich.

Let now G be a finite group of order $|G| = k$ which acts by homeomorphisms on a Z-cohomological n-sphere X. Let $k = \prod k_j$, $k_j = p_j^{\alpha_j}$, $j = 1, \cdots, J$ be a canonical decomposition and let G_j be Sylow subgroup in G corresponding to the divisor k_j and $\{G_{ji}\}_{i=1}^{\alpha_j}$ — a cyclic tower, $e \in G_{j1} \subset \cdots \subset G_{j\alpha_j} = G_j$; denote by F_{ji} the set of fixed points of the action G_{ji}, then F_{j1} is the greatest subset in the chain $\{F_{ji}\}_{i=1}^{\alpha_j}$ and it is a set of fixed points of the action of the group $G_{j1} = Z_{p_j}$; by Smith theory F_{j1} is a Z_{p_j}-cohomological m_j-sphere and it can be shown that F_{j1} is a Z_k-cohomological m_j-sphere.

Consider a G-equivariant mapping $f : X_1 \to X_2$ of a Z-cohomological n-sphere with action of the group G on X_1, X_2 and let $F_{j1}^{(i)} \subset X_i$ $(i = 1, 2)$ be the sets of fixed points of the action of the subgroup G_{j1} on X_i $(j = 1, 2, \cdots, J)$ and let $\gamma_j^i : F_{j1}^{(i)} \to X_i$ be inclusions. Then we have

Theorem VIII. The following equality holds

$$[\deg f]_k = \sum_{\langle j \rangle} \left[\frac{k}{k_j}\right]_{k_j}^{-1} \cdot \frac{k}{k_j} \left[I(\gamma_j^1)\right]_{k_j}^{-1} \cdot I(\gamma_j^2) \cdot \left[\deg f | F_{j1}^{(1)}\right]_{k_j},$$

where $I(\gamma_j^1), I(\gamma_j^2)$ are equivariance indices over the ring Z_{k_j}, $\langle j \rangle$ is a collection of indices such that $\dim F_{j1}^{(1)} = \dim F_{j1}^{(2)}$ and a square bracket $[\ \ ]_k$ denotes residue modulo k.

From theorem VIII we obtain a useful

Corollary. Under the conditions of theorem VIII let $f, g : X_1 \to X_2$ be equivariant mappings for which $I(f|F_{j1}^{(1)}) = I(g|F_{j1}^{(1)})$ $(mod\ k_j)$, $j = 1, \cdots, J$.

Then $\deg f \equiv \deg g$ $(mod\ k)$.

References.

1. Borisovich Yu.G., Zvyagin V.G., Sapronov Yu.I. Nonlinear Fredholm mappings and Leray-Schauder theory // Uspekhi mat. nauk. 32:4(1977), 3–54 (in Russian).

2. Krasnosel'skii M.A., Zabreiko P.P. Geometrical methods of nonlinear analysis, Berlin, Heidelberg, New York, Tokyo, Springer, 1984.

3. Borisovich Yu.G., Gel'man B.D., Myshkis A.D., Obukhovskii V.V. Multivalued mappings // Itogi nauki i tekhn.VINITI. Matematicheskii analiz. 19(1982), 127–230 (in Russian).

4. Borisovich Yu.G., Gel'man B.D., Myshkis A.D., Obukhovskii V.V. Multivalued analysis and operator inclusions // Itogi nauki i tekhn. VINITI. sovremennye problemy matematiki. Noveishie dostizheniya. 29(1986), 151–211 (in Russian).

5. Borisovich Yu.G., Gel'man B.D., Myshkis A.D., Obukhovskii V.V. On new results in the theory of multivalued mappings. I. Topological characteristics and solvability

of operator relations // Itogi nauki i tekhn. VINITI. Matematicheskii analiz. 25 1987, 121–195 (in Russian).

6. Borisovich Yu.G., Zvyagin V.G. On one topological principle of solvability of equations with Fredholm operators // Dokl. AN USSR, 1976, ser.A, No.3, 204–206 (in Russian).

7. Ratiner N.M. To the degree theory of Fredholm mappings of non-negative index, Voronezh University, Voronezh, 1981, dep. VINITI, No.1493-81, 31p. (in Russian).

8. Borisovich Yu.G., Sapronov Yu.I. To the theory of nonlinear Fredholm mappings // Trudy VIII letnei matematicheskoi shkoly, Kiev, Institut matematiki AN USSR, 1971, 128–163 (in Russian).

9. Skubachevskii A.L. Elliptic problems with non-local conditions in the neighbouhood of the boundary // Matem.sb., 129:2(1986), 279–302 (in Russian).

10. Zvyagin V.G., Dmitrienko V.T. Homotopic classification of one class of continuous mappings // Mat.zametki. 31:5(1982), 801–812 (in Russian).

11. Borisovich Yu.G. Topology and nonlinear functional analysis // Uspekhi mat.nauk, 34:6(1979), 14–22 (in Russian).

12. Zvyagin V.G. To the theory of generalized condensing perturbations of continuous mappings // Topological and geometrical methods in mathematical physics.-Voronezh: Izd-vo VGU, 1983, 42–62 (in Russian) (English translation in: L.N. in Math., vol.1108).

13. Borisovich Yu.G. On topological methods in the problem of solvability of nonlinear equations. Tr. Leningr. mezhdunar. topol. konf., 23–27 August, 1982, Leningrad: Nauka, 1983, 39–49 (in Russian).

14. Borisovich Yu.G. On the solvability of nonlinear equations with Fredholm operators // Geometry and topology in global nonlinear problems.-Voronezh, VGU, 1984, 3–22 (in Russian, English translation in: L.N. in Math. vol.1108).

15. Borisovich Yu.G. Modern approach to the theory of topological characteristics of nonlinear operators. I. // Geometry and the theory of singularities in nonlinear equations, Voronezh, VGU, 1987, 24–46 (in Russian, English translation to appear in L.N in Math.).

16. Borisovich Yu.G. Modern approach to the theory of topological characteristics of nonlinear operators, II // Global analysis and nonlinear equations.-Voronezh, VGU, 1988 (in Russian).

17. Lecture Notes in Math. 1108, 1214, Global Analysis —Studies and Applications, I, II, Ed. by Yu.G.Borisovich, Yu.E.Gliklikh, Springer-Verlag, 1984, 1986.

18. Borisovich Yu.G., Sapronov Yu.I. To the topological theory of compact contracted mappings // Trudy seminara po funkts. analizu, Voronezh, VGU, 1969, vyp.12, 43–68 (in Russian).

19. Sapronov Yu.I. Breach of spherical symmetry in nonlinear variational problems // Analiz na mnogoobraziyakh i differentisial'nye uraveniya, Voronezh, VGU, 1986, 88–111 (in Russian).

20. Sapronov Yu.I. On two-mode bifurcations of solutions to Karman equation // VI Republican conference "Nonlinear problems of mathematical physics", Donetsk, tez, 1987, 132 (in Russian).

21. Sapronov Yu.I. Multidimensional sleeping tops // Global analysis and mathematical physics, Voronezh, VGU, 1987, 95–109 (in Russian, English translation to appear in L.N. in Math.).

22. Borisovich A.Yu. One geometrical application of the thoerem on a simple bifurcation point // Application of topology in modern analysis, Voronezh, VGU, 1985, 172–174 (in Russian).

23. Borisovich A.Yu. Reduction of the problem on bifurcation of minimal surfaces to operator equations and the search of bifurcations from catenoid, helicoid, surfaces of Scherk and Enneper // Uspekhi mat. nauk, 41:5(1986), 165–166 (in Russian).

24. Borisovich A.Yu. Plateau operator and bifurcation of two-dimensional surfaces // Global analysis and mathematical physics, Voronezh, VGU, 1987, 142–154 (in Russian, English translation to appear in L.N. in Math.).

25. Tromba A.J. Degree theory on oriented infinite dimensional varieties and the Morse number of minimal surfaces spanning a curve in R^n. Part I: $n \geq 4$ // Transactions of the Amer. Math. Soc. 290:1(1985), 385–412.

26. Zvyagin V.G. On the number of solutions to some boundary problems // Global analysis and mathematical physics, Voronezh, VGU, 1987, 60–72 (in Russian, English translation to appear in L.N. in Math.).

27. Borisovich Yu.G., Gel'man B.D., Myshkis L.D., Obukhovskii V.V. Topological methods in the theory of fixed points of multivalued mappings // Uspekhi mat. nauk. 35:1(1980), 59–126 (in Russian).

28. Gel'man B.D. To the fixed points theorem of Kakutani type for multivalued mappings // Global analysis and nonlinear equations, Voronezh, VGU, 1988 (in Russian).

29. Bliznyakov N.M. Cauchy indices and singular point index of a vector field. // Application of topology in modern analysis-Voronezh, VGU, 1985, 3–21 (in Russian, English translation in Lecture Notes in Math., vol.1214).

30. Izrailevich Ya.A. To the singular points theory of iterative processes // Geometry and the theory of singularities in nonlinear equations, Voronezh, VGU, 1987, 146–151 (in Russian, English translation in Lecture Notes in Math., in print).

31. Borisovich Yu.G., Fomenko T.N. Homological methods in the theory of periodic and equivariant mappings // Global analysis and mathematical physics, Voronezh, VGU, 1987, 3–25 (in Russian, English translation to appear in L.N. in Math.).

32. Shchelokova T.N. To the calculation problem of the degree of an equivariant mapping // Sibirskii mat. zhurn, 19:2(1978), 426–435 (in Russian).

FIXED POINTS OF MAP EXTENSIONS

Robert F. Brown, Robert E. Greene[1] and Helga Schirmer[2]

University of California, Los Angeles, California 90024-1555
University of California, Los Angeles, California 90024-1555
Carleton University, Ottawa, K1S 5B6, Canada

1. Introduction

Let X be a connected finite polyhedron and A a subpolyhedron. For a map of pairs $f:(X,A)\to(X,A)$, the paper [11] presented a "relative" Nielsen number $N(f;X,A)$ with the property that every map $g:(X,A)\to(X,A)$ homotopic to f (as a map of pairs) has at least $N(f;X,A)$ fixed points. The paper [13] introduced a "Nielsen number of the complement" $\tilde{N}(f;X,A)$ with the property that every such map g has at least $\tilde{N}(f;X,A)$ fixed points in the closure $\mathrm{cl}(X-A)$ of $X-A$. Although [13] considerably extended the relative Nielsen fixed point theory in many cases, it furnishes no additional information when X and A are such that $\mathrm{cl}(X-A)=X$, for instance, if X is a compact triangulated n-manifold with boundary and A is the boundary of X.

Thus the purpose of this paper is to present a theory for maps of pairs $f:(X,A)\to(X,A)$ that concerns fixed points on $X-A$ rather than on its closure, focusing particularly on the important case where $X-A$ is the interior of a manifold with boundary. However, the "Nielsen number" we present will not be a lower bound for the number of fixed points on $X-A$ of all maps $f:(X,A)\to(X,A)$ homotopic to f as maps of pairs, but only those whose restriction to A is identical to the restriction of f, and homotopic to f through maps with the same property.

Although we are working in a more restricted setting than that of [11], [12], [13], the setting has the advantage of permitting us to address a natural fixed point question that has not been considered before. Given a map $\phi:A\to A$, an *extension* of ϕ is a map $f:(X,A)\to(X,A)$ such that $f\mid A=\phi$. If ϕ admits extensions, is it possible to extend ϕ without introducing any more fixed points? That is, among all extensions $f:X\to X$ of ϕ, is there one with no fixed points in $X-A$? We will see that when X is a smooth (that is, C^1) manifold with boundary A and $\phi:A\to A$ is a smooth map, the answer to that question can depend on whether we consider all continuous extensions of ϕ or only the smooth ones. That is, there are smooth maps $\phi:A\to A$ that admit continuous extensions to X with no fixed points on $X-A$, but with the property that every smooth extension of ϕ must have a fixed point in $X-A$. This difference in behavior between the continuous and smooth cases has not been observed in Nielsen fixed point theory previously. In fact, Jiang showed in [6] that, working with a smooth map $f:X\to X$ on a smooth manifold (of dimension at least 3), the basic constructions of "classical" Nielsen theory can all be carried out entirely in the smooth category. However, significant differences between the continuous and smooth categories have been observed in other parts of fixed point theory, see e.g. [14].

Before going on, it may help the reader keep track of all these "Nielsen numbers" if we describe the results for one simple example. Let X be the disc, represented as the complex numbers of modulus less than or equal to one, and let A be its boundary circle. Suppose $f:(X,A)\to(X,A)$ has the property: if $|z|=1$, then $f(z)=z^3$. In other words, f is an extension of the smooth map $\phi:A\to A$ defined by $\phi(z)=z^3$. Then $N(f)=1$, $N(f;X,A)=2$ and $\tilde{N}(f;X,A)=1$. The

[1]Research supported in part by NSF Grant DMS 88-01999
[2]Research supported in part by NSERC Grant A 7579

lower bound for the number of fixed points in $X - A$ of an extension f of ϕ will be denoted by $N(f \mid \phi)$ in the continuous case and by $N^1(f \mid \phi)$ when we consider only smooth extensions. We will show that $N(f \mid \phi) = 0$ for any extension of $\phi(z) = z^3$ and in fact that there is an extension of ϕ with fixed point set just that of ϕ, namely, $\{-1, +1\}$. However, if f is any smooth extension of this ϕ, we will compute that $N^1(f \mid \phi) = 1$, so f has at least three fixed points, the two just mentioned plus at least one on the interior of the disc.

In the next section, we define the "extension Nielsen number" $N(f \mid \phi)$ for extensions $f : (X, A) \to (X, A)$ of a map $\phi : A \to A$, where X is a connected finite polyhedron and A a subpolyhedron, and develop its properties in Sections 3 and 4. Restricting the setting in Sections 5 through 7 so that X is a smooth manifold and ϕ a smooth map, we present the theory of the "smooth extension Nielsen number" $N^1(f \mid \phi)$. We will illustrate the distinction between the continuous and smooth theories by studying the fixed points of extensions to the disc of a family of maps of the circle that includes the example $\phi(z) = z^3$ described above.

2. The extension Nielsen number

We assume the elements of Nielsen fixed point theory: the fixed point index for maps on finite polyhedra, the fixed point class concept, and the Nielsen number; see [2], [7].

Let X be a connected finite polyhedron, A a nonempty subpolyhedron, and $\phi : A \to A$ a map. An *extension* of ϕ is a map $f : X \to X$ such that $f \mid A = \phi$, and if such a map exists, we say that ϕ is *extendable*. Denote by $E(\phi)$ the set of all extensions $f : X \to X$ of ϕ.

Recall from [13] that a fixed point class F of $f : (X, A) \to (X, A)$ *does not assume its index in* A if $i(X, f, \mathsf{F}) \neq i(A, \phi, \mathsf{F} \cap A)$. We define the *extension Nielsen number* of f with respect of ϕ, denoted by $N(f \mid \phi)$, as the number of fixed point classes of $f : X \to X$ which do not assume their index in A and which do not intersect ∂A. To see that $N(f \mid \phi)$ is a lower bound, we use

LEMMA 2.1. *Let F be a fixed point class of $f : (X, A) \to (X, A)$ with $\mathsf{F} \cap \partial A = \emptyset$. Then F does not assume its index in A if and only if $i(X, f, \mathsf{F} \cap (X - A)) \neq 0$.*

Proof. Note first that $\mathsf{F} \cap \partial A = \emptyset$ implies that $i(X, f, \mathsf{F} \cap (X - A))$ is defined because $\mathsf{F} \cap (X - A) = \mathsf{F} \cap \mathrm{cl}(X - A)$ is compact since F is compact, and $\mathsf{F} \cap (X - A)$ is open in Fix f because F has that property. Hence $\mathsf{F} \cap (X - A)$ is an isolated set of fixed points in the sense of [7], page 16, and so the lemma follows from

$$\begin{aligned} i(X, f, \mathsf{F}) &= i(X, f, \mathsf{F} \cap A) + i(X, f, \mathsf{F} \cap (X - A)) \\ &= i(A, \phi, \mathsf{F} \cap A) + i(X, f, \mathsf{F} \cap (X - A)). \end{aligned}$$ □

THEOREM 2.2. *The map $f : (X, A) \to (X, A)$ has at least $N(f \mid \phi)$ fixed points in $X - A$.*

We will now discuss some relationships between $N(f \mid \phi)$ and the other "Nielsen numbers", and present some cases in which $N(f \mid \phi)$ is easy to compute. For $\tilde{N}(f; X, A)$ the Nielsen number of the complement of [13], it is evident from the definition of $N(f \mid \phi)$ that

THEOREM 2.3. $N(f \mid \phi) \leq \tilde{N}(f; X, A)$.

This result can be sharpened if Fix ϕ is contained in ∂A. This hypothesis is always satisfied in the case we will focus on later in the paper when X is a manifold and A is its boundary, so $A = \partial A$.

THEOREM 2.4. *If Fix ϕ is contained in ∂A, then*

$$N(f \mid \phi) \leq N(f; X, A) - N(\phi)$$

where $N(f; X, A)$ is the relative Nielsen number of [11]. *It follows that $N(f \mid \phi) \leq N(f)$.*

Proof. Let F be a fixed point class of $f : X \to X$. If Fix $\phi \cap \partial A = \emptyset$, then Fix ϕ contained in ∂A implies that F is not a common fixed point class of f and ϕ. If, in addition, F does not assume its index in A, then Lemma 2.1 implies that F is essential. So the theorem follows from the fact that, by the definition of the relative Nielsen number,

$$N(f; X, A) - N(\phi) = N(f) - N(f, \phi)$$

counts the number of essential fixed point classes of f which are not common fixed point classes of f and ϕ. □

Theorem 2.4 is a sharper result than 2.3 because, by Theorem 3.4 of [13],

$$N(f; X, A) - N(\phi) = N(f) - N(f, \phi) = \tilde{N}(f; X, A) - \tilde{n}(f; X, A).$$

We next state a simple condition under which the inequality of the previous theorem is reversed. This condition is satisfied, for example, by any map of nonzero Lefschetz number on a Jiang space (that is, a space in which the Jiang subgroup is the entire fundamental group). See [2] Chap. VII A or [7] Theorem 4.2, page 33.

THEOREM 2.5. *If all the fixed point classes of $\phi : A \to A$ are essential, then $N(f \mid \phi) \geq N(f; X, A) - N(\phi)$.*

Proof. Let F be an essential fixed point of $f : X \to X$ which is not a common fixed point class of f and ϕ. If all the fixed point classes of ϕ are essential, then F$\cap$ Fix $\phi = \emptyset$ and hence F is contained in $X - A$. Therefore,

$$i(X, f, \mathsf{F} \cap (X - A)) = i(X, f, \mathsf{F}) \neq 0$$

because F is essential. The theorem then follows from 2.1 and the definitions of $N(f; X, A)$ and $N(f \mid \phi)$. □

Combining Theorems 2.4 and 2.5, we have

COROLLARY 2.6. *If Fix ϕ is contained in ∂A and all the fixed point classes of ϕ are essential, then $N(f \mid \phi) = N(f; X, A) - N(\phi)$.*

We use the Corollary in proving

THEOREM 2.7. *Suppose that* int $A = \emptyset$ *and that all the fixed point classes of ϕ are essential. If $g : (X, A) \to (X, A)$ is any map which is homotopic to $f : (X, A) \to (X, A)$ by a homotopy of pairs of spaces, then $N(g \mid \psi) \leq N(f \mid \phi)$, where ψ denotes the restriction of g to A.*

Proof. We are given a homotopy of the form $H : (X \times I,\ A \times I) \to (X, A)$ and we know that $N(\phi)$ and $N(f; X, A)$ are invariant under such a homotopy. Now int $A = \emptyset$ implies that Fix ψ is contained in ∂A, so by 2.4 and 2.6 we have

$$N(g \mid \psi) \leq N(g; X, A) - N(\psi) = N(f; X, A) - N(\phi) = N(f \mid \phi). \qquad \square$$

The assumptions in Theorems 2.4 and 2.5 are necessary. To see that Theorem 2.4 is false without the assumption that Fix ϕ is contained in ∂A, let X be an annulus in $\mathbb{R}^2$ and let A be the area bounded by two rays starting at the center of the annulus and by the two boundary arcs of the annulus between them. Let $\phi : A \to A$ be the constant map to some $a \in$ int A. Let $f : X \to X$ be a deformation extending ϕ, that is f is homotopic to the identity map, then $N(f \mid \phi) = 1$ by its definition, but

$$N(f) = N(f; X, A) - N(\phi) = 0.$$

To see that Theorem 2.5 is false without the assumption that all the fixed point classes of ϕ are essential, let X be the 2-disc and A its bounding circle, and let $f : X \to X$ be the identity map. Then the definition tells us that $N(f \mid \phi) = 0$, but $N(f; X, A) - N(\phi) = 1$.

If Fix f is disjoint from ∂A rather than entirely contained in it, the behavior of $N(f \mid \phi)$ is quite different, for we have as a consequence of Theorem 4.4 of [13]:

THEOREM 2.8. *If Fix $\phi \cap \partial A = \emptyset$, then $N(f \mid \phi) = \tilde{N}(f; X, A)$, the Nielsen number of the complement. If, furthermore, $L(f) \neq L(\phi)$ where L denotes the Lefschetz number, then $N(f \mid \phi) > 0$.*

If Fix $\phi = \emptyset$, then both 2.6 and 2.8 reduce to the next result, which also follows immediately from the definition of $N(f \mid \phi)$ and Lemma 2.1.

THEOREM 2.9. *If Fix $\phi = \emptyset$, then $N(f \mid \phi) = N(f)$.*

There is a more general situation in which $N(f \mid \phi)$ can easily be computed.

THEOREM 2.10. *Suppose either that X is simply-connected or that $f : X \to X$ is a deformation then*

$$N(f \mid \phi) = \begin{cases} 1 & \text{if Fix } \phi \cap \partial A = \emptyset \text{ and } L(f) \neq L(\phi) \\ 0 & \text{otherwise.} \end{cases}$$

Proof. We consider the case where X is simply-connected. The argument when f is a deformation is very similar. Let $\phi : A \to A$ be any map and let $f \in E(\phi)$. The result is obvious if Fix $f = \emptyset$, so we assume f has fixed points, which must be a single fixed point class F. Thus Fix $\phi = \mathsf{F} \cap A$ and so Fix $\phi \cap \partial A \neq \emptyset$ implies $N(f \mid \phi) = 0$. If Fix $\phi \cap \partial A = \emptyset$, then $\mathsf{F} \cap \partial A = \emptyset$, and therefore

$$i(X, f, X) = i(X, f, \mathsf{F} \cap A) + i(X, f, \mathsf{F} \cap (X - A))$$

which gives us

$$L(f) = L(\phi) + i(X, f, \mathsf{F} \cap (X - A))$$

so Lemma 2.1 implies that the theorem is true in this case as well. $\square$

EXAMPLE. Let X be a closed n-ball, with $n \geq 2$, and let A be the disjoint union of finitely many balls contained in X. Thus f is a deformation and Theorem 2.10 shows that if k is the number of components of A which are mapped into themselves by a map $\phi : A \to A$, then for any extension $f : X \to X$ of ϕ

$$N(f \mid \phi) = \begin{cases} 1 & \text{if } \operatorname{Fix}\phi \cap \partial A = \emptyset \text{ and } k \neq 1 \\ 0 & \text{otherwise.} \end{cases}$$

(Compare [13], Example 3.8, where $\tilde{N}(f; X, A)$ is computed.)

3. Basic properties of the extension Nielsen number

We now discuss homotopy invariance and commutativity for the extension Nielsen number. First we show that $N(f \mid \phi)$ is invariant under relative homotopies and, more generally, under homotopies that are unchanged on ∂A throughout the homotopy.

THEOREM 3.1 (*Homotopy invariance*). *Let $f \in E(\phi)$ and $g \in E(\psi)$ and let $H : (X \times I, A \times I) \to (X, A)$ be a homotopy from $H(x,0) = f(x)$ to $H(x,1) = g(x)$ with $H(a,t) = \phi(a) = \psi(a)$ for all $a \in \partial A$. Then $N(f \mid \phi) = N(g \mid \psi)$.*

Proof. It follows from the proof of [13], Theorem 4.1, that H relates every fixed point class F of $f : X \to X$ which does not assume its index in A to a unique fixed point class G of $g : X \to X$ which does not assume its index in A. Hence it only remains to show that $\mathsf{F} \cap \partial A = \emptyset$ implies that $\mathsf{G} \cap \partial A = \emptyset$. But if there exists a point $a \in \mathsf{G} \cap \partial A$, then a is in $\operatorname{Fix}\psi \cap \partial A = \operatorname{Fix}\phi \cap \partial A$, and so there exists a unique fixed point class F' of $f : X \to X$ with $a \in \mathsf{F}'$. Using the path $a(t) = a$ for all $0 \leq t \leq 1$, we see that F' is H-related to G ([2], Chap. VI D and E, page 87 ff., or [7], Chap. 1.2, page 7 ff.). Since F is H-related to G also, we have $\mathsf{F} = \mathsf{F}'$ in contradiction to the assumption that $\mathsf{F} \cap \partial A = \emptyset$. As usual, the homotopy H^{-1} shows that H induces a bijection between those fixed point classes of f and g that do not assume their index in A and do not intersect ∂A. □

To establish commutativity, it is necessary to assume that boundaries of subspaces are mapped into boundaries of subspaces:

THEOREM 3.2 (*Commutativity*). *Let $\phi : (A, \partial A) \to (B, \partial B)$ and $\psi : (B, \partial B) \to (A, \partial A)$ be maps and let $f : (X, A) \to (Y, B)$ and $g : (Y, B) \to (X, A)$ be maps such that $f \mid A = \phi$ and $g \mid B = \psi$. Then $N(gf \mid \psi\phi) = N(fg \mid \phi\psi)$.*

Proof. Let F be a fixed point class of gf which does not assume its index in A and which does not intersect ∂A, and let $\mathsf{G} = f(\mathsf{F})$ be the fixed point class of fg which is the image of F under the homeomorphism $f : \operatorname{Fix}(gf) \to \operatorname{Fix}(fg)$ (see [7], p. 20, Lemma 5.1 and Theorem 5.2) which respects fixed point classes. Then since $\mathsf{F} \cap \partial A = \emptyset$ and ψ takes ∂B into ∂A it follows that $\mathsf{G} \cap \partial B = \emptyset$. Furthermore, $i(X, gf, \mathsf{F}) = i(Y, fg, \mathsf{G})$ and $i(A, \psi\phi, \mathsf{F} \cap A) = i(B, \phi\psi, \mathsf{G} \cap B)$ imply

$$i(X, gf, \mathsf{F} \cap (X - A)) = i(Y, fg, \mathsf{G} \cap (Y - B))$$

and the result is obtained as in [7], p. 20, Theorem 5.2. □

The assumptions that ϕ takes ∂A into ∂B and ψ takes ∂B into ∂A are necessary to insure that the comutativity property holds. For example, let

$$X = Y = D^2 = \{z \in \mathbf{C} : |z| \leq 1\}$$
$$A = \{z \in \mathbf{C} : 1 \geq |z| \geq 1/2\}$$
$$B = S^1 = \{z \in \mathbf{C} : |z| = 1\}.$$

Let $f = f_2 f_1 : X \to Y$ where $f_1(z) = \bar{z}$ and

$$f_2(z) = \begin{cases} z/|z| & \text{if } 1/2 \leq |z| \leq 1 \\ 2z & \text{if } |z| \leq 1/2. \end{cases}$$

Let $g : Y \to X$ be the identity map, and take $\phi = f \mid A$, $\psi = g \mid B$. Then $\mathrm{Fix}\, fg = \mathrm{Fix}\, gf = \{0, -1, +1\}$, but, by Theorem 2.10, $N(gf \mid \psi\phi) = 1$ while $N(fg \mid \phi\psi) = 0$.

The extension Nielsen number $N(f \mid \phi)$ is also a homotopy-type invariant if the definition of homotopy type is suitably restricted, that is, if the definition from [11], §3 and [13], §4 is modified by the requirement that all homotopies remain fixed on the boundaries of the subspaces. The proof is straightforward, and left to the reader.

4. The toplogical minimum theorem

We now show that $N(f \mid \phi)$ is optimal for a large class of pairs of polyhedra. As in [11], we say that a subpolyhedron A *can be by-passed* if every path in X with end points in $X - A$ is path-homotopic to a path in $X - A$. By a *homotopy* between two maps $f, g : (X, A) \to (X, A)$ we always mean a homotopy of the form $H : (X \times I, A \times I) \to (X, A)$ with $H(a, t) = H(a, 0)$ for all $a \in A$. As usual, maps with minimal fixed point sets are constructed from maps with finite fixed point sets. That such maps exist follows from the next theorem, which is easily proved by a modification of the proof of Theorem 4.4 of [11].

THEOREM 4.1. *Every map $f : (X, A) \to (X, A)$ is homotopic to a map $f' : (X, A) \to (X, A)$ such that f' is fix-finite on $X - A$, all fixed points on $X - A$ lie in maximal simplexes, and $f(a) = f'(a)$ for all $a \in A$.*

THEOREM 4.2 (*Minimum Theorem*). *If $X - A$ has no local cut-point and is not a 2-manifold and if A can be by-passed, then for any $f : (X, A) \to (X, A)$ there exists a homotopic map $g : (X, A) \to (X, A)$ that has exactly $N(f \mid \phi)$ fixed points on $X - A$.*

Proof. According to 4.1, we can assume that f is fixed finite on $X - A$, and that all fixed points in $X - A$ lie in maximal simplexes. Let F be a fixed point class of f such that $\mathsf{F} \cap (X - A) \neq \emptyset$. If $\mathsf{F} \cap (X - A)$ consists of more than one point, then we unite these points in $X - A$, using by-passing, as in the proof of [11], Theorem 6.2, to one fixed point, and we remove this fixed point in the usual way if it is of index zero. Hence we can homotope f to a map $f' : (X, A) \to (X, A)$ so that for each fixed point class F' of f', we have that $\mathsf{F}' \cap (X - A)$ is either empty or consists of one essential fixed point. Now let F' be a fixed point class of f' with $\mathsf{F}' \cap \partial A \neq \emptyset$ and $\mathsf{F}' \cap (X - A) \neq \emptyset$, then $\mathsf{F}' \cap (X - A) = \{x_0\}$ for some point x_0. We pick $x_1 \in \mathsf{F}' \cap \partial A$ and proceed as in the proof of Theorem 6.2 of [11] to homotope f' to a map g' with $\mathrm{Fix}\, g' = \mathrm{Fix}\, f' - \{x_0\}$. After dealing in this way with all fixed point classes which intersect both ∂A and $X - A$, we obtain a map $g : (X, A) \to (X, A)$ which is homotopic to f and such that for each fixed point classes G of g we have either that $\mathsf{G} \cap \partial A \neq \emptyset$ and $\mathsf{G} \cap (X - A) = \emptyset$ or that $\mathsf{G} \cap \partial A = \emptyset$ and $\mathsf{G} \cap (X - A)$ is either empty or consists of one fixed point of non-zero index. Since no changes are made to the given map on ∂A during these homotopies, it follows from Lemma 2.1 and Theorem 3.1 that g has exactly $N(g \mid \phi) = N(f \mid \phi)$ fixed points in $X - A$. □

The example of Jiang [9] demonstrates the necessity of the assumption that $X - A$ is not a 2-manifold. We take X to be the disc with two holes and A to be a single point, not on the boundary of the manifold X. We can take A to be one of the three fixed points of the map f constructed by Jiang. Jiang showed that $N(f) = 0$, so $N(f \mid \phi) = 0$ by Theorem 2.4. Jiang also proved that every map homotopic to f has at least two fixed points, so there is no map homotopic to f without fixed points in $X - A$.

As is usual in Nielsen fixed point theory, the assumptions on X and A in the Minimum Theorem can be relaxed if we consider only maps in the identity class. We use the terms 2-dimensionally connected and proximity map, and the symbol $\alpha(x, y, t)$, as in [12], §2 (see also [2]).

THEOREM 4.3. *If $X - A$ is 2-dimensionally connected, then every deformation $\phi : A \to A$ extends to a deformation $f : X \to X$ which has $N(f \mid \phi)$ fixed points on $X - A$.*

Proof. As in [15], Chap. 3.3, proof of Corollary 11, page 124, we find a subcomplex N of a triangulation K of X, where $|N|$ is disjoint from A, so that there exists a retraction $r : |K| - |N| \to A$ which maps each point x of $|K| - |N|$ to a point of the closed carrier simplex of x in $|K|$. If $\mathrm{Fix}\, \phi \cap \partial A \neq \emptyset$, we select a point $a_0 \in \mathrm{Fix}\, \phi \cap \partial A$. Note that then ϕ is a proximity map on a neighborhood of a_0. We use a star neighborhood of A with respect to a subdivision K' of K to find (as in [11], proof of Theorem 4.1) a compact polyhedron A_1 in $X - |N|$ which contains A in its interior and such that $B = X - \mathrm{int}(A_1)$ is a polyhedron in X. By choosing the subdivision K' fine enough, we can make sure that B is still 2-dimensionally connected and that there exists an open simplex σ of K' with $|\sigma|$ contained in $X - A$ so that a_0 lies in the closed simplex $|\bar{\sigma}|$ and $fr \mid |\bar{\sigma}| : |\bar{\sigma}| \to A$ is a proximity map with respect to K. It follows from the construction of A_1 with a star neighborhood that the boundary of $|\bar{\sigma}|$ intersects B. A map $f_1 : A_1 \to X$ which extends φ can be defined by $f_1 = \phi r \mid A_1$ and a homotopy $H_1 : (A \cup \partial A_1) \times I \to X$ by

$$H_1(x,t) = \begin{cases} \phi(x) & \text{if } x \in A \text{ and } 0 \leq t \leq 1 \\ \Phi(r(x), t) & \text{if } x \in \partial A_1 \text{ and } 0 \leq t \leq 1, \end{cases}$$

where $\Phi : A \times I \to A$ is a homotopy from $\Phi(a, 0) = \phi(a)$ to the identity $\Phi(a, 1) = a$. Because H_1 has no fixed points on $\partial A_1 \times I$, it is a special homotopy and hence extends to a special homotopy $H : A_1 \times I \to X$ of the map $H(x, 0) = f_1(x)$. (See [8], Lemma 2.1.) If $f_2 : A_1 \to X$ is given by $f_2(x) = H(x, 1)$, then f_2 extends ϕ, and since every $x \in \partial A_1$ lies in a closed simplex of $|K|$ which also contains $r(x)$, we see that f_2 is a proximity map on ∂A_1. Now let $H' : \partial A_1 \times I \to X$ be a homotopy from the identity map on ∂A_1 to f_2 which is a proximity map at each level. (The standard homotopy which uses $\alpha(x, f_2(x), t)$ has this property.) Then it follows from [12], Lemma 2.2 (i) that H' extends to a homotopy from the identity map on B to a proximity map $f_2' : B \to X$ with $f_2'(x) = f_2(x)$ for all $x \in \partial A_1$. If we define a map $f' : X \to X$ by

$$f'(x) = \begin{cases} f_2(x) & \text{for } x \in A_1 \\ f_2'(x) & \text{for } x \in B, \end{cases}$$

then f' extends ϕ, is fixed point free on $A_1 - A$, and is a proximity map on B. If $\mathrm{Fix}\, \phi \cap \partial A \neq \emptyset$, then f' is also a proximity map on a simplex $|\bar{\sigma}|$ with $|\bar{\sigma}| \cap (\mathrm{Fix}\, \phi \cap \partial A) \neq \emptyset$ and $|\bar{\sigma}| \cap B \neq \emptyset$. We next use an argument similar to the one in the proof of [12], Theorem 2.3, to obtain an extension f'' of ϕ with $f'' = f'$ on A_1 and such that f'' is a proximity map on B and has all its fixed points located in maximal simplexes. Since B is 2-dimensionally connected, we can unite the fixed points on B as usual to one fixed point x_0, and this can be carried out away from A_1. If x_0 has index zero, we remove it. If x_0 has non-zero index and Fix $\phi \cap \partial A \neq \emptyset$, we proceed as in the proof of [12], Lemma 3.2 and first move x_0 to $|\sigma|$ and then unite it with the selected point $a_0 \in$ Fix $\phi \cap \partial A$ to obtain a deformation which extends ϕ and is fixed point free on $X - A$. Theorem 2.10 shows that now the result holds in all cases. □

A class of maps $\phi : A \to A$ of special interest are those which have an extension $f : X \to X$ with Fix $f =$ Fix ϕ, that is, such that f has no fixed points on $X - A$. The previous results can be used to obtain conditions for the existence of such extensions. The next theorem follows from Theorems 2.7, 2.9, 2.10 and 4.2.

THEOREM 4.4. *Suppose that $X - A$ has no local cut-point and is not a 2-manifold and that A can be by-passed.*

(i) *A fixed point free map $\phi : A \to A$ has an extension with no fixed points on $X - A$ if and only if it has an extension $f : X \to X$ with $N(f) = 0$.*

(ii) *If X is simply-connected, then an extendable map $\phi : A \to A$ has an extension with no fixed points on $X - A$ if and only if either* Fix $\phi \cap \partial A \neq \emptyset$ *or ϕ has an extension $f : X \to X$ with $L(f) = L(\phi)$.*

(iii) *If* $\operatorname{int}(A) = \emptyset$ *and $\phi : A \to A$ is a map which has only essential fixed point classes and which has an extension without fixed points in $X - A$, then every map $\psi : A \to A$ homotopic to ϕ has an extension without fixed points on $X - A$.*

For deformations, we obtain from Theorems 2.10 and 4.3:

THEOREM 4.5. *Suppose $X - A$ is 2-dimensionally connected. Then a deformation $\phi : A \to A$ can be extended to a deformation $f : X \to X$ with no fixed points on $X - A$ if and only if either* Fix $\phi \cap \partial A \neq \emptyset$ *or $\chi(X) = \chi(A)$, where χ denotes the Euler characteristic.*

Taking X to be a 4-ball, A its bounding 3-sphere, and $\phi : A \to A$ a fixed point free rotation, every extension of ϕ must have a fixed point, so it is in $X - A$. This example illustrates the necessity of the hypothesis Fix $\phi \cap \partial A \neq \emptyset$ in Theorems 4.4 (ii) and 4.5. For an example of an extendable map ϕ that *has* fixed points yet every extension has fixed points on $X - A$, let X be the torus $S^1 \times S^1$, let A be the figure-eight $S^1 \times \{1\} \cup \{1\} \times S^1$, and define ϕ by $\phi(e^{i\theta}, 1) = (e^{-i\theta}, 1)$ and $\phi(1, e^{i\omega}) = (1, e^{-i\omega})$, so ϕ has three fixed points. Any extension $f : S^1 \times S^1 \to S^1 \times S^1$ of ϕ has Lefschetz number equal to 4, so $N(f) = 4$ by [1] and f has at least one fixed point not in A.

Both Theorems 4.4 and 4.5 can be simplified if $A = \partial X$, a case which will be considered (for manifolds) in the next section. Theorem 4.4 (ii) can be sharpened to yield the following result:

THEOREM 4.6. *Assume that X has no local cut-point. If X is simply-connected, then an extendable map $\phi : \partial X \to \partial X$ has an extension without fixed points in* $\operatorname{int}(X) = X - \partial X$ *if and only if either* Fix $\phi \neq \emptyset$ *or ϕ has an extension $f : X \to X$ with $L(f) = 0$.*

Proof. Because of Theorem 4.4 (ii), we need only consider the case where $X = D^2$ is a disc. Since $L(f) = 1$ for any map on the disc, we only have to prove that if Fix $\phi \neq \emptyset$, then ϕ has an extension without fixed points in $\operatorname{int}(D^2)$. We can perturb the map obtained by forming the cone over ϕ to obtain a self-map of D^2 with the center of D^2 as the only fixed point in the interior. Then the construction of [11], Lemma 6.1 produces the desired map. □

The final result of this section follows from Theorem 4.4:

THEOREM 4.7. *Assume that X is 2-dimensionally connected. Then a deformation $\phi : \partial X \to \partial X$ can be extended to a deformation $f : X \to X$ without fixed points on* $\mathrm{int}(X)$ *if and only if either* Fix $\phi \neq \emptyset$ *or* $\chi(X) = 0$.

Note that the last two results hold for X a compact triangulated manifold with boundary and $A = \partial X$ because they are trivially true for the interval. For an example of a map $\phi : \partial X \to \partial X$ on the boundary of a manifold with Fix $\phi \neq \emptyset$ and such that every extension to X has a fixed point on $\mathrm{int}(X)$, let $M = S^1 \times D^2$ and start with $\phi^* : \partial M \to \partial M = S^1 \times S^1$ defined by $\phi^*(e^{i\theta}, e^{i\omega}) = (e^{-i\theta}, e^{i\omega})$ so Fix $\phi^* = \{1, -1\} \times S^1$. Homotope ϕ^*, in the second factor only, to a map ϕ which is a rotation on $\{-1\} \times S^1$ and is a deformation with a single fixed point on $\{1\} \times S^1$. Let $f \in E(\phi)$, then $N(f) = 2$ so Fix $f \neq$ Fix ϕ.

5. The index of fixed points of smooth maps

The rest of this paper will be concerned with the fixed point theory of map extensions in the category of smooth, that is C^1, maps on smooth n-dimensional manifolds M with nonempty boundary ∂M. For a given smooth map $\phi : \partial M \to \partial M$, we will consider smooth maps $f : M \to M$ that are extensions of the map ϕ. Suppose that $p \in \partial M$ is an isolated fixed point of f, that is, $f(q) \neq q$ for all $q \neq p$ in a neighborhood of p in M. Then certainly p is an isolated fixed point of ϕ as well. In this section we will investigate the relationship between $i(\partial M, \phi, p)$, the index of ϕ at p, and $i(M, f, p)$, the index of f at the same point.

We will first show that, without further hypotheses, the values of $i(\partial M, \phi, p)$ and of $i(M, f, p)$ can be arbitrary integers, with the values unrelated to each other even for C^∞ maps. Since our discussion is entirely local, we may write f in a neighborhood of p in terms of a local coordinate system in which M is identified with the half-space

$$\mathbf{R}^n_- = \{(x_1, x_2, \dots, x_n) \in \mathbf{R}^n : x_n \leq 0\}$$

in such a way that p is the origin $\mathbf{O} = (0, 0, \dots, 0)$, and ∂M is identified with the subspace

$$\mathbf{R}^{n-1} = \{(x_1, x_2, \dots, x_{n-1}, 0) \in \mathbf{R}^n\}.$$

We will also make use of the half-space $\mathbf{R}^n_+$ of $\mathbf{R}^n$ for which $x_n \geq 0$, the unit sphere S^{n-1} in $\mathbf{R}^n$, the "lower hemisphere" $S^{n-1}_- = \mathbf{R}^n_- \cap S^{n-1}$, which we will write more compactly as S_-, the "upper hemisphere" $S_+ = S^{n-1}_+ = \mathbf{R}^n_+ \cap S^{n-1}$, and the "equator" $S^{n-2} = \mathbf{R}^{n-1} \cap S^{n-1}$.

Consider the map $G : \mathbf{R}^n_- \to \mathbf{R}^n$ defined by $G(x) = f(x) - x$, where the difference is defined in the sense of vector subtraction in the $(x_1, x_2, \dots, x_n)$ coordinates. The map G has the following properties: (i) $G(\mathbf{O}) = \mathbf{O}$, (ii) $G(q) \neq \mathbf{O}$ for $q \neq \mathbf{O}$ near $\mathbf{O}$, (iii) G takes $\mathbf{R}^{n-1}$ to itself and (iv) the x_n coordinate of $x + G(x)$ is nonpositive. Conversely, a map G satisfying (i) - (iv) produces a map $f : (\mathbf{R}^n_-, \mathbf{R}^{n-1}) \to (\mathbf{R}^n_-, \mathbf{R}^{n-1})$ with an isolated fixed point at the origin, by setting $f(x) = x + G(x)$.

Now suppose that $\Gamma : S_- \to \mathbf{R}^n - \mathbf{O}$ is a C^∞ function. Define a map $G : \mathbf{R}^n_- \to \mathbf{R}^n$ by setting $G(\mathbf{O}) = \mathbf{O}$ and

$$G(x) = h(|x|)\Gamma(x/|x|) \qquad \text{for } x \neq \mathbf{O},$$

where $h : \mathbf{R} \to \mathbf{R}$ is a C^∞ real function with $0 \leq h(t) \leq 1$ for all t, $h(t) = \exp(-t^{-2})$ for $0 < t < 1/3$, and $h(t) = 1$ for $t \geq 1$. Then G is a C^∞ function because the only problematic point, the origin, is smoothed by the $\exp(-t^{-2})$ factor. The associated map $f(x) = x + G(x)$ is of the required type mapping $\mathbf{R}^n_-$ to $\mathbf{R}^n_-$ with an isolated fixed point at p provided that the nth coordinate of $G(x_1, \dots, x_n)$ is less than or equal to $|x_n|$.

We will make use of the retraction $\rho : \mathbf{R}^n - \mathbf{O} \to S^{n-1}$ defined by $\rho(x) = x/|x|$. If $f = \phi \mid \partial M$, where f is the map of the previous paragraph, then the index $i(\partial M, \phi, p)$, when identified with $i(\mathbf{R}^{n-1}, \phi, \mathbf{O})$, is the degree of the map $-\rho\Gamma \mid S^{n-2} : S^{n-2} \to S^{n-2}$. On the other hand, identifying $i(M, f, p)$ with $i(\mathbf{R}^n_-, f, \mathbf{O})$, we can compute this index as the degree of the map $\rho E : S^{n-1} \to S^{n-1}$, where $E : \mathbf{R}^n \to \mathbf{R}^n - \mathbf{O}$ is defined by setting

$$E(x_1, x_2, \dots, x_n) = \begin{cases} (x_1, x_2, \dots, x_n) - f(x_1, x_2, \dots, x_{n-1}, 0) & \text{if } x_n > 0 \\ -G(x_1, x_2, \dots, x_n) & \text{if } x_n \leq 0 \end{cases}$$

$$= \begin{cases} (0, 0, \dots, 0, x_n) - G(x_1, x_2, \dots, x_{n-1}, 0) & \text{if } x_n > 0 \\ -G(x_1, x_2, \dots, x_n) & \text{if } x_n \leq 0. \end{cases}$$

Note that ρE maps the upper hemisphere S_+ to itself and that this map is determined by the values of G just on S^{n-2}. With G defined in terms of Γ as before, the map ρE on S_+ depends only on $\Gamma \mid S^{n-2}$, not on the values of Γ elsewhere on the lower hemisphere. In particular, it is straightforward to choose Γ so that $i(\partial M, \phi, p) = i(\mathbf{R}^{n-1}, \phi, \mathbf{O})$ and $i(M, f, p) = i(\mathbf{R}^n_-, f, \mathbf{O})$ each have arbitrary, unrelated values.

Specifically, we can do this as follows: Given two integers r and s, we shall construct Γ so that $(-1)^n i(\partial M, \phi, p) = r$ and $(-1)^{n-1} i(M, f, p) = s$. For this purpose, we first choose a smooth map $\alpha : S^{n-2} \to S^{n-2}$ of degree r. For J an interval in [-1,1], define

$$S_J = \{(x_1, \dots, x_n) \in S^{n-1} : x_n \in J\}.$$

Now extend the map α to be a smooth map of $S_{(-1/3,0]}$ to S_- in such a way that the image of $S_{(-1/3,-1/4)}$ is the single point $(0, \dots, 0, -1)$. This construction is accomplished by smoothing the suspension of α, using a construction similar to our definition of $G(x)$ above. In more detail, we set $\alpha(x_1, \dots, x_{n-1}, 0) = (\gamma(x_1, \dots, x_n), 0)$, where $\gamma(x_1, \dots, x_n) \in \mathbf{R}^{n-1}$. Then choose a C^∞ function $\delta : S_- \to [0,1]$ which is identically 0 on $S_{[-1,-1/4]}$ and identically 1 on $S_{(-1/10,0]}$. Then, for $x = (x_1, \dots, x_n)$ a point of $S_{(-1/3,0]}$, set

$$\Gamma(x_1, \dots, x_n) = \delta(x) \cdot \alpha(\rho(x_1, \dots, x_{n-1}, 0)) + (1 - \delta(x))(0, \dots, 0, -1)$$

where ρ is the retraction to S^{n-1} defined before. Now extend Γ smoothly to $S_{(-1/2,-1/3)}$ so that Γ on this region has x_n coordinate nonpositive and so that Γ maps $S_{(-1/2,-5/12)}$ on the point $(0, \dots, 0, -1/10)$. Finally, using a smoothed suspension process similar to the definition of Γ on $S_{(-1/3,0)}$, we can extend Γ to $S_{[-1,-1/2]}$ so that G as a whole is smooth and Γ on $S_{[-1,-1/2)}$ is generically $s - r$ to one onto the sphere of radius 1/10. By this we mean that the signed number of preimages of a noncritical value, counting orientation, is $s - r$ (which may be negative). (See Figure 1.) It is easy to see that $\Gamma(x_1, \dots, x_n)$ has its x_n coordinate less than or equal to $|x_n|$, for every point in S_-. In particular, the associated G maps $\mathbf{R}^n_-$ to itself. This will be the case automatically if $\rho(x)$ is in $S_{(-1/2,0]}$ since Γ has nonpositive x_n coordinate there. If $\rho(x)$ is in $S_{[-1,-1/2]}$, then $|\Gamma(x)| \leq 1/10$, so $|G(x)| \leq (1/10)h(|x|)$. On a neighborhood of $\mathbf{O}$, we have $h(t) = \exp(-t^{-2}) < t$. Hence it is easy to have chosen the original h so that $h(t) \leq 5t$. Therefore we have $|G(x)| \leq (1/10)(10|x|) = |x|$ on all x with $\rho(x) \in S_{[-1,-1/2]}$ and thus $x + G(x)$ has nonpositive x_n coordinate. Since the indices of ϕ and f arise from the degrees of $x - f(x) = -G(x)$, we have $i(\partial M, \phi, p) = (-1)^n \deg(\Gamma \mid \partial M) = (-1)^n r$ and

$i(M,f,p) = (-1)^{n-1}\deg\Gamma = (-1)^{n-1}s$. Consequently, we have an isolated fixed point of f at p, with the required properties, and f maps $\mathbf{R}^n_-$ to $\mathbf{R}^n_-$.

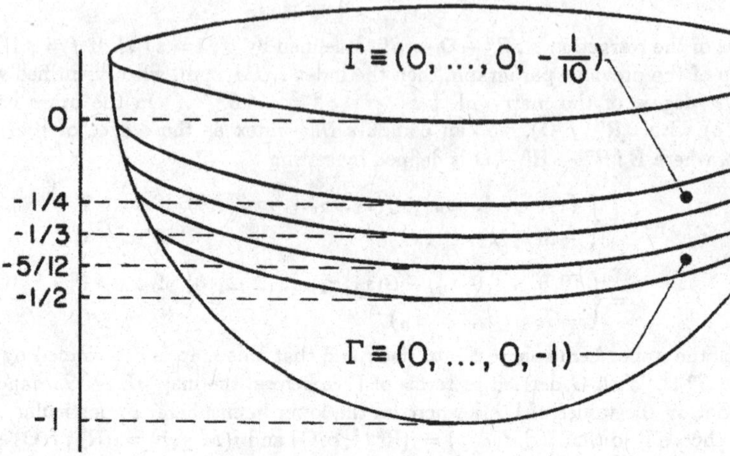

FIGURE 1

Careful consideration of the construction just given shows that it depends in some sense upon the fact that the map G we defined by $G(x) = f(x) - x$ has the properties that $G(\mathbf{O}) = \mathbf{O}$ and that its first derivatives vanish at $\mathbf{O}$. The next theorem makes this observation more precise. We still have the smooth map $f:(M,\partial M)\to(M,\partial M)$ with an isolated fixed point at $p\in\partial M$ and we continue to identify $(M,\partial M)$ with $(\mathbf{R}^n_-,\mathbf{R}^{n-1})$ in a neighborhood of $p=\mathbf{O}$. We write $T_p(\partial M)$ for the tangent space of ∂M at p, $d\phi_p : T_p(\partial M)\to T_p(\partial M)$ for the differential of ϕ at p, and $\mathbf{I}_\partial : T_p(\partial M)\to T_p(\partial M)$ for the identity map. If the linear transformation $d\phi_p - \mathbf{I}_\partial : T_p(\partial M)\to T_p(\partial M)$ is nonsingular, then $i(\partial M,\phi,p)$ can take on the values +1 or -1 only. This follows from the fact that

$$|G(\varepsilon x) - dG_p(\varepsilon x)| \le o(\varepsilon)$$

uniformly over $\{x\in\partial M : |x| = 1\}$ and that there exists a fixed constant $C>0$ such that $|dG_p(x)|\ge C|x|$ for all $x\in T_p(\partial M)$. The theorem will show that this hypothesis puts a restriction on the possible values of $i(M,f,p)$ as well, so the construction of a map with unrelated arbitrary indices, like that above, is impossible.

THEOREM 5.1. *Given a smooth map $\phi : \partial M \to \partial M$ and a smooth map $f:(M,\partial M)\to(M,\partial M)$ extending ϕ, suppose that $p\in\partial M$ is an isolated fixed point of f and that $d\phi_p - \mathbf{I}_\partial : T_p(\partial M)\to T_p(\partial M)$ is a nonsingular linear transformation. Then either $i(M,f,p)=0$ or $i(M,f,p)=i(\partial M,\phi,p)$.*

Proof. We divide the proof into cases: (1) the linear transformation $df_p - \mathbf{I} : T_p(M)\to T_p(M)$, where $\mathbf{I}$ denotes the identity on $T_p(M)$, is nonsingular and (2) $df_p-\mathbf{I}$ is a singular linear transformation. So we first assume that $df_p-\mathbf{I}$ is nonsingular. For $\varepsilon>0$ sufficiently small, we may define $E_\varepsilon : S^{n-1}\to\mathbf{R}^n - \mathbf{O}$ by

$$E_\varepsilon(x_1,x_2,\dots,x_n) = \begin{cases} (0,0,\dots,\varepsilon x_n) - G(\varepsilon x_1,\dots,\varepsilon x_{n-1},0) & \text{if } x_n>0 \\ -G(\varepsilon x_1,\dots,\varepsilon x_n) & \text{if } x_n\le 0,\end{cases}$$

where $G(x) = f(x) - x$ as usual. Then the index $i(M, f, p) = i(\mathbb{R}^n_-, f, O)$ is the degree of $\rho E_\varepsilon : S^{n-1} \to S^{n-1}$, where $\rho(x) = x/|x|$ for $x \in \mathbb{R}^n - O$ as before. Now E_ε takes S_+ into $\mathbb{R}^n_+$. Moreover, we have

(a) $|G(\varepsilon x_1, \ldots, \varepsilon x_n) - dG_p(\varepsilon x_1, \ldots, \varepsilon x_n)| \leq o(\varepsilon)$ and

(b) $|dG_p(\varepsilon x_1, \ldots, \varepsilon x_n)| \geq C\varepsilon$

for some $C > 0$ independent of ε and $(x_1, x_2, \ldots, x_n)$. Condition (b) holds because $dG_p = df_p - \mathrm{I}$ is nonsingular by hypothesis, while condition (a) follows from the fact that G is a C^1 function with value O at p. From (a) and (b) it follows that E_ε is homotopic, as a map of S^{n-1} into $\mathbb{R}^n - O$, to the following map

$$D_\varepsilon(x_1, x_2, \ldots, x_n) = \begin{cases} (0, 0, \ldots, \varepsilon x_n) - dG_p(\varepsilon x_1, \ldots, \varepsilon x_{n-1}, 0) & \text{if } x_n > 0 \\ -dG_p(\varepsilon x_1, \ldots, \varepsilon x_n) & \text{if } x_n \leq 0. \end{cases}$$

Hence $i(M, f, p)$ is the degree of $\rho D_\varepsilon : S^{n-1} \to S^{n-1}$. Since dG_p is a nonsingular linear map, this last degree is easily seen to be 0, +1, or -1 (because D_ε is linear on the upper and lower hemispheres). But we can say more. Since the image under D_ε of the lower hemisphere and on the upper hemisphere are each contained entirely in either the lower or the upper half-space, it follows immediately that either D_ε is of degree 0 or D_ε is homotopic in $\mathbb{R}^n - O$ to the suspension of $D_\varepsilon \mid S^{n-2}$. Hence either $i(M, f, p) = 0$ or $i(M, f, p) = i(\partial M, \phi, p)$ (which is either +1 or −1, as we remarked earlier).

The case that $df_p - \mathrm{I}$ is singular requires somewhat different reasoning. We define E_ε as in the first part of the proof and examine the behavior of E_ε on $\mathbb{R}^n_-$ in some detail. For this purpose, it is easiest to choose new x coordinates, if necessary, and a y coordinate system, linearly related to the x coordinates such that

(a) $\{(x_1, x_2, \ldots, x_{n-1}, 0) \in \mathbb{R}^n\} = \{(y_1, y_2, \ldots, y_{n-1}, 0) \in \mathbb{R}^n\}$

(b) $\{(x_1, x_2, \ldots, x_n) \in \mathbb{R}^n : x_n \leq 0\} = \{(y_1, y_2, \ldots, y_n) \in \mathbb{R}^n : y_n \leq 0\}$

so we may still associate M with $\mathbb{R}^n_-$ and ∂M with $\mathbb{R}^{n-1}$, in either coordinate system, and

(c) dG_p, as a map from the x coordinates to the y coordinates, has the form

$$dG_p(\alpha_1, \ldots, \alpha_{n-1}, \alpha_n) = (\alpha_1, \ldots, \alpha_{n-1}, 0) + \alpha_n(0, \ldots, 0, A, 0)$$

for some (possibly 0) constant A. These conditions can always be arranged because $G \mid \partial M$ is nonsingular onto ∂M at p and dG_p has image equal to

$$\{(x_1, x_2, \ldots, x_{n-1}, 0)\} = \{(y_1, y_2, \ldots, y_{n-1}, 0)\}.$$

Specifically, we first rotate the $x_1, \ldots, x_{n-1}$ coordinates in their coordinate plane so that the positive x_{n-1} direction points along the direction of the differential image $dG_p(0, \ldots, 0, 1)$. (The vector $(0, \ldots, 0, 1)$ is unchanged by this rotation, and hence so is its differential image.) Then we choose the $y_1, \ldots, y_{n-1}, y_n$ coordinates so that the hyperplane $y_n = 0$ is the same as the hyperplane $x_n = 0$ and, for each $j = 1, \ldots, n-1$, the transformation dG_p takes the x_j unit vector to the y_j unit vector.

We will next examine the map $dG_p \mid S_-$, expressed as a map from x to y coordinates. We claim that for each sufficiently small fixed δ with $0 < \delta < 1$, the following properties hold (only property (i) actually involves δ):

(i) $dG_p \mid S_{[-\delta,1]}$ is bounded away from $\mathbf{O}$, with the bound independent of δ

(ii) the point $x = (0, \dots, 0, A/(1+A^2)^{1/2}, -1/(1+A^2)^{1/2})$ is the unique point of S_- with the property that $dG_p(x) = \mathbf{O}$

(iii) there is a (small) neighborhood U of the point x in (ii) such that $dG_p \mid S_-$ is a (non-singular) diffeomorphism of U onto a neighborhood of $\mathbf{O}$ in the hyperplane $y_n = 0$ and $(dG_p \mid S_-)^{-1}(dG_p(U)) = U$.

Properties (i), (ii) and (iii) follow from elementary algebra and the properties of the x and y coordinates with respect to dG_p. (Note: $dG_p \mid S_-$ is not injective if $A \neq 0$; it is two-to-one at some points.) Now consider the maps A_ε defined by $A_\varepsilon(x) = \varepsilon^{-1} E_\varepsilon(x)$. It is an elementary application of the chain rule to show that the maps $A_\varepsilon \mid S_-$ converge in the C^1 topology to $dG_p \mid S_-$. In particular, for δ fixed and small enough so that (i) holds, we have, for $\varepsilon > 0$ sufficiently small, the following analogues of (i) and (ii) above:

(1) $A_\varepsilon \mid S_{[-\delta,1]}$ is bounded away from $\mathbf{O}$, with the bound independent of δ

(2) there is a unique point $x_\varepsilon \in S_-$ such that the $y_1, \dots, y_{n-1}$ coordinates of $A_\varepsilon(X_\varepsilon)$ are all 0.

(See Figure 2.) Note that $A_\varepsilon(S_-)$ can intersect both the upper and lower open half-spaces that form the complement of $y_n = 0$.

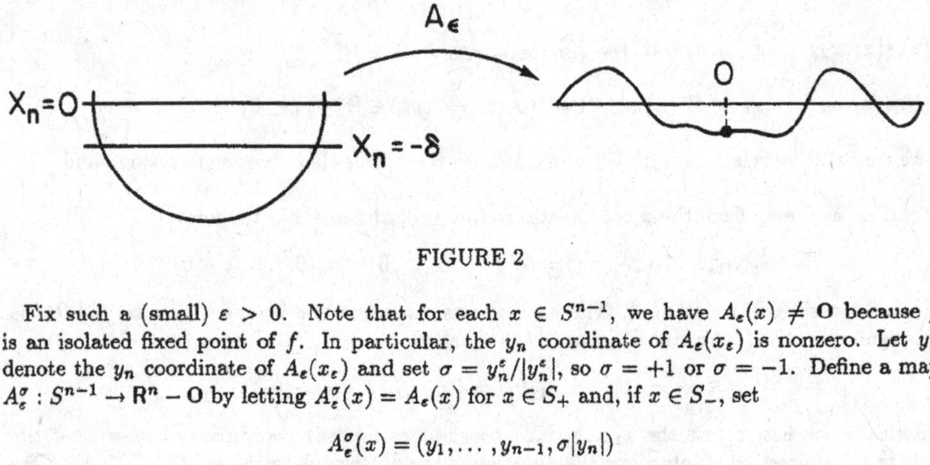

FIGURE 2

Fix such a (small) $\varepsilon > 0$. Note that for each $x \in S^{n-1}$, we have $A_\varepsilon(x) \neq \mathbf{O}$ because p is an isolated fixed point of f. In particular, the y_n coordinate of $A_\varepsilon(x_\varepsilon)$ is nonzero. Let y_n^ε denote the y_n coordinate of $A_\varepsilon(x_\varepsilon)$ and set $\sigma = y_n^\varepsilon/|y_n^\varepsilon|$, so $\sigma = +1$ or $\sigma = -1$. Define a map $A_\varepsilon^\sigma : S^{n-1} \to \mathbf{R}^n - \mathbf{O}$ by letting $A_\varepsilon^\sigma(x) = A_\varepsilon(x)$ for $x \in S_+$ and, if $x \in S_-$, set

$$A_\varepsilon^\sigma(x) = (y_1, \dots, y_{n-1}, \sigma|y_n|)$$

where $A_\varepsilon(x) = (y_1, \dots, y_{n-1}, y_n)$. Note that the values of $A_\varepsilon^\sigma \mid S_-$ lie entirely in either the lower half-space of $\mathbf{R}^n$ with respect to the y coordinates

$$\mathbf{R}_-(y) = \{y_1, \dots, y_{n-1}, y_n) : y_n \leq 0\}$$

or the upper half-space $\mathbf{R}_+(y)$, that is, with $y_n \geq 0$. Moreover, A_ε and A_ε^σ are homotopic as maps into $\mathbf{R}^n - \mathbf{O}$. A homotopy is given by

$$H(x,t) = \begin{cases} A_\varepsilon(x) & \text{if } x \in S_+ \\ tA_\varepsilon(x) + (1-t)A_\varepsilon^\sigma(x) & \text{if } x \in S_-. \end{cases}$$

It is obvious that $H(x,t) \neq \mathbf{O}$ if $x \in S_+$. For $x \in S_-$, we have *either* that $(y_1, \ldots, y_{n-1}) \neq (0, \ldots, 0)$ *or* that $x = x_\varepsilon$. But in the case that $(y_1, \ldots, y_{n-1}) \neq (0, \ldots, 0)$, it is clear that $H(x,t) \neq (0, \ldots, 0)$ because $H(x,t)$ has the form $(y_1, \ldots, y_{n-1}, \alpha)$ for $\alpha = ty_n + (1-t)\sigma|y_n|$. And for $x = x_\varepsilon$, we see that $H(x,t) = A_\varepsilon(x)$ since $y_n = \sigma|y_n|$ in this case by definition. (See Figure 3.) The map A_ε^σ takes S_+ into $\mathbf{R}_+(y)$ and

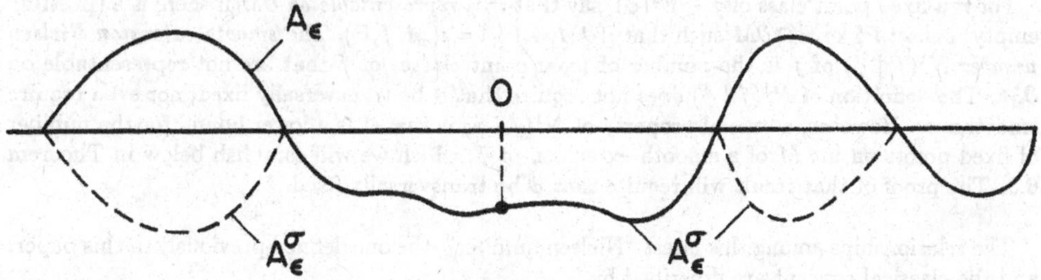

FIGURE 3

maps S_- either entirely into $\mathbf{R}_+(y)$ or entirely into $\mathbf{R}_-(y)$. Hence A_ε^σ is homotopic either to a constant or to the suspension of $A_\varepsilon^\sigma \mid S^{n-2} = A_\varepsilon \mid S^{n-2}$. The homotopy H then establishes that either A_ε is of degree 0 or $\deg(A_\varepsilon) = \deg(A_\varepsilon \mid S^{n-2})$. These degrees are computed as maps from x to y coordinates, but since the positive y_n axis is directed into the same half-space as the positive x_n axis, it follows that these equalities persist as maps from x coordinates to x coordinates. We have established that either $i(M,f,p) = 0$ or else $i(M,f,p) = i(\partial M, \phi, p) = \pm 1$, as we claimed. □

6. The smooth extension Nielsen number

Now let $(X,A) = (M, \partial M)$ where M is a compact connected smooth (that is, C^1) manifold. Given a smooth (C^1) map $\phi : \partial M \to \partial M$, we say that ϕ is *transversally fixed* if the linear transformation $d\phi_p - \mathbf{I}_\partial : T_p(\partial M) \to T_p(\partial M)$ is nonsingular for each fixed point p of ϕ. This condition implies that each fixed point of ϕ is isolated and of index either $+1$ or -1. To explain the terminology, let $\Gamma(\phi)$ denote the graph of ϕ, that is the smooth submanifold of $\partial M \times \partial M$ consisting of all points of the form $(x, \phi(x))$; then ϕ is transversally fixed if and only if $\Gamma(\phi)$ and the diagonal in $\partial M \times \partial M$ are transverse submanifolds of $\partial M \times \partial M$.

Define $E^1(\phi)$ to be the set of smooth extensions $f : M \to M$ of ϕ.

Remark. If ϕ admits a continuous extension, then it admits a smooth one as well, that is, $E(\phi)$ nonempty implies that $E^1(\phi)$ is also nonempty. To verify this, first note that ϕ always extends smoothly to a sufficiently small tubular neighborhood of ∂M in M: one can simply take the extension $\phi_1(p) = \phi(\pi(p))$ where π is a smooth projection of the tubular neighborhood of ∂M onto ∂M. Suppose that ϕ also extends to a continuous map $f : M \to M$. By standard techniques of smooth approximation (e.g., convolution smoothing, etc., cf. [10] and [4]), we can find a smooth map ϕ_2 on $M - U$, where U is a small tubular neighborhood of ∂M, arbitrarily close to f. In particular, by continuity, if U is small enough (and f and ϕ_2 close enough) then

ϕ_2 will be close to ϕ_1 on $(M-U)\cap V$, where V is a tubular neighborhood of ∂M in M, slightly larger than U, with ϕ_1 defined on V. Since ϕ_1 and ϕ_2 are close together on this intersection, they can be combined by a (smooth) partition of unity to yield a smooth extension F of ϕ to all of M. For such a partition of unity to apply, it suffices that ϕ_1 and ϕ_2 on $(M-U)\cap V$ have images sufficiently close so that there is a unique interior geodesic connection between them relative to some fixed Riemannian metric on M. Then, with $\{\rho, 1-\rho\}$ the partition unity functions, $F(p) = \rho(p)\phi_1(p) + (1-\rho(p))\phi_2(p)$ is defined to be the unique point which divides the geodesic segment from $\phi_1(p)$ to $\phi_2(p)$ in the ratio $\rho(p) : 1-\rho(p)$. For F to be well-defined (with image in M) here, we choose the Riemannian metric to be a product metric near the boundary of M. Then ∂M is a locally convex submanifold and $F(p)$ a well-defined point of M.

For F a fixed point class of $f \in E^1(\phi)$, say that F is *representable on* ∂M if there is a (possibly empty) subset F'_∂ of $\mathsf{F}\cap\partial M$ such that $i(\partial M, \phi, \mathsf{F}'_\partial) = i(M, f, \mathsf{F})$. The *smooth extension Nielsen number* $N^1(f \mid \phi)$ of f is the number of fixed point classes of f that are not representable on ∂M. The definition of $N^1(f \mid \phi)$ does not require that ϕ be transversally fixed, nor even require smoothness. However, a crucial property of $N^1(f \mid \phi)$ is that it is a lower bound for the number of fixed points on int M of a smooth extension of ϕ, which we will establish below in Theorem 6.3. The proof of that result will require that ϕ be transversally fixed.

The relationships among this latest "Nielsen number", the one defined previously in this paper, and the classical concept are described by

THEOREM 6.1. *If $f \in E^1(\phi)$, then $N(f \mid \phi) \le N^1(f \mid \phi) \le N(f)$.*

Proof. When (X, A) consists of a manifold and its boundary, the extension Nielsen number $N(f \mid \phi)$ is the number of essential fixed point classes F of f such that $\mathsf{F}\cap\partial M = \emptyset$. Thus, for such a class, $i(M, f, \mathsf{F}) \neq 0$ but $\mathsf{F}\cap\partial M = \emptyset$, so F cannot be representable on ∂M and we conclude that $N(f \mid \phi) \le N^1(f \mid \phi)$. If $i(M, f, \mathsf{F}) = 0$ then F is representable by taking $\mathsf{F}'_\partial = \emptyset$, so all the non-representable classes are essential, which proves that $N^1(f \mid \phi) \le N(f)$. □

We showed in 2.3 that $N(f \mid \phi) \le \tilde{N}(f; X, A)$. In the smooth manifold case, we also have the following relationship with the Nielsen number of the complement which, by the previous result, is stronger:

THEOREM 6.2. *If $f \in E^1(\phi)$, then $N^1(f \mid \phi) \le \tilde{N}(f; M, \partial M)$.*

Proof. If F assumes its index on ∂M, we can take $\mathsf{F}'_\partial = \mathsf{F}\cap\partial M$, so F is representable on ∂M. Thus every fixed point class which is not representable on ∂M fails to assume its index on ∂M. □

The Nielsen number $N^1(f \mid \phi)$ has the required properties.

THEOREM 6.3 (*Lower bound*). *If $\phi : \partial M \to \partial M$ is transversally fixed, then every map $f \in E^1(\phi)$ has at least $N^1(f \mid \phi)$ fixed points in int M.*

Proof. If F is a fixed point class of f entirely contained in ∂M then it must be representable in ∂M. The argument goes: F is in $\mathrm{Fix}(\phi)$, hence finite since ϕ is transversally fixed, so write

$\mathsf{F} = \{x_1, x_2, \ldots, x_r\}$, then

$$i(M,f,\mathsf{F}) = \sum_{i=1}^{r} i(M,f,x_i) = \sum_{i=1}^{r} \gamma(j) i(\partial M, \phi, x_i)$$

where, by Theorem 5.1, $\gamma(j) = 0$ or 1. Let $\mathsf{F}'_\partial = \{x_j : \gamma(j) = 1\}$. □

THEOREM 6.4 (*Homotopy invariance*). *If the maps $f, g : (M, \partial M) \to (M, \partial M)$ are homotopic, then $N^1(f \mid \phi) = N^1(g \mid \phi)$.*

Proof. By hypothesis, there is a homotopy $H : (M, \partial M) \times I \to (M, \partial M)$ such that $h_0 = f$, $h_1 = g$ and $H(y,t) = \phi(y)$ for all $y \in \partial M$ and all t. Define $\mathsf{H} : M \times I \to M \times I$ by $\mathsf{H}(x,t) = H((x,t),t)$. Let F be a fixed point class of f which is not representable on ∂M and let F^* be the fixed point class of H containing F. Since F must be essential, there is a fixed point class $\mathsf{G} = \mathsf{F}^* \cap (M \times \{1\})$ of g such that $i(M,g,\mathsf{G}) = i(M,f,\mathsf{F})$. If G is representable on ∂M, then there is a finite set G'_∂ in ∂M such that $i(\partial M, g, \mathsf{G}'_\partial) = i(M,g,\mathsf{G})$. Since $\mathsf{G}'_\partial \times I$ is in Fix H, then it must be in F^*, so G'_∂ is contained in F and thus F would be representable on ∂M, using the set G'_∂, contrary to assumption. Therefore, G is not representable on ∂M and we have the required one-to-one correspondence of non-representable fixed point classes. □

THEOREM 6.5 (*Commutativity*). *Let $\phi : \partial M \to \partial N$ and $\psi : \partial N \to \partial M$ be smooth maps. Suppose $f : M \to N$ and $g : N \to M$ are smooth extensions of ϕ and ψ, respectively, such that $gf \in E^1(\psi\phi)$ and $fg \in E^1(\phi\psi)$, then $N^1(gf \mid \psi\phi) = N^1(fg \mid \phi\psi)$.*

Proof. (Compare the proof of Theorem 3.2.) The restriction of f is a homeomorphism of Fix gf onto Fix fg, whose restriction to Fix $\psi\phi$ is a homeomorphism onto Fix $\phi\psi$. The inverse is the restriction of g. The homeomorphisms preserve fixed point classes and the index so, if F is a representable class of gf, represented by the finite subset F'_∂ of $\mathsf{F} \cap \partial M$, then $f(\mathsf{F}'_\partial)$ is contained in $f(\mathsf{F}) \cap \partial N$. The fixed point class $f(\mathsf{F})$ is representable because $i(M, gf, \mathsf{F}) = i(\partial M, \psi\phi, \mathsf{F}'_\partial)$ by assumption while preservation of the index implies $i(M, gf, \mathsf{F}) = (N, fg, f(\mathsf{F}))$ and $i(\partial M, \psi\phi, \mathsf{F}'_\partial) = i(\partial N, \phi\psi, \phi(\mathsf{F}'_\partial))$, so $i(N, fg, f(\mathsf{F})) = i(\partial N, \phi\psi, \phi(\mathsf{F}'_\partial))$ as required. Since we can use g in the same way for a representable class of fg, the one-to-one correspondence of fixed point classes preserves representability on the boundary of the manifold. □

If $\psi\phi$ is transversally fixed then $\phi\psi$ is transversally fixed as well (and conversely). This is a simple consequence of the following observation about linear algebra: If $A : V \to W$ and $B : W \to V$ are linear transformations of vector spaces, and if I_V and I_W are the identity maps on V and W, respectively, then $BA - \mathsf{I}_V$ is injective if and only if $AB - \mathsf{I}_W$ is injective. The proof is: If $BA - \mathsf{I}_V$ is injective, then $BA(v) = v$ (if and) only if $v = 0$. So if $AB(w) = w$, then $BA(Bw) = Bw$. Thus $Bw = 0$, hence $w = A(B(w)) = A(0) = 0$. Therefore $AB - \mathsf{I}_w$ is injective. Applying this to the differentials yields the conclusion about $\psi\phi$ and $\phi\psi$. It is actually this case, where $\psi\phi$ and $\phi\psi$ are transversally fixed, that is of the most interest in this paper.

7. The smooth minimum theorem

The main purpose of this final section is to establish the smooth version of Theorem 4.2. We will also complete the discussion of the example on the disc that we described in the introduction. Since we are concerned with extensions of a given map $\phi : \partial M \to \partial M$, by a *homotopy* between two such extensions $f, g : M \to M$ we continue to mean a map $H : (M \times I, \partial M \times I) \to (M, \partial M)$ such that $H(x,t) = \phi(x)$ for all $x \in \partial M$ and $t \in I$.

We begin the section by developing some techniques we will need to perform fixed point constructions in the setting of smooth maps.

Let $f : (M, \partial M) \to (M, \partial M)$ be a smooth extension of $\phi : \partial M \to \partial M$. Even if the map ϕ has only isolated fixed points, it is quite possible that those same points will not be isolated when they are considered as fixed points of the map $f : M \to M$. For such an example, consider the map f from the closed unit disc in the plane to itself defined (in complex-number polar coordinates) by setting $f(0) = 0$ and $f(re^{i\theta}) = \rho(r)e^{2i\theta}$ for $r > 0$, where $\rho : [0, 1] \to \mathbf{R}$ is a smooth map with $\rho(t) = 0$ for $0 \leq t \leq 1/4$, $\rho(t) = t$ for $3/4 \leq t \leq 1$, and ρ is monotone strictly increasing on $[1/4, 3/4]$.

In spite of such possibilities, all such smooth maps can be perturbed, while retaining the same values on ∂M, in such a way that f has no fixed points near ∂M except those of ϕ itself. Precisely, we have the following

LEMMA 7.1. *Let $f : (M, \partial M) \to (M, \partial M)$ be a smooth extension of $\phi : \partial M \to \partial M$ and let U be a neighborhood of ∂M in M. Then there is a smooth map $F : (M, \partial M) \to (M, \partial M)$ homotopic to f and an open set U_1 with $\partial M \subset U_1 \subset U$ such that*

(i) *$F(q) = f(q)$ for all $q \in M - U$*

(ii) *$F(q) \neq q$ for all $q \in U_1 - \partial M$*

(iii) *if $q \in \partial M$ is an isolated fixed point of F, then $i(M, F, q) = 0$*

(iv) *for each $q \in \text{Fix}\,\phi$ that is transversal, q is also a transversal fixed point of F itself.*

Proof. Choose a Riemannian metric on M and an $\varepsilon > 0$ so small that the exponential map of a 5ε neighborhood of 0 in the (inward) normal bundle of ∂M is a diffeomorphism onto its image and that its image is contained in U. We let T_λ be the λ-neighborhood of 0 in the inward normal bundle, that is

$$T_\lambda = \{v \in T_q(M) : q \in \partial M,\ v \perp T_q(\partial M),\ \|v\| < \lambda,\ v \text{ is a nonnegative multiple of the inward unit normal}\}.$$

If, for $q \in \partial M$, we write N_q for the inward unit normal, then

$$T_\lambda = \{\theta N_q : q \in \partial M \text{ and } 0 \leq \theta < \lambda\}.$$

We write the exponential map E of T_λ into M in the form $E(q, \theta)$ for $q \in \partial M$ and $\theta \in \mathbf{R}$, with $E(q, \theta)$ the Riemannian exponentiation of θN_q from q. Thus, letting $d(p, S)$ denote the distance from a point p to a set S, we see that E is a diffeomorphism of $T_{5\varepsilon}$ onto the set $\{q \in M : d(q, \partial M) < 5\varepsilon\}$. Now, for all $q \in M$ sufficiently close to ∂M, define $F_1(q) \in M$ by $F_1(q) = E(\pi_1(f(q)), 2d(q, \partial M))$ where $\pi_1(f(q))$ is the (unique) point q' in ∂M such that $f(q) = E(q', \theta N_{q'})$, $\theta = d(q, \partial M)$. Note that $\pi_1(f(q))$ is well-defined for q close enough to ∂M because $f(q)$ is close to ∂M by continuity (since f takes ∂M to itself). Also, if $q \notin \partial M$ then $F_1(q) \neq q$ because the distance from $F_1(q)$ to ∂M is twice the distance from q to ∂M. If $x \in \partial M$ is an isolated fixed point of F_1, then $i(M, F_1, x) = 0$ because this index is the degree of a map of a sphere into $\mathbf{R}^n - O$ whose image lies in a half-space. Now choose ε^* with $0 < \varepsilon^* < \varepsilon$ so that F_1 is well-defined on $T_{5\varepsilon*}$ and also small enough so that $F_1(T_{5\varepsilon*})$ is contained in T_ε. Then choose a smooth map $\rho : M \to [0,1]$ such that $\rho(q) = 1$ for all $q \in T_{\varepsilon *}$ and $\rho(q) = 0$ for all $q \in M - T_{3\varepsilon * /2}$. (The value $\rho(q)$ can in fact be chosen to depend on $d(q, \partial M)$). Then for each $q \in T_{2\varepsilon *}$, set

$$F(q) = E(\pi_1(f(q)),\ 2\rho(q)d(q, \partial M) + (1 - \rho(q))d(f(q),\ \partial M)).$$

Note that if $q \in T_{\varepsilon *}$, then $\rho(q) = 1$ and $F = F_1$. Also if $q \in T_{2\varepsilon *} - T_{3\varepsilon * /2}$, then $\rho(q) = 0$ and $F(q) = E(\pi_1(f(q)), d(f(q), \partial M)) = f(q)$. Thus if we extend F to the rest of M by letting $F(q) = f(q)$ for $q \in M - T_{2\varepsilon *}$, we obtain a smooth map with the required properties, with $U_1 = T_{\varepsilon *}$. □

Next we show that a smooth map from a convex set in Euclidean space to another convex set containing the first and with a compact set of fixed points on the interior of the set can be smoothly deformed to a mapping with at most one fixed point on the interior of the convex set, without changing the original map near the boundary. Here a map on a set C (with possibly non-smooth boundary) in $\mathbf{R}^n$ is called smooth if it extends to a smooth map on an open neighborhood of C in $\mathbf{R}^n$.

LEMMA 7.2. *Let C, C_1 be closed, bounded convex sets in $\mathbf{R}^n$ with nonempty interior, and with $C \subset C_1$. Suppose $f : C \to C_1$ is a smooth map such that* Fix $f \cap$ int C *is compact. Then there is a neighborhood U of ∂C in C and a smooth map $F : C \to C_1$ such that $F(q) = f(q)$ for all $q \in U$ and F has as most one fixed point in* int C*. The map F can be made fixed point free on* int C *if and only if $i(C, f,$ int $C) = 0$. If $i(C, f,$ int $C) = \pm 1$ then F can in addition be made transversally fixed at its unique fixed point in* int C.

Proof. Choose $\varepsilon > 0$ such that $d(p, \partial C) > 5\varepsilon$ for all p in the compact set Fix $f \cap$ int C. We shall write $C_\lambda = \{q \in C : d(q, \partial C) \geq \lambda\}$. Also, let β denote the diameter of C_1, that is, $\beta = \max\{||q_1 - q_2|| : q_1, q_2 \in C_1\}$. We consider the smooth vector field V on C defined by $V(q) = f(q) - q$, where the subtraction is in the sense of $\mathbf{R}^n$ as a vector space. Note that $||V(q)|| \leq \beta$ for every $q \in C$ since $q, f(q) \in C_1$. Now choose a smooth map $\rho : C \to \mathbf{R}$ such that $0 < \rho(q) \leq 1$ for all $q \in C, \rho(q) = 1$ for all $q \in C - C_\varepsilon$, and $\rho(q) = \varepsilon / 2\beta$ for all $q \in C_{4\varepsilon}$. We define a smooth map $F_1 : C \to C_1$ by

$$F_1(q) = q + \rho(q)(f(q) - q).$$

The image $F_1(C)$ lies in C_1 because C_1 is convex and $0 < \rho(q) \leq 1$ for all q. Moreover, $F_1(q) = f(q)$ for all $q \in C - C_\varepsilon$. Now consider the vector field V_1 defined by $V_1(q) = \rho(q)(f(q) - q) = \rho(q)V(q)$. This vector field has no zeros on the set $C_{4\varepsilon} - C_{5\varepsilon}$. The set $C_{4\varepsilon}$ is homeomorphic to an n-ball and its boundary S is homeomorphic to an $(n-1)$-sphere, because $C_{4\varepsilon}$ is a compact convex set with nonempty interior. Thus we can speak of the degree of the map $V_1 : S \to \mathbf{R}^n - \mathbf{O}$, which we denote by $\deg(V_1 \mid S)$. If V_1 has only isolated zeros in $C_{4\varepsilon}$, then the Poincare-Hopf Index Theorem (or, more precisely, one of the steps in its proof) yields the result that $\deg(V_1 \mid S)$ is the sum of the indices of the zeros of V_1 in $C_{4\varepsilon}$ (cf. [3] or [4]). Since the sum of the local Lefschetz indices equals the global index, we have that

$$\begin{aligned} \deg(V_1 \mid S) &= \sum (-1)^{n-1} i(C_{4\varepsilon}, f, p) \quad \text{(summed over all fixed points on } C_{4\varepsilon}) \\ &= (-1)^{n-1} i(C, f, \text{int } C). \end{aligned}$$

Here we have used the fact that the (local) index of the vector field at a zero is $(-1)^{n-1}$ times the Lefschetz index of the zero point as a fixed point of the map taking q to $q + V_1(q)$. The $(-1)^{n-1}$ arises from the fact that V_1 is a (positive) multiple of $f(q) - q$ rather than $q - f(q)$. In case V_1 does not have isolated zeros, we can nevertheless conclude that

$$\deg(V_1 \mid S) = (-1)^{n-1} i(C, f, \text{int } C).$$

For this, it suffices to note that we can slightly deform V_1 so it will be unaltered on a neighborhood of S but have only isolated zeros on $C_{4\varepsilon}$. This deformation can be produced by standard transversality arguments: we need only think of V_1 as a smooth section of the tangent bundle $TM \mid C_{4\varepsilon}$ and deform V_1 by a small deformation so that it becomes transversal to the zero section of $TM \mid C_{4\varepsilon}$. This implies that the deformed vector field V_1^* has only isolated zeros. Now if the deformation is taken small enough, $V_1^* \mid S$ is homotopic (as a map into $\mathbf{R}^n - \mathbf{O}$) to $V_1 \mid S$ so $\deg(V_1 \mid S) = \deg(V_1^* \mid S)$. On the other hand, the map taking q to $q + V_1^*(q)$ on $C_{4\varepsilon}$ is homotopic, relative to S, to the map taking q to $q + V_1(q)$ which is in turn homotopic to f. Thus $\deg(V_1^* \mid S) = (-1)^{n-1} i(C, f, \text{int } C)$ and the same statement for $\deg(V_1 \mid S)$ follows. Now fix a point $q_1 \in \text{int } C_{4\varepsilon}$, and let U_1 be a neighborhood of q_1 with $U_1 \subset \text{int } C_{4\varepsilon}$. Suppose W is a smooth vector field on U_1 with $\|W\| \leq \varepsilon/2$ on U_1 and with $\{q \in U_1 : W(q) = \mathbf{O}\} = \{q_1\}$. If the vector field index of W at q_1 equals $\deg(V_1^*|S)$, then, according to standard homotopy theory, there is a continuous vector field $V_2 : C_{4\varepsilon} \to \mathbf{R}^n$ with $V_2(q) = V_1(q)$ for all $q \in S$, with $\|V_2\| \leq \varepsilon/2$ everywhere on $C_{4\varepsilon}$, with $\{q \in C_{4\varepsilon} : V_2(q) = \mathbf{O}\} = \{q_1\}$, and with $V_2 = W$ on some neighborhood (contained in U_1) of q_1. In case $\deg(V_1^*|S) = \pm 1$, we can and shall take W to be of the form $W(q) = \Psi(q) - q$ for some smooth function Ψ on U_1 having q_1 as a transversal fixed point. By standard theorems on smooth approximations (see [4] and [10]), there is a smooth vector field $V_3 : C_{4\varepsilon} \to \mathbf{R}^n$ such that $V_3 = V_2$ in some neighborhood of q_1, that $\|V_3(q) - V_2(q)\| \leq 2\varepsilon/3$ for all $q \in C_{4\varepsilon}$, and that $\|V_3(q) - V_2(q)\| \leq (1/10)\|V_2(q)\|$ for all q in $C_{4\varepsilon}$. (Note that $V_2 = V_3$ near q_1, the only zero of V_2, so that the norm $\|V_2\|$ is not near 0 in the region where V_2 is being approximated by smoothing.) Because $V_2 = V_1$ on $S = \partial C_{4\varepsilon}$, there exists δ, with $0 < \delta \leq \varepsilon$, such that

$$\|V_2(q) - V_1(q)\| < (1/10)\|V_2(q)\| \quad \text{for all} \quad q \in C_{4\varepsilon} - C_{4\varepsilon+\delta}.$$

Chose a smooth non-negative partition of unity ρ_1, ρ_2 for C subordinate to the cover $\{\text{int } C_{4\varepsilon}, C - C_{4\varepsilon+\delta}\}$. Thus the support of ρ_1 lies in int $C_{4\varepsilon}$ and the support of ρ_2 lies in $C - C_{4\varepsilon+\delta}$, and of course $\rho_2 = 1 - \rho_1$. Define a vector field V_4 on C by setting

$$V_4(q) = \rho_1(q)V_3(q) + \rho_2(q)V_1(q),$$

then V_4 has the following properties:

(i) V_4 is nonvanishing on $C_{4\varepsilon} - C_{4\varepsilon+\delta}$ because if some $V_4(q*) = \mathbf{O}$, we would have

$$\|V_3(q*) - V_1(q*)\| = \|V_3(q*)\| + \|V_1(q*)\| \geq (9/10)\|V_2(q*)\|$$

since $\|V_2(q*) - V_1(q*)\| \leq (1/10)\|V_2(q*)\|$, but that is impossible: for all $q \in C_{4\varepsilon} - C_{4\varepsilon+\delta}$ we have that $V_2(q) \neq \mathbf{O}$ and that

$$\|V_3(q) - V_1(q)\| \leq \|V_3(q) - V_2(q)\| + \|V_2(q) - V_1(q)\| \leq (1/5)\|V_2(q)\|,$$

(ii) V_4 is nonvanishing on int $C - C_{4\varepsilon}$ because it equals V_1 on that set,

(iii) $V_4(q) = V_3(q)$ for $q \in C_{4\varepsilon+\delta}$ and hence can vanish only at q_1; and it vanishes at that point if and only if $i(C, f, \text{int } C) \neq 0$.

Now define the map $F : C \to \mathbf{R}^n$ by $F(q) = q + V_4(q)$. Then, for $q \in C - C_\varepsilon = U$, we have $F(q) = q + V_1(q) = f(q) \in C_1$. Note that $F(q) = F_1(q) \in C_1$ if $q \in C - C_{4\varepsilon}$. If $q \in C_{4\varepsilon+\delta}$, then $F(q)$ is in C and hence C_1 because $d(q, \partial C) > 4\varepsilon$ while we know that $\|V_4(q)\| = \|V_3(q)\| \leq 2\varepsilon/3$. Similarly, $\|V_4(q)\| < 4\varepsilon$ for $q \in C_{4\varepsilon} - C_{4\varepsilon+\delta}$ and thus $F(q) \in C \subset C_1$. The map F is now easily seen to have all the required properties. □

The next result will furnish us with a method for smoothly "cancelling" a fixed point in the interior of a manifold by means of a suitably related fixed point on the boundary. The construction in the proof consists of a smooth local modification of the map near the boundary fixed point to produce an interior fixed point that we will use subsequently to cancel the given interior fixed point.

THEOREM 7.3. *Let $f : (M, \partial M) \to (M, \partial M)$ be a smooth map which is an extension of $\phi : \partial M \to \partial M$. Suppose $p \in \partial M$ is an isolated fixed point of f, with $i(M, f, p) = 0$, such that ϕ is transversely fixed at p. We write $i(\partial M, \phi, p) = \alpha (= \pm 1)$. Let U be a given neighborhood of p in M containing no other fixed point of f. Then f may be homotoped moving only points on U to a smooth map $F : (M, \partial M) \to (M, \partial M)$ so that the map F has one fixed point q_1 in $U - \partial M$, F is transversally fixed at q_1, and $i(M, F, p) = \alpha$ while $i(M, F, q_1) = -\alpha$. Furthermore, p and q_1 are in the same fixed point class of F.*

Proof. Since all constructions will be local, in a neighborhood of p contained in U we can choose coordinates $(\mathbf{x}, x_n)$ where $\mathbf{x} \in \mathbf{R}^{n-1}$ and $x_n \in \mathbf{R}$ so that $\partial M = \{(\mathbf{x}, 0)\}$, $M = \{(\mathbf{x}, x_n) : x_n \leq 0\}$, and p corresponds to $(\mathbf{0}, 0)$. We write $(\mathbf{x}(q), x_n(q))$ for the coordinates of a point $q \in M$. Define a map F_1 in a neighborhood of p, small enough so that its image under f still lies in the coordinate neighborhood, by $F_1(q) = (\mathbf{x}(f(q)), 0)$. Clearly p is an isolated fixed point of F_1. Moreover, $F_1(q) = f(q)$ for $q \in \partial M$ in a neighborhood of p. It is easy to see that $i(M, F_1, p) = \alpha$, since the restriction of F_1 to the lower hemisphere around p is homotopic to the suspension of the restriction of f to the boundary of the hemisphere. Now choose neighborhoods V and W of p in M, with V contained in W, where W is contained in the lower half of the unit ball in the $(\mathbf{x}, x_n)$ coordinate neighborhood and it contains no fixed point of f other than p. We choose also a smooth map $\rho : M \to [0, 1]$ such that $\rho(q) = 1$ for all $q \in V$ and $\rho(q) = 0$ for all $q \in M - W$. Then we define, for all $q \in M$,

$$F_2(q) = (1 - \rho(q))f(q) + \rho(q)F_1(q)$$

where the multiplications and additions are in the sense of $(\mathbf{x}, x_n)$ vector operations. Then, because M is (locally) convex in these coordinates, $F_2 : M \to M$. Also, $F_2(q) = F_1(q)$ for $q \in V$. Now $F_2(q) = f(q)$ for all $q \in \partial M$ near p since $F_1(q) = f(q)$ there. We have $F_2(q) = f(q)$ as well outside the unit ball in the $(\mathbf{x}, x_n)$ coordinates because then $\rho(q) = 0$. Combining these observations, we see that $F_2(q) = f(q)$ for all $q \in \partial M$. Let H denote the lower half of the unit ball in the $(\mathbf{x}, x_n)$ coordinate neighborhood of p. The homotopy property of the fixed point index implies that

$$i(M, F_2, H) = i(M, f, p) = 0$$

(by hypothesis). On the other hand,

$$i(M, F_2, V) = i(M, F_1, V) = i(M, F_1, p) = \alpha.$$

Noting that ∂H is contained in $U - W$, we see that F_2 has no fixed points on $\partial(H - V)$, so the additivity property of the index implies that $i(M, F_2, H - V) = -\alpha$. Finally, since Fix $F_2 \cap \operatorname{int}(H - V)$ is compact and F_2 has no fixed points on V except p, by Lemma 7.2 we can modify F_2 on the convex set H to a map $F : (M, \partial M) \to (M, \partial M)$ that agrees with f on ∂M and has a single transversal fixed point q_1 in int H with $i(M, F, q_1) = -\alpha$. Let ω be an arc in H from p to q_1, then $F(\omega)$ still lies in the $(\mathbf{x}, x_n)$ coordinate neighborhood of p, so p and q_1 are in the same fixed point class. □

COROLLARY 7.4. *Let M be the unit disc D^2. Define $\phi_k : S^1 \to S^1$ by $\phi_k(e^{i\theta}) = e^{ki\theta}$. If $k \leq 0$, then there is a smooth extension $f_k : D^2 \to D^2$ of ϕ_k that has no fixed points on $\operatorname{int}(D^2) = D^2 - S^1$.*

Proof. We will write $re^{i\theta} \in D^2 - O$ in the form $\exp(\ln r + i\theta)$ and define $\Phi_k : D^2 \to D^2$ by setting $\Phi_k(O) = O$ and

$$\Phi_k(\exp(\ln r + i\theta)) = \exp(|k|r^2 \ln r + ki\theta)$$

for $r > 0$, so $\Phi_k \in E^1(\phi_k)$. Since $i(D^2, \Phi_k, q) = 0$ for all $q \in \text{Fix } \phi_k$, it must be that $i(D^2, \Phi_k, O) = 1$. For $k \leq 0$, we have that $i(S^1, \phi_k, 1) = \alpha = 1$. Let U be a neighborhood of 1 in D^2 that contains no other fixed point of Φ_k. By Theorem 7.3 there exists $F_k \in E^1(\phi_k)$ such that Fix $F_k \cap \text{int } D^2 = \{O, q_1\}$ for some $q_1 \in U$ with $i(D^2, F_k, q_1) = -\alpha = -1$. Since from 7.3 we also have that $i(D^2, F_k, O) = i(D^2, \Phi_k, O) = 1$, the additivity property of the index tells us that $i(D^2, F_k, \text{int } D^2) = 0$ and the required map $f_k \in E^1(\phi_k)$, with no fixed points on int D^2, is constructed by Lemma 7.2. □

REMARK. If $k \geq 2$, then the fixed point behavior of smooth extensions of $\phi_k : S^1 \to S^1$ is quite different. In this case, $i(S^1, \phi_k, x) = -1$ for any $x \in \text{Fix } \phi_k$. Let $f \in E^1(\phi_k)$, then f can have only one fixed point class and its index must therefore be $L(f) = 1$. But if the class were representable on S^1, then Theorem 5.1 would imply $L(f) \leq 0$, so $N^1(f \mid \phi) = 1$. Thus we see that when $k \geq 2$, the smooth extensions of the maps ϕ_k must have fixed points on int(D^2). On the other hand, there are continuous extensions without fixed points on int(D^2) by Theorem 4.6 since Fix $\phi_k \neq \emptyset$. This illustrates the fact that the topological and smooth extension fixed point theories are really different and in particular that Theorem 4.6 does not hold in the smooth category. It also furnishes on example, for Theorem 6.2, in which $N(f \mid \phi) \neq N^1(f \mid \phi)$. If $k \leq 0$, we also have an example for Theorem 6.2 where $N^1(f \mid \phi) \neq N(f)$ since then $N^1(f \mid \phi) = 0$ while $N(f) = 1$.

THEOREM 7.5 (*Minimum Theorem*). *If* dim $M \geq 3$ *and* ϕ *is transversally fixed, then for any* $f \in E^1(\phi)$ *there exists a homotopic map* $g \in E^1(\phi)$ *such that* g *has exactly* $N^1(f \mid \phi)$ *fixed points on* int M.

Proof. By Lemma 7.1, we may assume that f has no interior fixed points near ∂M, that $i(M, f, p) = 0$ for all p in the finite set Fix ϕ and that each $p \in \text{Fix } \phi$ is a transversal fixed point of f. In particular, there is a neighborhood U in M of ∂M such that the graph $\Gamma(f \mid U) = \{(q, f(q)) : q \in U\}$ is transversal to the diagonal in $M \times M$. By transversality theory, we can replace f by a perturbation identical to f in a (possibly smaller) neighborhood of ∂M, homotopic to f and with its graph everywhere transversal to the diagonal in $M \times M$. In particular, f is then transversally fixed. Since dim $M \geq 3$, given a fixed point class F, by [6] we can homotope f to replace $\mathsf{F} \cap \text{int } M$ by a single fixed point. And we can do it without changing f near ∂M; see Remark 1 on page 167 of [6]. Thus we now assume that f has the further property that for each of its fixed point classes F, either $\mathsf{F} \cap \text{int } M$ is empty or it consists of a single point p_{F} with the property that $i(M, f, p_{\mathsf{F}}) = i(\mathsf{F})$. If F is representable on ∂M, either $i(\mathsf{F}) = 0$ so $\mathsf{F} = \{p_{\mathsf{F}}\}$ can be removed as in [6], or there exists $\mathsf{F}'_\partial = \{p_1, \dots, p_s\}$ in $\mathsf{F} \cap \partial M$ such that

$$\sum_{j=1}^{s} i(\partial M, \phi, p_j) = i(\mathsf{F}).$$

Applying Theorem 7.3 once for each $p_j \in \mathsf{F}'_\partial$, we obtain $f_1 \in E^1(\phi)$ homotopic to f and a fixed point class $\mathsf{F}_1 = \mathsf{F} \cup \{q_1, \dots, q_s\}$ corresponding to F with $i(M, f_1, p_{\mathsf{F}}) = i(M, f, p_{\mathsf{F}})$ and $i(M, f_1, q_j) = -i(\partial M, \phi, p_j)$ for all $p_j \in \mathsf{F}'_\partial$. Thus we have

$$i(M, f_1, p_{\mathsf{F}}) + \sum_{j=1}^{s} i(M, f_1, q_j) = 0.$$

By the method of [6] we can homotope f_1 to $f_2 \in E^1(\phi)$ in which the fixed point class F_2 corresponding to F_1 has no fixed points in int M. Repeating this step for each fixed point class that is representable on ∂M, we obtain the required map $g \in E^1(f)$ because it has one fixed point in int M for each fixed point class that is not representable on ∂M while every representable class lies entirely in ∂M. □

REMARK. If the map $f \in E^1(\phi)$ given in Theorem 7.5 is of class C^k, for any $k = 1, 2, \ldots, \infty$, then all the constructions in this section can be carried out using C^k maps and so the conclusion of the Minimum Theorem is that the map $g \in E^1(\phi)$ with exactly $N^1(f \mid \phi)$ fixed points is C^k as well.

References

[1] R. Brooks, R. Brown, J. Pak and D. Taylor, *Nielsen numbers of maps of tori*, Proc. Amer. Soc. **52** (1975), 398-400.

[2] R. Brown, *The Lefschetz Fixed Point Theorem*, Scott, Foresman and Co., Glenview, Il., 1971.

[3] T. Gamelin and R. Greene, *Introduction to Topology*, Saunders, 1983.

[4] R. Greene and H. Wu, *C^∞ approximations of convex, subharmonic, and plurisubharmonic functions*, Ann. Sci. École Norm. Sup. (4) **12** (1979), 47-84.

[5] V. Guillemin and A. Pollack, *Differential Topology*, Prentice Hall, 1974.

[6] B. Jiang, *Fixed point classes from a differentiable viewpoint*, in *Fixed Point Theory* (Proceedings, Sherbrooke, Quebec, 1980), Springer-Verlag, Berlin, 1981. (Lecture Notes in Mathematics, vol. 886.)

[7] ____________, *Lectures on Nielsen Fixed Point Theory*, Contemporary Mathematics vol. 14, Amer. Math. Soc., Providence, RI, 1983.

[8] ____________, *On the least number of fixed points*, Amer. J. Math. **102** (1960), 749-763.

[9] ____________, *Fixed points on braids*, Inv. Math. **75** (1984), 69-74.

[10] J. Munkres, *Elementary Differential Topology*, Princeton Univ., 1966.

[11] H. Schirmer, *A relative Nielsen number*, Pacific J. Math. **122** (1986), 459-473.

[12] ____________, *Fixed point sets of deformations of pairs of spaces*, Topology and its Applications, **23** (1986), 193-205.

[13] ____________, *On the location of fixed points on pairs of spaces*, Topology and its Applications, to appear.

[14] M. Shub and D. Sullivan, *A remark on the Lefschetz fixed point formula for differentiable maps*, Topology **13** (1974), 189-191.

[15] E. Spanier, *Algebraic Topology*, McGraw-Hill, 1966.

TWO VIGNETTES IN FIXED POINT THEORY

E. Fadell*

1. THE FIXED POINT PROPERTY AND THEOREMS OF THE BORSUK–ULAM TYPE.

As anyone who has worked on both fixed point theorems and Borsuk–Ulam type theorems will agree, there is a striking similarity in both subjects as to strategy and technique. The object of this section is to give precise evidence of this admittedly vague assertion. We will show that a compact homogeneous space $M = G/K$ has the fixed point property (FPP), i.e., every self-map of M has fixed points if, and only if a certain Borsuk–Ulam theorem holds, which of course varies with M. As usual G is a compact, connected Lie group and K a closed subgroup. $M = G/K$ is the space of left cosets gK and is naturally a left K-space via the action $x(gK) = (xg)K$, $x \in K$. G is also a left K-space using the action $x \circ g = gx^{-1}$, $g \in G$, $x \in K$. Let $x_0 \in M$ denote the point corresponding to the coset K.

Consider the following statements.

<u>Property BU</u>. If $\phi : G \longrightarrow M$ is any K-map, then $Z = \phi^{-1}(x_0)$ is non-empty.

<u>Property FPP</u>. If $f : M \longrightarrow M$ is any map, then $\mathrm{Fix} f = \{x \mid f(x) = x\}$ is non-empty.

We may formulate our result as follows.

1.1 <u>Theorem</u>. M has Property BU if, and only if, M has Property FFP.

*Supported in part by the National Science Foundation under NSF Grant No. DMS-8722295.

Before giving the proof of theorem 1.1 we develop some preliminary material.

G is a free right K-space using right multiplication by K and the projection $p: G \longrightarrow G/K = M$ is a principal K-bundle. $G \times_K M$, where (gk^{-1}, kx) and (g,x), $g \in G$, $x \in M$, $k \in K$, are identified, yields the associated bundle with fiber M and group K. The map $q: G \times_K M \longrightarrow M$ is induced by projection. We note that $M - x_0$ is a K-subset of M and hence $(M, M - x_0)$ is a K-pair. We then have a (locally trivial) bundle pair $G \times_K (M, M - x_0)$.

Now consider the (locally trivial) fibered pair $(M \times M,\ M \times M - \Delta,\ \pi,\ M)$ where Δ is the diagonal in $M \times M$ and π is projection on the first factor [1]. The fiber pair over x_0 is $(M, M - x_0)$.

1.2 Proposition. There is a fiber-preserving homeomorphism of pairs

$$\begin{array}{ccc} G \times_K (M, M - x_0) & \xrightarrow{\ \alpha\ } & (M \times M,\ M \times M - \Delta) \\ \Big\downarrow q & & \Big\downarrow \pi \\ M & \xrightarrow{\ id\ } & M \end{array}$$

Proof. Define α by $\alpha[g,y] = (gK, gy)$, $g \in G$, $y \in M$. Note that $\alpha[gk^{-1}, ky] = [gk^{-1}K, gk^{-1}ky] = [gK, gy]$ so that α is well-defined. If $y = vK \neq x_0$, $gK \neq gvK$ since $v \notin K$. Thus $\alpha(G \times_K (M - x_0)) \subset M \times M - \Delta$. α is clearly fiber-preserving and if we define $\beta(x,y) = [g, g^{-1}y]$, where $x = gK$, then β serves as a well-defined inverse for α and α is a homeomorphism.

Since sections σ in the bundle $M \times M \longrightarrow M$ are of the form $\sigma(x) = (x, f(x))$, where $f: M \longrightarrow M$ is a map, we have the following corollary.

1.3 Corollary. α induces a bijection of the set of sections of the bundle $G \times_K M \longrightarrow M$ to the set of maps $f: M \longrightarrow M$. In particular, α also induces a bijection from the sections of the subbundle $G \times_K (M - x_0)$ to the set of fixed point free maps $f: M \longrightarrow M$, i.e. $f(x) \neq x$ for all $x \in M$.

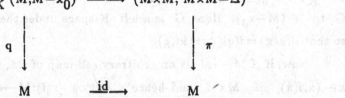

The sections of $G \times_K M$ are well understood as follows. Given a section $\sigma: M \longrightarrow G \times_K M$, and $g \in G$, $\sigma(gK) = [g,y]$, $y \in M$, i.e. $\sigma(gK)$ has a unique representative of the form $[g,y]$. Set $y = f(x)$ thus defining a map $\phi: G \longrightarrow M$. Note that

$$\sigma(gk^{-1}K) = [gk^{-1}, ky]$$

so that $\phi(gk^{-1}) = k\phi(g)$. Thus, $\phi: G \longrightarrow M$ is a K-map where K acts on the left of G by $k \circ g = gk^{-1}$, $k \in K$, $g \in G$. Conversely, given a K-map $\phi: G \longrightarrow M$, define a section $\sigma_\phi: M \longrightarrow G \times_K M$ by

$$\sigma_\phi(gK) = [g, \phi(g)] .$$

The correspondence $\phi \mapsto \sigma_\phi$ defines a bijection from the set of K-maps $\phi: G \longrightarrow M$ to the set of sections in $G \times_K M \longrightarrow M$. We observe also that σ_ϕ is a section in the subbundle $G \times_K M - x_0 \longrightarrow M$ if, and only if, $\phi(G) \subset M - x_0$. Summarizing:

1.4 Proposition. The sections of $G \times_K M \longrightarrow M$ $(G \times_K (M - x_0) \longrightarrow M)$ are in bijective correspondence with the K-maps of G to M $(M - x_0)$. Here G is a left K-space under the action $k \circ g = gk^{-1}$ so that $f(k \circ g) = f(gk^{-1}) = kf(g)$.

Now, if $f: M \longrightarrow M$ is an arbitrary self-map of M, f defines a section $x \mapsto (x, f(x))$ in $M \times M$ and hence a section $\sigma(f): M \longrightarrow G \times_K M$. $\sigma(f)$ in turn determines a K-map $\phi(f): G \longrightarrow M$.

1.5 Theorem. (a) $f: M \longrightarrow M$ is fixed point free if, and only if, $\phi(f)$ takes G into $M - x_0$.

(b) f is deformable into a fixed point free map if, and only if, the K-map $\phi(f): G \longrightarrow M$ is K-deformable into $M - x_0$.

(c) M has the fixed point property if, and only if every K-map $G \longrightarrow M$ covers $x_0 \in M$, i.e., M has Property FPP if, and only if, M has Property BU.

1.6 Remark. Theorem 1.1 is (c) of theorem 1.5.

A study of some consequences of this result will be the objective of a future work. We content ourselves here with some elementary consequences. First, it is useful to describe the correspondence Maps(M,M) to K-maps(G,M) explicitly.

The map $x \mapsto f(x)$ first corresponds to the graph $x \mapsto (x, f(x))$;

which corresponds to the section $x = gK \mapsto [g, g^{-1}f(gK)]$; which in turn corresponds to the K-map $\phi(f): G \longrightarrow M$ defined by

$$\phi(f)(g) = g^{-1}f(gK), \quad g \in G.$$

On the other hand, given the K-map $\phi: G \longrightarrow M$ the corresponding map $f: M \longrightarrow M$ is easily seen to be given by

$$f(x) = f(gK) = g\phi(g), \quad x = gK \in M.$$

We observe that the only K-maps $\phi: G \longrightarrow M$ which are constant must require $\phi(G) \in M^K$. Let us identify M^K, the fixed points under the K-action.

$kgK = gK$ for all $k \in K$ if, and only if, $g^{-1}Kg \subset K$. Thus, $M^K = \{gK: g \in NK\}$, where NK is the normalizer of K in G.

1.7 <u>Proposition</u>. A necessary condition that M have property FPP is that $NK = K$, i.e. K is self-normalizing.

<u>Proof</u>. If $NK - K \neq \phi$, take $g_0 \in NK - K$ and the constant K-map $\phi: G \longrightarrow M$ which takes G to g_0K. Then, $x_0 \notin \phi(G)$ and hence the corresponding map (given by theorem 1.5) $f: M \longrightarrow M$ is fixed point free, namely $f(gK) = gg_0K$.

1.8 <u>Corollary</u>. If $W(K) = NK/K$ is non-trivial, M admits a fixed point free map.

2. A UNIVERSAL FORMULA FOR RELATIVE LEFSCHETZ NUMBERS.

Let M denote an ANR (metric) and $\mathbb{F}$ a field. Let us assume that $H_q(M;\mathbb{F})$ is finitely generated for each q and vanishes for q sufficiently large. Let $P_n(M)$ denote the n-fold product of M with itself and $\overline{\Delta}$ the "fat" diagonal in $P_n(M)$, i.e.,

$$\overline{\Delta} = \{(x_1, \ldots, x_n) \in P_n(M) | x_i = x_j \text{ for some } i \neq j\}.$$

If $f: M \longrightarrow M$ is a map, f induces an n-fold product map $f^n = f \times f \times \cdots \times f: P_n(M) \longrightarrow P_n(M)$. We let $f^n_{\overline{\Delta}}$ denote $f^n|\overline{\Delta}$ and $(f^n, f^n_{\overline{\Delta}})$ the map of pairs $(f^n, f^n_{\overline{\Delta}}): (P_n(M), \overline{\Delta}) \to (P_n(M), \overline{\Delta})$. Our objective is to prove the following result.

2.1 <u>Theorem</u>. For any M (as above) and any self-map $f: M \longrightarrow M$, the relative Lefschetz number of the pair $(f^n, f^n_{\overline{\Delta}})$ is given by

$$L(f^n, f^n_{\overline{\Delta}}) = L(f)(L(f)-1)(L(f)-2)\cdots(L(f)-n+1)$$

where $L(f)$ is the Lefschetz number of f.

2.2 Remark. This result is also valid for metric spaces if one employs Alexander–Spanier or Čech cohomology over F and this cohomology is finitely generated so that $L(f)$ is defined. We will employ singular homology in the proof of theorem 2.1.

We first consider some preliminaries.

Let X denote an ANR (metric) where $X = X_1 \cup X_2 \cup \cdots \cup X_n$ and each X_i is a closed subset and all intersections of the form $X_{i_1} \cap X_{i_2} \cap \cdots \cap X_{i_k}$ are ANR (metric). Let $f: X \longrightarrow X$ denote a self map such that $f(X_i) \subset X_i$. For indices $i_1 < i_2 < \cdots < i_k$, set $f(i_1 \cdots i_k) = f|(X_{i_1} \cap X_{i_2} \cap \cdots \cap X_{i_k})$.

2.3 Lemma. $L(f) = \sum_{k=1}^{n} \sum_{i_1 < \cdots < i_k} (-1)^{k+1} L(f(i_1 \cdots i_k))$.

Proof. The case $n = 2$ yields

$$L(f) = L(f(1)) + L(f(2)) - L(f(12))$$

which starts the induction and the inductive step is a simple exercise.

2.4 Proposition. Let $f: M \longrightarrow M$ denote the map in theorem 2.1 and $f^n_{\overline{\Delta}}: \overline{\Delta} \longrightarrow \overline{\Delta}$ the induced map on the fat diagonal $\overline{\Delta}$. Then, the Lefschetz number $L(f^n_{\overline{\Delta}})$ is a polynomial in $L(f)$ of degree $n-1$ with coefficients independent of the space M.

Proof. Let $g = f^n_{\overline{\Delta}}: \overline{\Delta} \longrightarrow \overline{\Delta}$. If we set

$$\Delta_{ij} = \{(x_1, \ldots, x_n) \in P_n(M) \mid x_i = x_j\}, \quad i < j,$$

then, $\overline{\Delta} = \bigcup_{i<j} \Delta_{ij}$. Let K denote the set of all ordered pairs ij, $i<j$. Order K in an arbitrary way, say $\alpha_1, \ldots, \alpha_m$, $m = n(n-1)/2$ and set $X_j = \Delta_{\alpha_j}$. Then, $\overline{\Delta} = \bigcup_{j=1}^{m} X_j$ and applying Lemma 2.3,

$$L(g) = \sum_{k=1}^{m} \sum_{i_1 < \cdots < i_k} (-1)^{k+1} L(g(i_1 \cdots i_k)).$$

The set $X_{i_1} \cap \cdots \cap X_{i_k}$ is homeomorphic to an q-fold product $(q = q(i_1 \cdots i_k))$ of spaces M, $q < n$, and hence $L(g(i_1 \cdots i_k)) = [L(f)]^q$. Thus, our assertion follows.

2.5 Corollary. $L(f^n, f^n_{\Delta}) = L(f)^n - L(f^n_{\bar{\Delta}})$ and hence is a monic polynomial of degree n in $L(f)$, with coefficients independent of M.

Proof of theorem 2.1. Let M denote a finite set of cardinal greater than $n+1$. Let $f_j : M \longrightarrow M$ denote a permutation which fixes exactly j points, $j = 0, \ldots, n-1$. Then, recall that the n-th configuration space $F_n(M)$ is the set of n-tuples $(x_1, \ldots, x_n) \in P_n(M)$ such that $x_i \neq x_j$ for $i \neq j$. f^n_j has no fixed points in $F_n(M) = P_n(M) - \bar{\Delta}$ and $L(f_j) = j$. Let $Q(x)$ denote the polynomial of degree n obtained in Corollary 2.5 where $Q(L(f)) = L(f^n, f^n_{\Delta})$. Since, there are no fixed points in $F_n(M)$, $L(f^n, f^n_{\Delta}) = 0$. Hence $x = 0, \ldots, n-1$ are roots of $Q(x)$ so that

$$Q(x) = x(x-1) \cdots (x-n+1)$$

and

$$L(f^n, f^n_{\Delta}) = L(f)(L(f)-1) \cdots (L(f)-n+1)\ .$$

BIBLIOGRAPHY

1. E. Fadell, Generalized Normal Bundles for Locally-Flat Imbeddings, Trans. Amer. Math. Soc. 114 (1965), 488–513.

University of Wisconsin, Madison, U.S.A.

INDEX THEORY FOR NONCOMPACT GROUP ACTIONS WITH APPLICATIONS TO BORSUK–ULAM THEOREMS

E. Fadell* and S. Y. Husseini*

1. INTRODUCTION.

The study of Borsuk–Ulam theorems, originally limited to $\mathbb{Z}_2$ actions, has received considerable attention recently for more general compact Lie groups G (see, e.g., [1], [2], [3]). The classifying space BG plays a crucial role, especially when one employs the Borel cohomology of G-spaces. More specifically, the cohomology of BG plays a crucial role in the cohomological index theory of G-spaces where, for example the index of a G-space X is an ideal in $H^*(BG)$. If one is to consider Borsuk–Ulam type theorems for G-spaces X when G is not compact numerous difficulties occur. One of them is the fact that BG may in fact be acyclic. However, in this case, one may employ the de Rham model for equivariant cohomology for smooth G-manifolds based on the work of H. Cartan, A. Weil, et al. (see, e.g. [4]). Our objective here is to begin the study of such noncompact situations, develop the necessary "infinitesimal" index theory and apply it to a Borsuk–Ulam theorem with symmetry group $G = \mathbf{C}$, the additive group of complex numbers.

2. PRELIMINARIES ON G–DE RHAM COHOMOLOGY.

We set down in this section the material we will require in subsequent sections concerning the de Rham model for equivariant cohomology of smooth G-manifolds based on the work of H. Cartan, C. Chevalley, J.-L. Koszul and A. Weil. An excellent introduction to the subject may be found in Atiyah–Bott [5]. A useful original source is also Cartan [4]. We will also make use of references to the 3 volume work of Greub, Halperin and Vanstone [6].

In this section M will denote a smooth manifold and G a Lie group (not necessarily compact) which acts (on the right) on M so that M is a smooth G-manifold. $\mathfrak{G}$ will denote the Lie algebra of G, $\mathfrak{G}^*$ its dual. $\Lambda\mathfrak{G}^*$ is the exterior algebra on the dual $\mathfrak{G}^*$ and $S\mathfrak{G}^*$ is the symmetric algebra generated by $\mathfrak{G}^*$. $S\mathfrak{G}^*$ is graded by assigning degree 2 to elements of $\mathfrak{G}^*$. The Weil algebra

$$W(\mathfrak{G}) = \Lambda\mathfrak{G}^* \otimes S\mathfrak{G}^*$$

serves as a model for the contractible universal G-space EG. Furthermore, the so-called "basic subcomplex" $B\mathfrak{G}$ serves as a model for the classifying space BG. In fact, when G is compact and connected $B\mathfrak{G}$ and $H^*(BG;\mathbb{R})$ are naturally isomorphic. However, $B\mathfrak{G}$ and $H^*(BG;\mathbb{R})$ may differ in the noncompact case. In fact, when G is contractible and non-compact, BG is contractible and $B\mathfrak{G}$ must be employed in order to obtain the results in this paper.

It is useful to describe "basic complex" as well as the model for equivariant de Rham cohomology which employs $W(\mathfrak{G})$ by recalling the notion of an <u>operation</u> of a Lie algebra in a graded differential algebra [6]. The base field of scalars will be the real field $\mathbb{R}$.

2.1 <u>Definition</u>. An operation of a Lie algebra E in a graded differential algebra R is a system $(E, R, \delta, i, \theta)$ where:

1) E is a finite dimensional Lie algebra and (R, δ) is a graded commutative algebra, $R = \sum_{p \geq 0} R^p$, with differential δ, degree $\delta = +1$.

2) θ is a representation of E in R, i.e., for each $x \in R$, $\theta(x)\colon R \to R$ is a derivation of degree 0 $[\theta(x)(ab) = (\theta(x)(a)b + a(\theta(x)b)]$, and

$$\theta[x,y] = \theta(x)\theta(y) - \theta(y)\theta(x), \qquad x, y \in E.$$

3) i is a linear map from E to the space of antiderivations of R of degree -1, i.e. for $x \in E$, $i(x)\colon R^q \longrightarrow R^{q-1}$ and

$$i(x)(ab) = (i(x)a)b + (-1)^{|a|}a(i(x)b).$$

4) The following relations hold for x,y in E

a) $i(x)^2 = 0$

b) $i[x,y] = \theta(x)i(y) - i(y)\theta(x)$

c) $\theta(x) = i(x)\delta + \delta i(x)$.

Note that 4c) implies $\delta\theta(x) = \theta(x)\delta$ for all $x \in E$.

Given any operation $(E, R, \delta, i, \theta)$ we have certain subalgebras of R as follows. The horizontal subalgebra of R, denoted by $R_{i=0}$, where

$$R_{i=0} = \bigcap_{x \in E} \ker i(x);$$

the invariant subalgebra, denoted by $R_{\theta=0}$, where

$$R_{\theta=0} = \bigcap_{x \in E} \ker \theta(x);$$

and the basic subalgebra,

$$R_{i=0,\theta=0} = R_{i=0} \cap R_{\theta=0}.$$

It should be noted that while the invariant and basic subalgebras are stable with respect to δ, the horizontal algebra is not, in general.

Here are some examples which we will encounter.

1. The exterior algebra operation $(E, \Lambda E^*, \delta, i, \theta)$ where E is a finite dimensional Lie algebra and ΛE^* is the exterior algebra on its dual E^*. The "substitution operator" $i(x)$, $x \in E$, is the extension of $i(x): E^* \to \mathbf{R}$ given by $i(x)\phi = \langle x, \phi\rangle$. The "infinitesimal transformation" $\theta(x)$ is the extension of $\theta(x): E^* \to E^*$ defined by $\theta(x) = -(\mathrm{ad}\, x)^*$ where $\mathrm{ad}\, x: E \to E$ is given by $(\mathrm{ad}\, x)y = [x,y]$, $x,y \in E$. The coboundary $\delta: \Lambda E^* \to \Lambda E^*$ is defined by $\delta: E^* \to \Lambda^2 E^*$ where $(\delta\phi)(x \wedge y) = \phi[x,y]$, $x,y \in E$. δ also is given by the "Kozul formula"

$$\delta = \frac{1}{2} \sum_k \mu(x_k')\theta(x_k)$$

where $\{x_k\}$ and $\{x_k'\}$ are dual bases for E and E^* and $\mu(x_k')$ is left multiplication by x_k'.

2. The Weil algebra of the Lie algebra E, $W(E) = (E, \Lambda E^* \otimes SE^*, \delta, i, \theta)$. SE^* denotes symmetric algebra on E^*, with generators of degree 2, so that SE^* is evenly graded. Take for $i(x)$ the unique extension to $\Lambda E^* \otimes SE^*$ which is $i(x)$ from example 1 on ΛE^* and 0 on SE^*. $\theta(x)$ will also be an extension of $\theta(x)$ on ΛE^*. Let u_k

be a basis for SE^*, $\dim u_k = 2$, and $h: E^* \to S^2E^*$ defined by $h: x'_k \mapsto u_k$. Set $\theta(x)u_k = h(\theta(x)x'_k)$. This defines $\theta(x)$ on S^*E and hence on $\Lambda E^* \otimes S^*E$. The differential operator is defined explicitly by

$$\delta = \delta_E + \delta_S + h$$

where

$$h = \sum_k i(x_k)\mu(h(x'_k))$$

$$\delta_E = \frac{1}{2} \sum_k \mu(x'_k)\theta_E(x_k)$$

$$\delta_S = \sum_k \mu(x'_k)\theta_S(x_k)$$

and $\theta(x) = \theta_E + \theta_S$, with $\theta_E = \theta(x)$ on ΛE^* and 0 on SE^* and vice versa for θ_S.

If we also use the notation $W(E) = \Lambda E^* \otimes SE^*$, then $H^0(W(E)) = \mathbb{R}$ and $H^q(W(E)) = 0$ for $q \neq 0$ ([4]). Furthermore,

$$(\Lambda E^* \otimes SE^*)_{i=0,\theta=0} \simeq (SE^*)_{\theta=0}$$

and both serve as the algebraic analogue of BG, when $E = \mathfrak{G}$ the Lie algebra of G and we denote both by $B\mathfrak{G}$.

When E is abelian, $\theta(x) = 0$ for $x \in E$, δ reduces to h and $[SE^*]_{\theta=0} = SE^*$.

3. Let G denote a Lie group, M a smooth manifold and $M \times G \to M$ a smooth action. $\Omega(M)$ will denote the differential forms on M. This situation gives rise to an operation $(\mathfrak{G}, \Omega(M), \delta, i, \theta)$, where $\mathfrak{G}$ is the Lie algebra of G. δ is the usual derivation for the de Rham complex $\{\Omega(M), \delta\}$, $\theta(x)\colon \Omega(M) \to \Omega(M)$ is the Lie derivative restricted to vector fields on M induced by $x \in \mathfrak{G}$ and $i(x)(\omega) = \omega(h)$, where h is the vector field on M induced by $x \in \mathfrak{G}$. This is the classical diffeogeometric situation.

The invariant differential forms in $\Omega(M)$ are those $\omega \in \Omega(M)$ such that $T_a^*\omega = \omega$ where $T_a\colon M \to M$ is given by $T_a x = xa$, $a \in G$. If we let $\Omega_I(M)$ denote the subalgebra of invariant forms then $\Omega_I(M) \subset \Omega(M)_{\theta=0}$ and when G is connected $\Omega_I(M) = \Omega(M)_{\theta=0}$ (see [6]).

When a connected G acts freely on M, and the natural projection $p\colon M \longrightarrow M/G$ is locally trivial, i.e. we have a locally trivial principal bundle, then

$$p^*:\Omega(M/G) \longrightarrow \Omega(M)_{i=0,\theta=0}$$

is an isomorphism. The forms in $\Omega(M)_{i=0}$ are called horizontal forms, so that the "basic subalgebra" of $\Omega(M)$ consists of invariant, horizontal forms.

4. The infinitesimal de Rham complex, $\{\mathfrak{G},\Omega(M)\otimes W(G),\delta,\theta,i\}$. the operators δ,θ,i have already been defined in examples 2 and 3 on the subalgebras $\Omega(M)$ and $W(G)$. They have unique extensions to $\Omega(M)\otimes W(G)$. The basic subcomplex of $\Omega(M)\otimes W(G)$ is denoted by $\Omega_{\mathfrak{G}}(M)$, i.e.

$$\Omega_{\mathfrak{G}}(M) = [\Omega(M)\otimes W(G)]_{i=0,\theta=0}$$

and we set

$$H^*_{\mathfrak{G}}(M) = H^*(\Omega_{\mathfrak{G}}(M)) \ .$$

When, G is compact and connected $H^*_{\mathfrak{G}}(M)$ may be identified with the Borel cohomology $H^*_G(M) = H^*_G(EG\times_G M)$ of M [see [5]). $H^*_{\mathfrak{G}}$ is an equivariant cohomology theory for the category of smooth G-manifolds and maps. If $f:M\longrightarrow N$ is a smooth G-map, then

$$H^*_{\mathfrak{G}}(f): H^*_{\mathfrak{G}}(N) \longrightarrow H^*_{\mathfrak{G}}(M)$$

is induced by $f^*:\Omega(N)\longrightarrow \Omega(N)$. We will refer to $H^*_{\mathfrak{G}}$ as "infinitesimal" equivariant cohomology.

Subsequent calculations will require a slightly more general naturality property of the functor $H^*_{\mathfrak{G}}$. Let $\phi: G\longrightarrow \overline{G}$ denote a homomorphism of Lie groups, M and N smooth G and $\overline{G}$-manifolds respectively. Suppose $f: M\longrightarrow N$ is an equivariant map in the sense that $f(xg) = f(x)\phi(g)$. Then ϕ induces $\phi^*:\overline{\mathfrak{G}}^*\longrightarrow \mathfrak{G}^*$, a homomorphism of the duals of the corresponding Lie algebras and, as usual, f induces $f^*:\Omega(N)\longrightarrow\Omega(M)$. Thus, we have induced

$$(f,\phi)_{\#}:\Omega(N)\otimes W(\overline{\mathfrak{G}}) \longrightarrow \Omega(M)\otimes W(\mathfrak{G})$$

and

$$H^*_{\mathfrak{G}}(f,\phi): H^*_{\overline{\mathfrak{G}}}(N) \longrightarrow H^*_{\mathfrak{G}}(M) \ .$$

5. Examples 3 and 4 have their relative counterparts, for smooth manifold pairs (M,A).

3. INFINITESIMAL INDEX THEORY.

If M is a smooth G-manifold and

$$\Omega_{\mathfrak{G}}(M) = [\Omega(M)\otimes W(G)]_{i=0,\theta=0}$$

is the basic subcomplex of $\Omega(M)\otimes W(G)$, as in Section 2, the inclusion map $j_M: W(G)\longrightarrow\Omega(M)\otimes W(G)$ $(u\to 1\otimes u)$ induces

$$\tilde{j}_M: B\mathfrak{G} \longrightarrow \Omega_{\mathfrak{G}}(M),$$

called the classifying map for the G-space M and this in turn induces

$$j_M^*: B\mathfrak{G} \longrightarrow H_{\mathfrak{G}}^*(M),$$

since $\delta = 0$ on $S(\mathfrak{G}^*)_{\theta=0}$.

3.1 Definition. The infinitesimal G-index of M, $\mathrm{Index}_{\mathfrak{G}}M$, is the kernel of j_M^*. In particular, $\mathrm{Index}_{\mathfrak{G}}M$, is ideal in the ring $B\mathfrak{G}$. Alternatively, $H_{\mathfrak{G}}^*(M)$ is a module over $B\mathfrak{G}$, using j_M^*, and $\mathrm{Index}_{\mathfrak{G}}M = \mathrm{Annih}\, H_{\mathfrak{G}}^*(M)$.

We may extend this concept to any G-subset X as follows. Let U denote the directed set of all open G-sets U, $M\supset U\supset X$.

3.2 Definition. Set $\mathrm{Index}_{\mathfrak{G}}X = \varinjlim_{U} \mathrm{Index}_{\mathfrak{G}}U$.

3.3 Remark. At this stage $\mathrm{Index}_{\mathfrak{G}}X$ depends on the ambient space M.

3.4 Proposition. If $B\mathfrak{G}$ is Noetherian, there is an open G-set V_0 such that, $V_0 \supset X$ and for every open G-set U, $X\subset U\subset V_0$

$$\mathrm{Index}_{\mathfrak{G}}X = \mathrm{Index}_{\mathfrak{G}}U.$$

Proof. Let Σ denote the ideals $\mathrm{Index}_{\mathfrak{G}}W$ for any open G-set $W\supset X$. Then Σ contains a maximal element $\mathrm{Index}_{\mathfrak{G}}V_0$, i.e. $\mathrm{Index}_{\mathfrak{G}}V_0 \supset \mathrm{Index}_{\mathfrak{G}}W$ for arbitrary W. If U is an open G-set contained in V_0, then $\mathrm{Index}_{\mathfrak{G}}U \supset \mathrm{Index}_{\mathfrak{G}}V_0$ and hence $\mathrm{Index}_{\mathfrak{G}}U = \mathrm{Index}_{\mathfrak{G}}V_0$ for all U, $X\subset U\subset V_0$. Hence

$$\mathrm{Index}_{\mathfrak{G}}X = \varinjlim_{W} \mathrm{Index}_{\mathfrak{G}}W = \mathrm{Index}_{\mathfrak{G}}V_0 = \mathrm{Index}_{\mathfrak{G}}U$$

for $X\subset U\subset V_0$.

3.5 Proposition. If $f: M\longrightarrow N$ is a smooth G-map, $X\subset M$ and $Y\subset N$ are G-subsets such that $f(X)\subset Y$, and $B\mathfrak{G}$ is Noetherian, then

$$\mathrm{Index}_{\mathfrak{G}}X \supset \mathrm{Index}_{\mathfrak{G}}Y.$$

Proof. Choose open G-sets V_0 and V'_0 such that $X \subset V_0$, $Y \subset V'_0$ satisfying Proposition 3.4 so that

$$\mathrm{Index}_{\mathfrak{G}} V_0 = \mathrm{Index}_{\mathfrak{G}} X, \ \mathrm{Index}_{\mathfrak{G}} V'_0 = \mathrm{Index}_{\mathfrak{G}} Y .$$

Then, there is an open G-set U, $X \subset U \subset V_0$ such that $f(U) \subset V'_0$. Then, the diagram

$$\begin{array}{ccc} H^*_{\mathfrak{G}}(U) & \xleftarrow{f^*} & H^*_{\mathfrak{G}}(V'_0) \\ {\scriptstyle j^*_U}\uparrow & & \uparrow{\scriptstyle j^*_{V'_0}} \\ B\mathfrak{G} & \xleftarrow{id} & B\mathfrak{G} \end{array}$$

implies $\mathrm{Index}_{\mathfrak{G}} V'_0 \supset \mathrm{Index}_{\mathfrak{G}} U$, which gives the desired result.

3.6 Proposition. If $B\mathfrak{G}$ is Noetherian and $A \cup B \subset M$ where A and B are G-sets, then

$$(\mathrm{Index}_{\mathfrak{G}} A) \cdot (\mathrm{Index}_{\mathfrak{G}} B) \subset \mathrm{Index}_{\mathfrak{G}} A \cup B$$

or, in terms of ideal quotients

$$\mathrm{Index}_{\mathfrak{G}} B \subset (\mathrm{Index}_{\mathfrak{G}} A \cup B : \mathrm{Index}_{\mathfrak{G}} A).$$

Proof. Consider the maps $j^*_A : B\mathfrak{G} \longrightarrow H^*_{\mathfrak{G}}(A)$, $j^*_B : B\mathfrak{G} \longrightarrow H^*_{\mathfrak{G}}(B)$, $j^*_{A\cup B} : B\mathfrak{G} \longrightarrow H^*_{\mathfrak{G}}(A \cup B)$ induced by the classifying maps $\bar{j}_A, \bar{j}_B, \bar{j}_{A\cup B}$. If $j^*_A(\lambda_1) = 0$, $j^*_B(\lambda_2) = 0$. $j^*_{A\cup B}(\lambda_1)$ pulls back to $\lambda'_2 \in H^*_{\mathfrak{G}}(A \cup B, B)$. $\lambda'_1 \lambda'_2 = 0$ easily implies $j^*_{A\cup B}(\lambda_1 \lambda_2) = 0$.

We state, without proof, the following consistency result.

3.7 Proposition. If A is a smooth G-submanifold of M, then if $\mathfrak{G}$ is Noetherian, $\mathrm{Index}_{\mathfrak{G}} A$ is the same whether we use $\Omega_{\mathfrak{G}}(A)$ or Definition 3.2.

3.8 Remark. Propositions 3.4, 3.5, and 3.6 are called, respectively, continuity, monotonicity, and additivity.

3.9 Remark. Since the requirement that $B\mathfrak{G}$ be Noetherian is important in the above propositions, a few remarks are in order. First, when $\mathfrak{G}$ is abelian $B\mathfrak{G} = S(\mathfrak{G}^*)_{\theta=0} = S(\mathfrak{G}^*)$ and $B\mathfrak{G}$, being a polynomial algebra on k generators ($k = \dim \mathfrak{G}$) is Noetherian. More generally, when $\mathfrak{G}$ is reductive, it follows from results in [6] (Chapter IV, §4), that $B\mathfrak{G}$ is Noetherian in this case. However, $B\mathfrak{G}$ may be Noetherian without $\mathfrak{G}$

being reductive. For example, let $\mathfrak{G}$ denote the "so-called" Heisenberg algebra with generators $X_1,\ldots,X_n,Y$ with Lie products $[X_i,X_j]=Y$ for $i<j$ and $[X_i,Y]=0$ for all i. Then, the center $Z(\mathfrak{G})$ of $\mathfrak{G}$ is one dimensional, namely, $Z(\mathfrak{G})=\langle Y\rangle$ and $\mathfrak{G}$ is one step nilpotent. Thus, $\mathfrak{G}$ is not reductive and since $S(\mathfrak{G}^*)_{\theta=0}$ is a polynomial algebra on n generators, $B\mathfrak{G}$ is Noetherian.

We next consider in this setting general theorems of the Borsuk–Ulam and Bourgin–Yang type. The setting is as follows: M and N are smooth G-manifolds; G a Lie group, and $f: M \longrightarrow N$ is a smooth G-map. Z' is a closed G-set in N and $Z=f^{-1}(Z')$. The following theorem is immediate using Proposition 3.5.

3.10 Theorem. (BU-type). If $B\mathfrak{G}$ is Noetherian and $\mathrm{Index}_{\mathfrak{G}}(M)\not\supset \mathrm{Index}_{\mathfrak{G}}(N-Z')$, then $Z\neq\phi$.

3.11 Theorem. (BY–type) If $B\mathfrak{G}$ is Noetherian, then

$$\mathrm{Index}_{\mathfrak{G}}Z \subset (\mathrm{Index}_{\mathfrak{G}}M\colon \mathrm{Index}_{\mathfrak{G}}(N-Z')).$$

Proof. Let U denote an open G-set such that $\mathrm{Index}_{\mathfrak{G}}U = \mathrm{Index}_{\mathfrak{G}}Z$, employing Proposition 3.4 (continuity). Let V denote an open G-set such that $Z\subset V\subset \bar{V}\subset U$. Then $M=U\cup W$, $W=M-\bar{V}$ and by Proposition 3.6 (additivity),

$$\mathrm{Index}_{\mathfrak{G}}U \subset (\mathrm{Index}_{\mathfrak{G}}M\colon \mathrm{Index}_{\mathfrak{G}}W)\,.$$

But by Proposition 3.5 (monotonicity),

$$\mathrm{Index}_{\mathfrak{G}}W \supset \mathrm{Index}_{\mathfrak{G}}(N-Z').$$

Therefore,

$$\mathrm{Index}_{\mathfrak{G}}Z \subset (\mathrm{Index}_{\mathfrak{G}}M\colon \mathrm{Index}_{\mathfrak{G}}(N-Z'))\,.$$

The computation of $\mathrm{Index}_{\mathfrak{G}}M$ invariably requires the following spectral sequence. Recall [6] that

$$\Omega_{\mathfrak{G}}(M) = [\Omega(M)\otimes W(G)]_{i=0,\theta=0} \cong [\Omega(M)\otimes S\mathfrak{G}^*]_{\theta=0}$$

and the natural inclusion $S\mathfrak{G}^* \longrightarrow \Omega(M)\otimes S\mathfrak{G}^*$ induces

$$B\mathfrak{G} = S\mathfrak{G}^*_{\theta=0} \longrightarrow [\Omega(M)\otimes S\mathfrak{G}^*]_{\theta=0}$$

which is identified with the "classifying map"

$$j_M\colon B\mathfrak{G} \longrightarrow \Omega_{\mathfrak{G}}(M).$$

$[\Omega(M)\otimes S\mathfrak{G}^*]_{\theta=0}$ can be filtered on the second factor as follows (regard

$S\mathfrak{G}^*$ as the "base" and $\Omega(M)$ as the "fiber"):

$$F^p = [\Omega(M) \otimes \sum_{n \geq p} S^n \mathfrak{G}^*]_{\theta=0} \,.$$

3.12 Proposition. When G is abelian, the above filtration induces a natural spectral sequence $\{E_r^{p,q}, d_r\}$ such that

$$E_2^{p,q} = H^q(\Omega(M)_{\theta=0}) \otimes S^p \mathfrak{G}^*$$

and $E_r^{*,*}$ converges to $H^*_{\mathfrak{G}}(M)$ additively, in the usual sense. There is also an associated "edge" homomorphism with the following identifications

$$\begin{array}{ccc}
B^p\mathfrak{G}^* = S^p\mathfrak{G}^* = H^0(\Omega(M)_{\theta=0}) \otimes S^p(\mathfrak{G}^*) & = & E_2^{p,0} \\
 & & \downarrow \alpha_2 \\
\Big\downarrow j_M^* & & E_3^{p,0} \\
 & & \downarrow \alpha_3 \\
 & & \vdots \\
H^p_{\mathfrak{G}}(M) & \xleftarrow[\beta]{\qquad\qquad} & E_\infty^{p,0}
\end{array}$$

where the α's surject and β injects, and $\ker j_M^* = \mathrm{Index}_{\mathfrak{G}} M$.

Proof. The proof is a simple exercise using the fact that in this case $[\Omega(M) \otimes S\mathfrak{G}^*]_{\theta=0} = \Omega(M)_{\theta=0} \otimes S\mathfrak{G}^*$.

3.13 Remark. The above spectral sequence exists for an arbitrary smooth G-space M but the E_1 and E_2 are more complicated.

4. AN APPLICATION.

We will need a preliminary result before giving a specific application of the general Borsuk–Ulam and Bourgin–Yang type theorems of the previous section.

4.1 Proposition. Let G denote a connected, simply connected, nilpotent Lie group which acts freely on the smooth manifold M. Suppose further that M admits a finite cover of tubular neighborhoods around orbits of M. Let $\Gamma \subset G$ denote a discrete subgroup of G such that G/Γ

is compact. Then, there is a natural chain equivalence

$$\nu\colon \Omega(M)_{\theta=0} \longrightarrow \Omega(M/\Gamma)\,.$$

Proof. a) We first consider the special case $M=G$. Here Nomizu's theorem (theorem 1 in [7]) applies. $\Omega(M)_{\theta=0}$ consists of differential forms right invariant by G and is a subcomplex of $\Omega^{\Gamma}(M)$, the forms invariant under Γ. Identifying, $\Omega^{\Gamma}(M)$ with $\Omega(M/\Gamma)$ we have

$$\nu\colon \Omega(M)_{\theta=0} \longrightarrow \Omega(M/\Gamma)$$

which induces isomorphisms in cohomology [7] and hence ν is a chain equivalence.

b) Next, let M denote an invariant tubular neighborhood of an orbit G. Then, we have

$$\begin{array}{ccc} \Omega(M)_{\theta=0} & \xrightarrow{\ \nu'\ } & \Omega(M/\Gamma) \\ \downarrow & & \downarrow \\ \Omega(G)_{\theta=0} & \xrightarrow{\ \nu'\ } & \Omega(G/\Gamma) \end{array}$$

where, since the vertical maps are chain equivalences induced by smooth retractions and ν is a chain equivalence, ν' is a chain equivalence.

c) The final step is a Mayer–Vietoris argument using b) and the hypothesis that M is covered by a finite number of such tubular neighborhoods of orbits. This argument is left for the reader.

We now consider the following example, $n \geq 1$. Let $\mathbf{C}^n$ denote complex n-space and $M=\mathbf{C}^n-0$. Let $G=\mathbf{C}$ and define a right action $M\times\mathbf{C}\longrightarrow M$ by

$$(z_1,\ldots,z_n)\zeta = (z_1,\ldots,z_n)A(\zeta)$$

where $A(\zeta) = \mathrm{diag}\left[e^{\lambda_1\zeta_1}, e^{\lambda_2\zeta_2},\ldots,e^{\lambda_n\zeta_n}\right]$, where $(\lambda_1,\ldots,\lambda_n)$ is a fixed n-tuple of non-zero real numbers, i.e.

$$(z_1,\ldots,z_n)\zeta = (e^{\lambda_1\zeta} z_1,\ldots,e^{\lambda_n\zeta} z_n)\,.$$

Since $e^{\lambda_i\zeta} z_i = z_i$ if, and only if, $\zeta = iy$ where $\lambda_i y = k2\pi$, $k\in Z$, the discrete subgroup

$$\Gamma_i = \left\{\left(\frac{2k\pi}{\lambda_i}\right)\right\}, \quad k \in \mathbf{Z},$$

appears as non-trivial isotropy. If we require that λ_i/λ_j be irrational for $i \neq j$, then these would be the only non-trivial isotropy subgroups. We note also that the representation

$$y \mapsto \operatorname{diag}\left[e^{i\lambda_1 y}, \ldots, e^{i\lambda_n y}\right]$$

has compact image $\mathbf{R}/D = S^1$, D discrete, in $U(n)$ if all ratios λ_i/λ_j are rational. Otherwise, this imbedding of $\mathbf{R}$ in $U(n)$ has closure which is a torus of dimension ≥ 2. We will refer to this action as an <u>exponential action</u> with <u>parameters</u> $\lambda_1, \ldots, \lambda_n$.

Our next objective is to compute the infinitesimal equivariant cohomology $H^*_{\mathfrak{G}}(M)$ and $\mathrm{Index}_{\mathfrak{G}} M$.

We first write $\mathbf{C} = G = G' + G''i$. Then G' acts freely on $\mathbf{C}^n - 0$ and $\mathbf{C}^n - 0 = \mathbf{R}^+ \times S^{2n-1}$, with orbit orbit space $(\mathbf{C}^n - 0)/G' = S^{2n-1}$. If Γ is a discrete subgroup of G', note that $G'/\Gamma = S^1$, a compact group and $M/\Gamma = (\mathbf{C}^n - 0)/\Gamma = S^1 \times S^{2n-1}$. Then by Proposition 4.1, we have the following.

4.2 <u>Lemma</u>. $\nu: \Omega(M)_{\theta'=0} \longrightarrow \Omega(S^1 \times S^{2n-1})$ is a chain equivalence, where $\theta' = 0$ is short for $\theta(x) = 0$ for $x \in \mathfrak{G}'$.

Next, we observe the following.

4.3 <u>Lemma</u>. $\Omega(M)_{\theta=0} = \Omega(M)_{\theta'=0, \theta''=0}$ where again $\theta'' = 0$ is short for $\theta'' = 0$ is short for $\theta''(x) = 0$ is short for $\theta'' = 0$ for $x \in \mathfrak{G}''$.

Consider the representation $G'' \longrightarrow U(n)$ given by

$$iy \mapsto \operatorname{diag}\left[e^{i\lambda_1 y}, \ldots, e^{i\lambda_n y}\right]$$

and let T denote the closure of the image of G''. T is a torus and the action of G'' on the second factor of $S^1 \times S^{2n-1}$ naturally extends to T. Then, by continuity

$$\Omega(S^1 \times S^{2n-1})_{\theta''=0} = \Omega(S^1 \times S^{2n-1})_{\theta_T=0}$$

where $\theta_T = 0$ stands for $\theta_T(x) = 0$ for all x in the Lie algebra of T.

4.4 <u>Lemma</u>. Since T is compact, there is a chain equivalence (inclusion induced)

$$\Omega(S^1\times S^{2n-1})_{\theta_T=0} \longrightarrow \Omega(S^1\times S^{2n-1})\,.$$

Proof. This is classical using integration on T to symmetrize forms ([4], [5], [6]).

Combining lemmas 4.2, 4.3, 4.4 we obtain

4.5 Proposition. There is a chain equivalence,

$$\Omega(M)_{\theta=0} \longrightarrow \Omega(S^1\times S^{2n-1})\,.$$

Now, let us again consider $C^n-0=M$ as a G'-space by restricting the G-action. Then the natural projection $M\longrightarrow M/G'=S^{2n-1}$ is a locally trivial principal G-bundle (using the results in [8]).

4.6 Proposition. There is a chain equivalence

$$\gamma\colon \Omega_{\mathfrak{G}'}(M) \longrightarrow \Omega(S^{2n-1})\,.$$

Proof. First, there is an isomorphism [6; II, p. 241] $\Omega(S^{n-1})\longrightarrow\Omega(M)_{i'=0,\theta'=0}$. Then, the injection

$$\Omega(M)_{i'=0,\theta'=0} \longrightarrow [\Omega(M)\otimes S(\mathfrak{G}'^*)]_{\theta=0}$$

given by $x\mapsto x\otimes 1$ is also a chain equivalence [6; III, p. 344]. The composition of these is the required chain equivalence γ.

At this point we will need a result of Husseini [9] which we recall as follows. Since T is compact, it follows that the Borel cohomology $H^*_T(S^{2n-1},\mathbb{R})$ is naturally isomorphic to the infinitesimal cohomology $H^*_{\mathfrak{T}}(S^{2n-1})$ ([5]). In fact, it is an exercise to verify that $\operatorname{Index}_T S^{2n-1}$ in the sense of [3] and $\operatorname{Index}_{\mathfrak{T}} S^{2n-1}$ coincide when $H^*(BT;\mathbb{R})$ and $B\mathfrak{T}$ are naturally identified. We summarize the results in [9] in the context of the infinitesimal theory as applied to the present situation.

The inclusion map $T\subset T^n\subset U(n)$ induces homomorphisms $\lambda_i\colon T\longrightarrow S^1$, $i=1,\dots,n$, and if $\mathfrak{s}$ is the Lie algebra of S^1,λ_i induces $\lambda_i^*\colon B\mathfrak{s}\longrightarrow B\mathfrak{T}$. If σ is the generator of $B\mathfrak{s}$, set $\lambda_i'=\lambda_i^*(\sigma)$. Then, the natural inclusion

$$B\mathfrak{T}\longrightarrow \Omega(M)_{\theta_T=0}\otimes B\mathfrak{T}$$

induces a surjection

$$B\mathfrak{T}\longrightarrow H_{\mathfrak{T}}(S^{2n-1})$$

with kernel P_T a principal ideal generated by

$$\gamma = \lambda'_1 \lambda'_2 \cdots \lambda'_n .$$

Note that each $\lambda_i: T \longrightarrow S^1$ is non-trivial because $(S^{2n-1})^{G''} = (S^{2n-1})^T = 0$. This also implies that if $\alpha: B\mathfrak{T} \longrightarrow B\mathfrak{G}''$ is induced by inclusion $\alpha(\lambda'_i) \neq 0$ for $i = 1,\dots,n$.

4.7 <u>Lemma</u>. $\alpha(\gamma) = ct^n$, $c \neq 0$, t a generator of $B\mathfrak{G}''$.

<u>Proof</u>. Suppose $t'_1,\dots,t'_p$ is a basis for T $(\dim T = p)$. Then

$$\lambda'_i = \sum_{j=1}^{p} \lambda_{ij} t_j$$

and $\alpha(t'_j) = \mu_j t$ then $\alpha(\lambda'_i) = c_j t$, $c_j \neq 0$ and hence

$$\alpha(\lambda'_1 \lambda'_2 \cdots \lambda'_n) = ct^n,$$

with $c = c_1 c_2 \cdots c_n \neq 0$.

4.8 <u>Theorem</u>. Let $M = \mathbb{C}^n - 0$ and $G = \mathbb{C}$ as above. Then the inclusion $j_M: W(G) \longrightarrow \Omega_{\mathfrak{G}}(M)$ induces a surjection $j_M^*: B\mathfrak{G} \longrightarrow H^*_{\mathfrak{G}}(M)$ with kernel $\mathrm{Index}_{\mathfrak{G}} M = \langle s, t^n \rangle$, where $B\mathfrak{G} = \mathbb{R}[s,t]$.

<u>Proof</u>. a) We first compare the spectral sequences of $\Omega(M)_{\theta'=0} \otimes S(\mathfrak{G}'^*)$ and $\Omega(M)_{\theta=0} \otimes S(\mathfrak{G}^*)$ via the filtration preserving map $\Omega(M)_{\theta=0} \otimes S(\mathfrak{G}^*) \longrightarrow \Omega(M)_{\theta'=0} \otimes S(\mathfrak{G}'^*)$ induced by $G' \subset G$. The induced map on "fibers" $\Omega(M)_{\theta=0} \longrightarrow \Omega(M)_{\theta'=0}$ is a chain equivalence and on the "base" $S(\mathfrak{G}^*) = \mathbb{R}[s,t] \longrightarrow \mathbb{R}[s] = S(\mathfrak{G}'^*)$ suppresses the variable t. At the E_2-level we obtain the diagram

$$\begin{array}{ccc} H^*(\Omega_{\theta=0}) \otimes S(\mathfrak{G}^*) & \longrightarrow & H^*(\Omega_{\theta'=0}) \otimes S(\mathfrak{G}'^*) \\ \downarrow d_2 & & \downarrow d'_2 \\ H^*(\Omega_{\theta=0}) \otimes S(\mathfrak{G}^*) & \longrightarrow & H^*(\Omega_{\theta'=0}) \otimes S(\mathfrak{G}'^*) . \end{array}$$

Let $u' \in H^1(\Omega(M)_{\theta'=0}) = H^1(S^1 \times S^{2n-1})$ denote a generator corresponding to the S^1 factor. Then, since $H^*_{\mathfrak{G}'}(M) = H^*(S^{2n-1})$, $d'_2 u \neq 0$ and we may assume without loss that $d_2 u = s$. Thus, if $u \in H^1(\Omega(M)_{\theta=0})$ is the generator corresponding to u', $d_2 u = s$.

b) Let $v \in H^{2n-1}(\Omega(M)_{\theta=0}) = H^1(S^1 \times S^{2n-1})$ denote a generator corresponding to the S^{2n-1} factor. We will show $d_n v = ct^n$, $c \neq 0$ by comparison with spectral sequence for $\Omega(S^{2n-1})_{\theta''=0} \otimes S(\mathfrak{G}'^*)$ via the natural map $\Omega(M)_{\theta=0} \otimes S(\mathfrak{G}^*) \longrightarrow \Omega(S^{2n-1})_{\theta''=0} \otimes S(\mathfrak{G}''^*)$ induced by $M = C^n - 0 \longrightarrow S^{2n-1}$, and $G'' \subset G$. but if we take a generator $v'' \in H^*(\Omega(S^{2n-1})_{\theta''=0}) = H^*(S^{2n-1})$, and apply Lemma 4.7, we see that $d_n'' v'' = ct^n$, $c \neq 0$. Since v'' may be chosen as the image of v, we have $d_n v = ct^n$, $c \neq 0$ and we may assume $c = 1$. Combining a) and b) and employing the edge homomorphism we see that $j_M^*: B\mathfrak{G} \longrightarrow H^*_{\mathfrak{G}}(M)$ surjects with kernel $\langle s, t^n \rangle$ as asserted.

While we are about the business of computing, let us consider the $\text{Index}_{\mathfrak{G}}$ of orbits in the G-space M. There are only two possibilities for $x \in M$. Either the isotropy $G_x = 0$ or G_x is a discrete subgroup Γ of G''; i.e. orbits are of the form G or G/Γ, $\Gamma \subset G''$.

4.9 Proposition. $H^*_{\mathfrak{G}}(G)$ is acyclic.

Proof. This is a general phenomenon, since $[\Omega(G) \otimes S\mathfrak{G}^*]_{\theta=0} = W(\mathfrak{G})_{\theta=0}$ is acyclic, [4], [6].

4.10 Proposition. $H^*_{\mathfrak{G}}(G/\Gamma)$ is acyclic.

Proof. The proof is an easy exercise, modelling the techniques used to prove theorem 4.8. In fact, if $N = G/\Gamma$, $N = R \times S^1$ and we apply theorem 4.8 in the case $n = 1$.

We may now state the following theorem of Borsuk–Ulam type.

4.11 Theorem. Let $G = C$ act on C^n and C^m with exponential actions with parameters $\lambda_1, \dots, \lambda_n$ and $\mu_1, \dots, \mu_m$, respectively, with $m < n$. Then every G-map $f: C^n \longrightarrow C^m$ has a non-trivial zero. Alternatively, there does not exist a G-map $f: C^n - 0 \longrightarrow C^m - 0$.

Proof. If $f: C^n - 0 \longrightarrow C^m - 0$ is a G-map $\text{Index}_{\mathfrak{G}}(C^n - 0) = \langle s, t^n \rangle \supset \langle s, t^m \rangle = \text{Index}_{\mathfrak{G}}(C^m - 0)$, which is impossible.

5. INDEX$_{\mathfrak{G}}$ AND G-CATEGORY.

We explore briefly in this section the Bourgin–Yang type analogue of theorem 4.11. If $G = C$ acts on C^n and C^m as in theorem 4.11, $m < n$, and $f: C^n - 0 \longrightarrow C^m$ is a G-map, we wish to get some information on the "size" of the zero set $Z \subset C^n - 0$.

If $n-m=1$, we know $Z \neq \phi$. On the other hand if $n-m=1$, Z must be infinite. This crude result follows from the fact that

$$\mathrm{Index}_{\mathfrak{G}} Z \subset (\mathrm{Index}_{\mathfrak{G}}(C^n - 0) : \mathrm{Index}_{\mathfrak{G}}(C^m - 0))$$

and hence

$$\mathrm{Index}_{\mathfrak{G}} Z \subset (\langle s, t^n \rangle : \langle s, t^m \rangle) = \langle s, t^{m-n} \rangle .$$

If Z is finite, $\mathrm{Index}_{\mathfrak{G}} Z = \langle s, t \rangle$ which is impossible.

To obtain a finer result we employ the concept of G-category. We recall [10] that if M is a G-space and $U \subset M$ is a G-subset of M, then U is G-categorical in M if there is a G-homotopy $F: U \times I \longrightarrow M$ such that F_0 is inclusion and $F_1: U \longrightarrow M$ has image in a single orbit. Then, if X is a G-subset of M, $G\text{-cat}_M X = n$ if X can be covered by n-categorical open G-sets U and n is minimal with this property.

5.1 Definition. Let W denote an open subset of a smooth G-manifold M. The $\mathfrak{G}$-cup length of W in M is k if there are elements

$$\alpha_i \in H^{n_i}_{\mathfrak{G}}(M), \ n_i \geq 1, \ i = 1, \ldots, k$$

such that $H^*_{\mathfrak{G}}(i_W)(\alpha_1 \alpha_2 \cdots \alpha_k) \neq 0$ and k is maximal with this property, where $i_W: W \subset M$.

5.2 Proposition. Let M denote a smooth G-manifold such that all orbits are $\mathfrak{G}$-acyclic, i.e. $H^q_{\mathfrak{G}}(Y) = 0$ for $q \geq 1$ and all orbits Y. Then if $W \subset M$ is an open G-set,

$$G\text{-cat}_M W > \mathfrak{G}\text{-cup length } W \text{ in } M .$$

Proof. Let $U_1, \ldots, U_n$ denote n open G-sets which are categorical in M and cover W. Then, the diagram

$$\begin{array}{ccc} U_i & \longrightarrow & M \\ \big\downarrow & & \big\uparrow \\ Y & \xrightarrow{\ id\ } & Y \end{array}$$

where Y is an orbit, induces

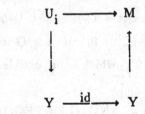

$$\begin{array}{ccc} H^*_{\mathfrak{G}}(U_i) & \longrightarrow & H^*_{\mathfrak{G}}(M) \\ \uparrow & & \downarrow \\ H^*_{\mathfrak{G}}(Y) & \xrightarrow{\ id\ } & H^*_{\mathfrak{G}}(Y) \end{array}$$

and hence if $\alpha_1,\ldots,\alpha_n \in H^*_{\mathfrak{G}}(M)$, $\dim \alpha_i \geq 1$, each α_i pulls back to β_i in $H^*_{\mathfrak{G}}(M,U_i)$ because Y is $\mathfrak{G}$-acyclic. Therefore, $\alpha_1\alpha_2\cdots\alpha_n$ is the image of $\beta_1\beta_2\cdots\beta_n \in H^*_{\mathfrak{G}}(M,W)$. Hence, $H^*_{\mathfrak{G}}(i_W)(\alpha_1\alpha_2\cdots\alpha_n)=0$.

5.3 <u>Theorem</u>. Let $f: C^n-0 \longrightarrow C^m$, $m<n$ denote a $G=C$ map as in theorem 4.12, with zero set Z. Then,

$$G\text{-cat}_M Z \geq m-n\ .$$

<u>Proof</u>. Choose an open G-set W such that $W \supset Z$, $\text{Index}_{\mathfrak{G}} W = \text{Index}_{\mathfrak{G}} Z$, and $G\text{-cat}_M W = G\text{-cat}_M Z$. Now,

$$\text{Index}_{\mathfrak{G}} W \subset (\langle s,t^n\rangle : \langle s,t^m\rangle)$$

and $t^{n-m-1} \not\in \text{Index}_{\mathfrak{G}} W$. Hence under the composition

$$B\mathfrak{G} \longrightarrow H^*_{\mathfrak{G}}(C^n-0) \longrightarrow H^*_{\mathfrak{G}}(W)$$

and t^{n-m-1} does not go to zero, so that

$$G\text{-cat}_M Z > \mathfrak{G}\text{-cup length } W \text{ in } M \geq m-n-1\ .$$

BIBLIOGRAPHY

1. E. Fadell and S. Husseini, Relative cohomological index theories, Advances in Mathematics, 64 (1987), 1–31.

2. E. Fadell and S. Husseini, Index theory for G-bundle pairs with application to Borsuk–Ulam type theorems for G-sphere bundles, Nonlinear Anaylsis, World Scientific, New Jersey, 1987, 307–325.

3. E. Fadell and S. Husseini, An ideal-valued cohomological index theory with applictions to Borsuk–Ulam and Bourgin–Yang theorems, to appear in Ergodic Theory and Dynamical Systems.

4. H. Cartan, Colloque de Topologie (espaces fibrés), Bruxelles, 1950, Liege et Paris, 1951, pp. 15-27 and pp. 57-71.

5. M. Atiyah and R. Bott, The moment map and equivariant cohomology, Topology, vol. 23, No. 1 (1984), 1-28.

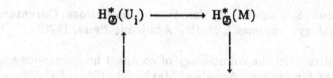

6. W. Greub, S. Halperin, R. Vanstone, Connections, Curvature, and Cohomology, volumes I, II, III, Academic Press, 1976.

7. K. Nomizu, On the cohomology of compact homogeneous spaces of nilpotent Lie groups, Annals of Math., 59 (1954), 531-538.

8. R. Palais, On the existence of slices for actions of noncompact Lie groups, Annals of Math., 73 (1961), 295-323.

9. S. Husseini, Applications of equivariant characteristic classes to problems of the Borsuk–Ulam and degree theory type, to appear in Topology and its Applications.

10. E. Fadell, The equivariant Ljusternik–Schnirelmann method for invariant functionals and relative cohomological index theories, Methods Topologiques en analyse Non Lineaire, edited by A. Granas, University of Montreal Press 1985, 41–71.

University of Wisconsin–Madison

*Supported in part by the National Science Foundation under NSF Grant No. DMS-8722295.

FULLER'S INDEX FOR PERIODIC SOLUTIONS OF FUNCTIONAL DIFFERENTIAL EQUATIONS

Christian C. Fenske
Mathematisches Institut der Justus-Liebig-Universität Gießen
Arndtstraße 2, D-6300 Gießen

F.B. Fuller [6] was the first to define an index of fixed point type counting periodic orbits of flows generated by smooth vector fields on smooth finite-dimensional manifolds. In [2], S.N. Chow and J. Mallet-Paret described a completely different approach using a complicated bifurcation argument. Their approach had the advantage that it generalizes to certain infinite-dimensional situations, viz., the local semiflow generated by a functional differential equation. The details (involving an adaptation of a result by Brunovský [1] which does not seem straightforward) are, however, not worked out. Here, we sketch a method which directly extends Fuller's idea to the case of functional differential equations. The main ingredient is a variant of an ingenious trick which is due to Fuller [6]. In addition, we use the author's earlier generalization [4] of the "homological part" of [6] to the case of semiflows on infinite-dimensional ANR. (We refer the reader to [8] for information on ANR spaces.)

Let $R > 0$ and denote by $C := C([-R,0], \mathbb{R}^n)$ the space of continuous $\mathbb{R}^n$-valued functions on $[-R,0]$. We shall consider a C^2-function $f: C \to \mathbb{R}^n$ which maps bounded sets in C into bounded sets in $\mathbb{R}^n$. If, for some $a > 0$, y is a continuous function from $[-R,a)$ to $\mathbb{R}^n$ then for $t \in [0,a)$ we denote by y_t the element in C defined by $y_t(s) := y(t+s)$. Let $x \in C$ and $a > 0$. A *solution on* $[0,a)$ *of*

$$(1) \qquad y'(t) = f(y_t)$$

with initial value x is a function $y: [-R,a) \to \mathbb{R}^n$ which solves (1) for $t \in [0,a)$ and satisfies $y_0 = x$. Letting $\phi_t x := y_t$ for $0 \le t < a$ we obtain a local semiflow on C.

We recall that a *local semiflow* on a topological space X is a continuous mapping Φ which maps an open set $\mathcal{D}(\Phi) \subset X \times [0,\infty)$ into X such that (with $\phi_t x := \Phi(x,t)$)

1) $X \times \{0\} \subset \mathcal{D}(\Phi)$

2) For every $x \in X$ there is an $\omega_x \in (0,\infty]$ such that $\mathcal{D}(\Phi) = \{(x,t) \in X \times [0,\infty) \mid 0 \le t < \omega_x\}$.

3) $\phi_0 = \mathrm{id}$.

4) $(x,t) \in \mathcal{D}(\Phi)$ and $(\phi_t x, s) \in \mathcal{D}(\Phi)$ imply $(x,t+s) \in \mathcal{D}(\Phi)$ and $\phi_{s+t}x = \phi_s\phi_t x$.

If $\phi_t x = x$ for some $t > 0$, then we have that $\omega_x = \infty$ and x is called a *periodic point* with *period* t. If x is not a rest point then we denote by $p(x)$ the *minimal period* of x and call $m(x,t) := t/p(x)$ the *multiplicity* of (x,t).

Consider now the local semiflow Φ on C defined by (1) and let $\Omega \subset C \times [0,\infty)$ be open and bounded with $\mathrm{cl}\,\Omega \subset \mathcal{D}(\Phi)$. Call $P := \{(x,t) \in \mathrm{cl}\,\Omega \mid \phi_t x = x\}$ and assume that

(B) $\qquad P \subset \Omega$.

In particular, there are then no rest points in Ω. Moreover,

(C) $\qquad P$ is compact, hence the projection $\mathrm{pr}_2 P$ into $(0,\infty)$ is compact.

In this situation, the *index* $i(\Phi,\Omega)$ assigned to Φ and Ω will be a rational number satisfying

(A) "Additivity": If Ω is a disjoint union of open sets Ω_1 and Ω_2, then $i(\Phi,\Omega) = i(\Phi,\Omega_1) + i(\Phi,\Omega_2)$.

(H) "Homotopy invariance": Let $F\colon C \times [0,1] \to \mathbb{R}^n$ be a continuous map sending bounded sets into bounded sets and assume that $F(\cdot,\alpha)\colon C \to \mathbb{R}^n$ is C^2 for every $\alpha \in [0,1]$. For $\alpha \in [0,1]$, call Φ^α the local semiflow on C generated by $y'(t) = F(y_t,\alpha)$. Assume that $\mathrm{cl}\,\Omega \subset \bigcap_{\alpha\in[0,1]} \mathcal{D}(\Phi^\alpha)$ and that $\bigcup_{\alpha\in[0,1]} \{(x,t) \in \mathrm{cl}\,\Omega \mid \Phi^\alpha(x,t) = x\} \subset \Omega$.
Then $i(\Phi^0,\Omega) = i(\Phi^1,\Omega)$.
(The assumptions on the homotopy in (H) can be weakened considerably - cf., [9].)

(N) "Normalization": If P consists of a single periodic orbit γ of multiplicity m, then $i(\Phi,\Omega) = i/m$ where i is the fixed point index of the Poincaré mapping associated with γ.

In the notation of (N), let x be a periodic point on γ and assume that $D_1\Phi(x,p(x))$ has precisely one eigenvalue of modulus one (which must then be the eigenvalue 1 corresponding to the eigenvector $D_2\Phi(x,p(x))$). Let $\mu_1,\dots,\mu_k$ be the eigenvalues of $D_1\Phi(x,p(x))$ of modulus greater than one and σ the number of $\mu_1^m,\dots,\mu_k^m$ which are greater than 1. As remarked in [2], in this situation $i = (-1)^\sigma$ (this is a consequence of the Leray-Schauder formula).

From [6] and [2] it is obvious how to define the index: By [9] (cf., also [7]) we find arbitrarily close to f a C^2-mapping g such that the local semiflow Ψ generated by $y'(t) = g(y_t)$ has only hyperbolic periodic and rest points. We can find g so close to f that there are no periodic points of Ψ on $\partial\Omega$. Since Ω is bounded there are only finitely many periodic orbits $\gamma_1,\dots,\gamma_r$, say, of Ψ in cl Ω. Choose disjoint neighbourhoods $\Omega_1,\dots,\Omega_r$ of $\gamma_1,\dots,\gamma_r$ in Ω and define $i(\Phi,\Omega) := i(\Psi,\Omega) := i(\Psi,\Omega_1) + \dots + i(\Psi,\Omega_r)$. Observe that the sum on the right hand side is defined by (N) and that it can be computed as described above.

As a matter of fact, it is not the definition itself that causes problems but rather the verification of (H) and of the fact that the index is well-defined (i.e., does not depend on the approximation g). We therefore sketch a proof of the following

Theorem. *The index $i(\Phi,\Omega)$ for periodic orbits in a bounded open set $\Omega \subset C \times [0,\infty)$ of the local semiflow Φ generated by a functional differential equation $y'(t) = f(y_t)$ with C^2 right hand side $f: C \to \mathbb{R}^n$ mapping bounded sets into bounded sets as defined above is well-defined and satisfies properties (A), (H), and (N).*

Proof. Property (A) is obvious. Once we have established (H), property (N) will be obvious as well. In order to show that the index is well-defined and satisfies (H) we have to consider the following situation: $\Omega \subset C \times [0,\infty)$ is a bounded open set, $F: C \times [0,1] \to \mathbb{R}^n$ is continuous and maps bounded sets into bounded sets, and each $F(\cdot,\alpha)$ is a C^1-map. We call Φ^α the local semiflow generated by the functional differential equation $y'(t) = F(y_t,\alpha)$ and we assume that $\mathrm{cl}\,\Omega \subset \bigcap_{\alpha\in[0,1]} \mathcal{D}(\Phi^\alpha)$ and that $P := \{(x,t,\alpha) \in \mathrm{cl}\,\Omega \times [0,1] \mid \Phi^\alpha(x,t) = x\} \subset \Omega \times [0,1]$. Finally, we assume that all periodic points of Φ^0 and Φ^1 in Ω are hyperbolic. We have to show that $i(\Phi^0,\Omega) = i(\Phi^1,\Omega)$. The proof is split in several steps.

1. Call $\gamma_1 \times \{t_1\},\dots,\gamma_r \times \{t_r\}$ and $\gamma_1' \times \{t_1'\},\dots,\gamma_s' \times \{t_s'\}$ the periodic orbits of Φ^0 and Φ^1 in Ω. Choose periodic points x_j and x_j' on γ_j and γ_j'. Write $m_j := m(x_j,t_j)$, $m_j' := m(x_j',t_j')$. We then have to prove

$$\sum_{j=1}^{r} \frac{i_j}{m_j} = \sum_{j=1}^{s} \frac{i_j'}{m_j'}$$

where i_j and i_j' are the fixed point indices of the Poincaré map corresponding to the orbit γ_j and the period t_j (to γ_j' and t_j').

Fix a prime k which is larger than max $\{m(x,t) \mid (x,t,\alpha) \in P\}$. Since the local semiflow associated with a functional differential equation of type (1) in C gets locally compact for $t > R$ we may choose an even integer q and a neighbourhood U of P in $\Omega \times [0,1]$ such that cl $\{\Phi^\alpha(x,t) \mid (x,t,\alpha) \in U\}$ is compact. Since $q+1$ is odd and the γ_j are hyperbolic by the remark following (N) we see that

(*) the index of the Poincaré mapping for the orbit γ_j and the period t_j of Φ^α equals the index of the Poincaré mapping of the same orbit for the period $(q+1)t_j$.

2. This is the key step and the main argument is a modification of an idea which is due to Fuller [6].

By $C^{(k)}$ we denote the set of all $(y_1,\ldots,y_k)$ with distinct elements $y_j \in C$. So $C^{(k)}$ is an open subset of a Banach space. The cyclic group $\mathbb{Z}_k$ acts on $C^{(k)}$ via cyclic permutations. By C_k we denote the quotient $C^{(k)}/\mathbb{Z}_k$ and by q_k the projection $q_k: C^{(k)} \to C_k$. We define $Q_k: C^{(k)} \times [0,\infty) \times [0,1] \to C_k \times [0,\infty) \times [0,1]$ by $Q_k(\xi,t,\alpha) = (q_k\xi,t,\alpha)$. Since C_k is locally homeomorphic to $C^{(k)}$ it is a Banach manifold. For $(x,t,\alpha) \in \Omega \times [0,1]$ we let $g_k(x,t,\alpha) := (x,\Phi^\alpha(x,t/k),\ldots,\Phi^\alpha(x,\frac{k-1}{k}t))$. By rather elementary arguments we find a $\delta > 0$ and open sets U' and U'' in U such that

i) $P \subset U'' \subset \mathrm{cl}\, U'' \subset U'$

ii) $(\Phi^\alpha(x,\tau),t,\alpha) \in U$ whenever $(x,t,\alpha) \in U'$ and $0 \le \tau \le (q+2)t$

iii) $g_k(\Phi^\alpha(x,s),t,\alpha) \in C^{(k)}$ whenever $(x,t,\alpha) \in U'$ and $0 \le s \le (q+2)t$

iv) $q_k g_k(x,t,\alpha) = q_k g_k(\Phi^\alpha(x,s),t,\alpha)$ for $(x,t,\alpha) \in U'$ and $|s - \frac{qk+1}{k}t| < \delta$ (if and) only if $(x,t,\alpha) \in P$ and $s = \frac{qk+1}{k}t$

v) $(\Phi^\alpha(x,\tau),t,\alpha) \in U'$ whenever $(x,t,\alpha) \in U''$ and $0 \le \tau \le (q+2)t$.

Define then

$$Z^{(k)} := \{(g_k(x,t,\alpha),t) \mid (x,t,\alpha) \in U'\}$$

$$O^{(k)} := \{(g_k(x,t,\alpha),t,s,\alpha) \mid (x,t,\alpha) \in U'',\ |s - \tfrac{qk+1}{k}t| < \delta\}$$

$$Z_k := \{(q_k g_k(x,t,\alpha),t) \mid (x,t,\alpha) \in U'\}$$

$$O_k := \{(q_k g_k(x,t,\alpha),t,s,\alpha) \mid (x,t,\alpha) \in U'',\ |s - \tfrac{qk+1}{k}t| < \delta\}.$$

For $i \in \{0,1\}$ let

$$O_k^i := \{(q_k g_k(x,t,i),t,s,i) \mid (x,t,i) \in U'',\ |s - \tfrac{qk+1}{k}t| < \delta\}$$

and define $\Psi^i: O_k^i \to Z_k$ by $\Psi^i(q_k g_k(x,t,i),t,s,i) := (q_k g_k(\Phi^i(x,s),t,i),t)$.

Now $Z^{(k)}$ is an ANR since it is homeomorphic to U', and so is $O^{(k)}$ since it is an open subset of $Z^{(k)} \times [0,\infty) \times [0,1]$. Since q_k is a local homeomorphism both Z_k and O_k are ANR. Finally, Ψ^0 and Ψ^1 are local semiflows which are homotopic, the homotopy being induced by the family Φ^α.

3. We now invoke the index $I(X,\Phi,\Omega) \in H_1\Omega$ (the first homology group of Ω) defined in [4]. This index is defined for local semiflows Φ on an ANR X with $\Omega \subset X \times [0,\infty)$ open and $D(\Phi) \subset \mathrm{cl}\,\Omega$ provided $\mathrm{cl}\,\Phi(\Omega)$ is compact and there are no periodic points on $\partial\Omega$. Call $i_0 : O_k^0 \to O_k$ and $i_1 : O_k^1 \to O_k$ the inclusion. Since the index I is homotopy invariant we have that $i_{0*}I(Z_k,\Psi^0,O_k^0) = i_{1*}I(Z_k,\Psi^1,O_k^1)$. By our assumption on Φ^0 and Φ^1 and by property iv) above we have that Ψ^0 has only finitely many periodic orbits in Z_k, viz., $\gamma_j^k : [0,t_j/k] \to Z_k$

$$\tau \to (q_k g_k(\Phi^0(x_j,\tau),t_j,0),t_j)$$

and the multiplicity of this orbit is m_j.

In order to compute $I(Z_k,\Psi^0,O_k^0)$ we have to know the fixed point index of the Poincaré map of $\tilde{\gamma}_j : [0,t_j] \to Z_k$

$$\tau \to (q_k g_k(\Phi^0(x_j,\tfrac{qk+1}{k}\tau),t_j,0),t_j).$$

(We refer the reader to [3] for the definition of the Poincaré map of a semiflow on a metric space.) Using the commutativity property of the fixed point index one can show that this index equals the fixed point index of the Poincaré mapping of γ_j for the period $(qk+1)t_j$ which in turn (by (*)) equals i_j. So we have that

$$I(Z_k,\Psi^0,O_k^0) = \sum_{j=1}^{r} \frac{1}{m_j} i_j [\gamma_j^k]$$

where $[\gamma_j^k]$ is the homology class represented by the 1-cycle γ_j^k in O_k^0. Similarly,

$$I(Z_k,\Psi^1,O_k^1) = \sum_{j=1}^{s} \frac{1}{m_j'} i_j' \cdot [\gamma_j'^k].$$

We thus obtain

$$(**)\qquad \sum_{j=1}^{r} \frac{1}{m_j} \cdot i_j i_{0*}[\gamma_j^k] = \sum_{j=1}^{s} \frac{1}{m_j'} \cdot i_j' i_{1*}[\gamma_j'^k]$$

Now $q_k : C^{(k)} \to C_k$ is a regular k-sheeted covering, and $\mathbb{Z}_k$ is its group of covering transformations. So there is a natural homomorphism $\mu_k : \pi_1 C_k \to \mathbb{Z}_k$. Since $\mathbb{Z}_k$ is abelian this induces a homomorphism $\mu_k : H_1 C_k \to \mathbb{Z}_k$.

Call $\iota : O_k \to C_k \times [0,\infty) \times [0,1] =: X_k$ the inclusion. Since X_k is homotopy equivalent to C_k we still have a homomorphism $\mu_k : H_1 X_k \to \mathbb{Z}_k$. Fuller's argument [6] shows that $\mu_k \iota_* i_{0*}[\gamma_j^k] = 1 \in \mathbb{Z}_k \quad (j = 1,\dots,r)$

and $\mu_k \iota_* i_{1*}[\gamma_j'^k] = 1 \in \mathbb{Z}_k \quad (j = 1,\dots,s)$.

We now multiply both sides of (**) by the least common multiple, ℓ, of the m_j and m_j' and apply $\mu_k \iota_*$ to both sides of the resulting equality. This yields

$$\sum_{j=1}^{r} \frac{\ell}{m_j}\cdot i_j = \sum_{j=1}^{s} \frac{\ell}{m_j'}\cdot i_j' \quad \text{mod } k.$$

Now observe that neither side of this equation depends on k. Since k can be taken arbitrarily large we have in fact that

$$\sum_{j=1}^{r} \frac{\ell}{m_j}\cdot i_j = \sum_{j=1}^{s} \frac{\ell}{m_j'}\cdot i_j' .$$

We finally divide both sides by ℓ and obtain the desired equality. This completes the argument.

Details and generalizations will be published elsewhere ([5]).

REFERENCES

[1] P. Brunovský: On one parameter families of diffeomorphisms. CMUC 11 (1970) 559-582

[2] S.N. Chow, J. Mallet-Paret: The Fuller index and global Hopf bifurcation. J. Diff. Eq. 39 (1978) 66-84

[3] C.C. Fenske: Periodic orbits of semiflows - Local indices and sections. In "Selected topics in Operations Research and Mathematical Economics" (Lecture Notes in Economics and Mathematical Systems 226) pp. 348-360. Springer-Verlag Berlin/Heidelberg/New York/Tokyo 1984

[4] --- : A simple-minded approach to the index of periodic orbits. J. Math. Analysis Appl. 129 (1988) 517-532

[5] --- : An index for periodic orbits of retarded functional differential equations (to appear)

[6] F.B. Fuller: An index of fixed point type for periodic orbits. Amer. J. Math. 89 (1967) 133-148

[7] J. Hale: Theory of functional differential equations. (Applied Mathematical Sciences 3). Springer-Verlag New York/Heidelberg/Berlin 1977

[8] S.T. Hu: Theory of retracts. Wayne State University Press Detroit 1965

[9] J. Mallet-Paret: Generic periodic solutions of functional differential equations. J. Diff. Eq. 25 (1977) 163-183

Simple variational methods for unbounded potentials.

by

G. Fournier and M. Willem

0. INTRODUCTION

When Fournier-Willem [6] defined the relative topological category (for a slightly different definition see Fadell [5]) an objectives was to find critical points of unbounded potentials. The main result in this direction is the well known Ambrosetti-Rabinowitz [1] mountain pass theorem.

The purpose of this paper is to show that the above can be attained, using the relative category, in the simple cases of some potentials having a disconnected level surface. We also give another proof of the mountain pass theorem and provide a very simple example of application to a differential system of equations.

1. Lusternik-Schnirelman relative category.

Let A be a subset of a topological space X.

The Lusternik-Schnirelman category of A in X, $cat_X(A)$, is the least integer n such that A can be covered by n closed subsets of X each of which is contractible in X[A is contractible in X if there exists h: $A \times I \to X$, where I=[0,1], such that (1) $h(x, 0) = x \ \forall x \in A$ and (2) $\exists x_0 \in X$ such that $h(x, 1) = x_0$]. If no such integer exists, we put catX(A) = $+\infty$.

We shall not list here the properties of the topological category which the reader can deduce from those of the topological relative category.

The following is due to Fournier-Willem [6] (see also [5]).

Definition 1.1: Let X be a topological space and Y be a closed subset of X. A closed subset A of X is of the k-th (weak) category relative to Y (we write $cat_{X,Y}(A) = k$) if and only if k is the

least positive integer n such that

$$A = \bigcup_{i=0}^{n} A_i$$

Work supported by the NSERC and the FCAR (Québec)

where, for each $i \geq 1$, A_i is closed and is contractible in X, and A_0 is strongly deformable into Y in X. [A_0 is strongly deformable into Y in X if and only if there exists $h_0: A_0 \times I \to X$, where $I=[0,1]$, such that (1) $h_0(x, 0) = x \quad \forall x \in A_0$, (2) $h_0(x, 1) \in Y \quad \forall x \in A_0$ and (3) $h_0(x, t) = x \quad \forall x \in Y \; \forall t \in I$.]

__Remark 1.2__: (1) If $Y = \emptyset$ then $A_0 = \emptyset$ and $cat_{X,\emptyset}(A) = Cat_X(A)$.

(2) If one such k does not exist, then $cat_{X,Y}(A) = \infty$.

(6) There exists an homeomorphism $\varphi: X \to X'$ such that $Y' = \varphi(Y)$ and $A' = \varphi(A)$ imply that $cat_{X',Y'}(A') = cat_{X,Y}(A)$.

We have the following properties see Fournier-Willem [6] (see also Fadell [5]).

__Proposition 1.3__: Let A, B, Y be closed subsets of X.

i) if $B \supset A$ then $cat_{X,Y}(A) \leq cat_{X,Y}(B)$

ii) $cat_{X,Y}(A \cup B) \leq cat_{X,Y}(A) + Cat_X(B)$

iii) $cat_{X,Y}(A) = 0$ if and only if there exists $h: A \times I \to X$ such that

(1) $h_0 = i_A: A \to X$ is the inclusion

(2) $Y \supset h_1(A)$

(3) $h(y, t) = y \quad \forall y \in Y \quad \forall t \in I$.

Let M be a complete C^2 Finsler manifold i.e. a C^2 Banach manifold with a Finsler structure on its tangent bundle. (Important examples are complete Riemannian manifolds and Banach spaces.) Let $\varphi \in C^1(M, R)$. Set

$$\varphi^c = \{ u \in M \mid \varphi(u) \leq c \}$$

$$K_c = \{ u \in M \mid \varphi(u) = c, \text{ and } \varphi'(u) = 0 \}.$$

__Definition 1.4__: (Palais-Smale condition; see [7])

A map $\varphi: M \to R$, where M is a C^1 Finsler manifold and φ is C^1, satisfies P-S if for every sequence $\{s_n\}$ of elements of M such that $\{\varphi(s_n)\}$ is bounded and $\|\varphi'(s_n)\| \to 0$ as $n \to \infty$ then there exists a subsequence

$$s_{n_k} \to s \text{ with } \varphi'(s) = 0.$$

The following variation of the deformation lemma due to Clark [3] for Banach spaces and to Ni [4] for Finsler manifolds, is the essential element in the proof of the fundamental theorem.

__Lemma 1.5__: If $\varphi \in C^1(M, R)$ satisfies the Palais-Smale condition and if U is an open neighbourhood of K_c, then, for every $\varepsilon' > 0$ there exists $\varepsilon \in]0, \varepsilon'[$ and a map $f: M \to M$ isotopic to id_M such that for all $d \in [0, \varepsilon]$, $\varphi^{c-d} \supset f(\varphi^{c+\varepsilon} \setminus U)$.

The following fundamental theorem generalises a result due to Palais [6] and is due to Fournier-Willem [6]. See also [5].

Theorem 1.6: If $\varphi \in C^1(M, R)$ satisfies the Palais-Smale condition and if $-\infty < a < b < +\infty$ and $K_a = K_b = \emptyset$, then

$$\# \left\{ u \in \varphi^{-1}([a,b]) \mid \varphi'(u) = 0 \right\} \geq \operatorname{cat}_{\varphi^b, \varphi^a}(\varphi^b) \geq \operatorname{cat}_{M, \varphi^a}(\varphi^b).$$

2. Connectedness theorems

We can immediately proye our main results. Note that they can also be proved using the Minimax principle (Palais [8]). Note also that the homological approach to the mountain pass theorem was pointed out in Tian [9] and in Chang [2].

Theorem 2.1

Let $\varphi \in C^1(M, R)$ satisfy the Palais-Smale condition on $\varphi^{-1}[a,b]$, when M is a complete C^2 Finsler manifold, with $-\infty < a < b < \infty$. If φ^a is disconnedted and φ^b is a connected set, then φ has at least one critical point in $\varphi^b \setminus \varphi^a$.

Proof

It is sufficient to prove that $\operatorname{cat}_{\varphi^b \varphi^a}(\varphi^b) > 0$, since then, by Th. 1.6, φ has a critical point in

$\varphi^b \setminus \varphi^a$.

If not the relative category is zero, and by proposition 1.3 iii), there exists $H: \varphi^b \times I \to \varphi^b$ continuous such that $H_1(\varphi^b) = \varphi^a$ contradiction (since φ^a is disconnedted and φ^b is a connected set).

Theorem 2.2

Let $\varphi \in C^1(M, R)$ satisfy the Palais-Smale condition on $\varphi^{-1}[a,b]$, where M is a complete C^2 Finsler manifold, with $-\infty < a < b < \infty$. If φ^b contains a connected set J intersecting two different components of φ^a, then φ has a critical point in $\varphi^{-1}(a,b]$.

Proof

Let A and B be two different connected components of φ^a such that $A \cap J \neq \emptyset$ and $B \cap J \neq \emptyset$.

Now copy the proof of theorem 2.1 to get $H_1: \varphi^b \to \varphi^b$ continuous with $H_1(x) = x \ \forall x \in \varphi^a$

and $H_1(\varphi^b) = \varphi^a$. Then, since J is connected, so is $H_1(J)$ which must then be contained in only one connected component of φ^a, contradicting $H_1(J \cap A) = J \cap A$ and $H_1(J \cap B) = J \cap B$.

Corollary 2.3

Let $\varphi \in C^1(M, R)$ satisfy the Palais-Smale condition on $\varphi^{-1}[a,\infty)$, where M is a complete C^2 Finsler manifold, with $-\infty < a < \infty$. If M is connected and φ^a is disconnected for some a then φ has a critical point in $\varphi^{-1}(a, \infty)$.

Proof

Let A and B be two different connected components of φ^a . Choose $y \in A$ and $z \in B$. If M is connected and, being a manifold, is locally path connected, it is path connected. So there is a path in M going from y to z. Denote J the image of that path, then J is compact connected and intersects both A and B. Set $b = \sup_{x \in J} \varphi(x)$, then $J \subset \varphi^b$ and, by theorem 2.2, we get the conclusion.

Corollary 2.4 (The Ambrosetti-Rabinowitz Mountain pass theorem)

Let $\varphi \in C^1(B, R)$ satisfy the Palais-Smale condition on $\varphi^{-1}[a, \infty)$, where B is a Banach space and $a \in R$. If there exists $x, y \in B$ and $r \in R$ such that $\|x-y\| > r$ and,

$$\varphi(z) > b \qquad \text{for all } z \text{ with } \|x-z\| = r, \text{ for some } b > a = \max\{\varphi(x), \varphi(y)\},$$

then φ has a critical point in $B \backslash \varphi^a$.

Proof

Evident from 2.3 and the fact that φ^b is disconnected: in fact φ^b is partitioned into the two closed non empty closed sets $\varphi^b \cap c^{-1}[0,r]$ and $\varphi^b \cap c^{-1}[r, \infty)$ where $c(z) = \|x-z\|$.

3. Example with a non bounded φ

Consider the following system of equations

$$\text{II} \quad \begin{cases} x'' = y + F_x(x, y) \\ y'' = x + F_y(x, y) \\ x(0) = x(T),\ y(0) = y(T),\ x'(0) = x'(T) \text{ and } y'(0) = y'(T) \end{cases}$$

For simplicity let us assume that F, F_x, F_y are continuous and bounded by M. Then the solution of II are the critical points of $\varphi: H \to R$ defined by

$$\varphi(x, y) = \int_0^T \Big[\frac{x'^2}{2} + \frac{y'^2}{2} + xy + F(x, y)\Big]dt$$

where φ is defined on the space $H = H'_T \times H'_T$ and

$$H_T^1 = \{y: [0,T] \to R \mid y' \in L_2, y \text{ is T-periodic}\}$$

is given the scalar products

$$«x, y» = \frac{1}{T} \int_0^T (xy + x'y')\, dt.$$

Let us define $\bar{x} = \frac{1}{T}\int_0^T x\,dt$, then $x = \bar{x} + \tilde{x}$ where $\bar{\tilde{x}} = 0$. Thus we can write

$$(3.1) \qquad \varphi(x, y) = \int_0^T \left[\frac{x'^2}{2} + \frac{y'^2}{2} + \tilde{x}\,\tilde{y} + \bar{x}\,\bar{y} + F(x, y)\right]dt.$$

In order to apply 2.3 we need the following two propositions.

Proposition 3.1

If $T < 2\pi$ then φ satisfies the Palais-Smale condition on H.

Proof

Assume that $S_n = (x_n, y_n) \in H$ where $\varphi(S_n)$ is bounded and $\|\nabla\varphi(S_n)\| \to 0$.

a) $\bar{x}_n\,\bar{y}_n$ is bounded.

In fact

$$(3.2) \qquad «\nabla\varphi(x, y), (a, b)» = \langle x'a'\rangle + \langle y'b'\rangle + \langle y + F_x(x, y), a\rangle + \langle x + F_y(x,y), b\rangle$$

where $\langle x, a\rangle = \frac{1}{T}\int_0^T xa\,dt$.

So choose m such that $\|\nabla\varphi(S_n)\| \le 1$ for all $n>m$ and choose $(a, b) = (1,0)$ the constant map, then

a', b', b = 0 and so

$$1 \ge \left| «\nabla\varphi(S_n), (1,0)» \right| = \left|\langle\tilde{y}_n, 1\rangle + \langle\bar{y}_n, 1\rangle + \langle F_x(x_n, y_n), 1\rangle\right|$$

but $\langle\tilde{y}_n, 1\rangle = 0$ and $\langle F_x(x_n, y_n), 1\rangle$ is bounded by M so $|\bar{y}_n| = |\langle\bar{y}_n, 1\rangle| \le 1+M$

and $\bar{y}_n$ is bounded.

Similarly $\bar{x}_n$ is bounded by 1+M.

b) Let us show that x_n, y_n, x'_n and y'_n are bounded in L_2.

In fact, since by a the last two terms of the integrant of 3.1 and $\varphi(x_n, y_n)$ are bounded, we get that

$$\int_0^T \left[\frac{x_n'^2}{2} + \frac{y_n'^2}{2} + \tilde{x}_n\,\tilde{y}_n\right]dt$$

is bounded but by the Wirtinger inequality ($\omega\|\tilde{x}\|_2 \le \|x'\|_2$ where $\omega = 2\pi T^{-1}$)

it dominates $\frac{1}{2}\int_0^T [(1-\frac{1}{\omega^2})(x_n'^2 + y_n'^2) + (\tilde{x}_n^2 + \tilde{y}_n^2 + 2\tilde{x}_n\tilde{y}_n)]dt$ wich in turn dominates

$$\frac{\omega^2-1}{2\omega^2}\int_0^T [x_n'^2 + y_n'^2]dt.$$

Thus x'_n and y'_n are bounded in L_2.

By the Wirtinger inequality $\tilde{x}_n$ and $\tilde{y}_n$ are also bounded in L_2. By a, so are x_n and y_n.

c) Let us prove that $(x_n, y_n) \to (x_0, y_0)$ for some $(x_0, y_0) \in H$.

By passing to a subsequence, we may assume that $(x_n, y_n) \to (x_0, y_0)$ weakly in H for some

$(x_0, y_0) \in H$. And so $(x_n, y_n) \to (x_0, y_0)$ strongly in C_0, thus strongly in L_2. It remains so show that

$\|x'_n - x'_0\|_2 \to 0$ and $\|y'_n - y'_0\|_2 \to 0$, in fact

$$\langle\langle\nabla\varphi(x_n, y_n), (x_n - x_0, y_n - y_0)\rangle\rangle - \langle\langle\nabla\varphi(x_0, y_0), (x_n - x_0, y_n - y_0)\rangle\rangle$$
$$= \langle x'_n - x'_0, x'_n - x'_0\rangle + \langle y'_n - y'_0, y'_n - y'_0\rangle + \langle y_n - y_0 + F_x(x_n, y_n) - F_x(x_0, y_0), x_n - x_0\rangle$$
$$+ \langle x_n - x_0 + F_y(x_n, y_n) - F_y(x_0, y_0), y_n - y_0\rangle.$$

Now since $\nabla\varphi(x_n, y_n) \to 0$ strongly and $(x_n - x_0, y_n - y_0) \to 0$ weakly and so is bounded, the first term of the left side goes to 0; evidently the second one also goes to zero. The last two terms of the right side also go to zero since for those the convergences are strong. We are left with

$$\|x'_n - x'_0\|_2^2 + \|y'_n - y'_0\|_2^2 \to 0 \quad \text{as } n \to \infty.$$

Proposition 3.2

If $T < 2\pi$ and $|F(x, y)| \leq M$ for all $x, y \in R$ then $\varphi^{-T(M+1)}$ is disconnected.

Proof

a) if $x = \bar{x}$, $y = \bar{y}$ and $\bar{x}\,\bar{y} \leq -(2M+1)$ then $\varphi(x, y) \leq -T(M+1)$.

In fact $x', y', \tilde{x}, \tilde{y} = 0$ and so

$$\varphi(x, y) = \varphi(\bar{x}, \bar{y}) = \int_0^T [\bar{x}\,\bar{y} + F(x, y)]dt$$

$$\leq T\bar{x}\,\bar{y} + TM \leq -T(2M+1) + TM.$$

b) If $\bar{x}\,\bar{y} = 0$ then $\varphi(x, y) \geq -TM$.

In fact, as in b of the proof of 3.1 we have that

$$\varphi(x, y) \geq \frac{\omega^2-1}{2\omega^2} \int_0^T [x'^2_n + y'^2_n]dt + \int_0^T \bar{x}\,\bar{y}\,dt + \int_0^T F(x, y)$$

$$\geq 0 + 0 - TM.$$

c) We have the conclusion.

In fact, since $\psi: H'_T \to R$ defined by $\psi(x) = \bar{x}$ is continuous, and we get that

$$U_1 = \varphi^{-T(M+1)} \cap [\psi^{-1}(-\infty, 0) \times H'_T]$$

and

$$U_2 = \varphi^{-T(M+1)} \cap [\psi^{-1}(0, \infty) \times H'_T]$$

are open disjoint containing respectively the constant maps $(-1, 2M+1)$ and $(1, -2M-1)$; more over, since

by b) $\bar{x} \neq 0$ for any $(x, y) \in \varphi^{-T(M+1)}$, their union is $\varphi^{-T(M+1)}$.

Using Theorem 2.1, from 3.1 and 3.2, we get trivialy the existence of a T periodic solution of II.

<u>Corollary 3.3</u>

If F, F_x, F_y are continuous and bounded, then II has a T-periodic solution for all $T < 2\pi$.

7. BIBLIOGRAPHY

[1] A. Ambrosetti and P.H.Rabinowitz, Dual variational methods in critical point theory and applications, J. Funct. Anal. 14 (1973), 349-381.

[2] K.-C.Chang, Infinite Dimentional Morse Theory and its Applications, S.M.S. les presses de l'Université de Montréal, (1985).

[3] D.C. Clark, A variant of the Lusternick-Schnirelman Theory, Indiana J. Math 22, (1972), 65-74.

[4] W.M. Ni, Some Minimax Principles and their Applications in nonlinear Elliptic Equations, Journal d'analyse mathématiques 37(1980), 248-275.

[5] E. Fadell, Cohomological methods in non-free G-spaces with applications to general Borsuk-Ulam theorems and critical point theorems for invariant functionals, Nonlinear Functional Analisis and its Applications, S.P. Singh (ed.) D. Reidel Publishing Company, (1986), 1-45.

[6] G. Fournier and M. Willem, Multiple Solutions of the Forced Double Pendulum Equation, Analyse non linéaire, Contributions en l'honneur de J.-J. Moreau, eds. H. Attouch, J.-P. Aubin, F. Clarke, I. Ekeland, CRM - Gauthier-Villars, Paris (1989), 259-281.

[7] R.S. Palais, The Lusternik-Schnirelman theory on Banach manifolds, Topology 5 (1966), 115-132.

[8] R.S. Palais, Critical point theory and the minimax principle, Global Analysis, Proc. Symp. Pure Math.15 (ed. S.S. Chern), A.M.S., Providence, (1970), 185-202.
[9] G. Tian, On the mountain pass theorem, Kexue Tongbao 29(1984),1150-1154.

G. Fournier and M.Willem
University of Sherbrooke
Sherbrooke
Canada

University of Louvain
L.L.N.
Belgium

The Lefschetz Function of a Point

BY JOHN FRANKS AND DAVID FRIED

Abstract

Associated to an isolated periodic point of a continuous self map on a manifold is an invariant called the Lefschetz function. It is a formal power series which encapsulates the values of the Lefschetz index of the fixed point for all iterates of the map. We prove that for a homeomorphism of a manifold of dimension at least three any formal power series with integer coefficients and leading term 1 can be realized as the Lefschetz function of a fixed point.

In this note we consider the Lefschetz function of an isolated periodic point of a homeomorphism. We will show that on manifolds of dimension three or greater any formal power series with integer coefficients and leading term 1 can occur as the Lefschetz function of an isolated fixed point. This should be compared with results which show that for a C^1 self map of a manifold the Lefschetz function of a point must be rational (see [Fr] for results and references). Results of M. Brown [B] deal with the case of two dimensional manifolds where the outcome is quite different.

§1 BACKGROUND AND DEFINITIONS

Suppose $f : M \to M$ is a homeomorphism of a manifold M, and $p \in M$ is a fixed point which is an isolated fixed point of f^n for all positive n. We will say that p is an *isolated periodic point* of f (of period 1). We will denote by $L(f^n, p)$ the Lefschetz index of the point p with respect to the map f^n. If Λ is a compact f invariant subset of M and $\Lambda \cap Fix(f^n)$ is open in $Fix(f^n)$, we will denote by $L(f^n, \Lambda)$ the Lefschetz index of Λ with respect to f^n (see [D]). If f^n has finitely many fixed points in Λ and they are isolated in M, then $L(f^n, \Lambda)$ is equal to the sum of $L(f^n, p)$ for all $p \in \Lambda$ fixed under f^n.

(1.1) Definition. The Lefschetz function $Z(f, \Lambda)$ of the set Λ, with respect to the map f is defined to be the formal power series in t given by

$$Z(f, \Lambda) = exp\,(\sum_{n=1}^{\infty} \frac{L(f^n, \Lambda)}{n} t^n).$$

We have chosen to use the term Lefschetz function for this power series rather than the more standard term *homology zeta function* because we feel it is less cumbersome and avoids confusion with cases (considered elsewhere) where one puts homology classes into the function as coefficients.

It is important to have an expression for this Lefschetz function as an infinite product with each term representing the contribution of a single periodic orbit. Recall that a map $f : M \to M$ is said to have a *transversal periodic point* p of period n

provided $f^n(p) = p$ and Df_p^n has no eigenvalues of absolute value 1. Let E^u be the space spanned by the generalized eigenspaces of Df_p^n corresponding to eigenvalues with modulus greater than 1. The dimension of E^u is called the *Morse index* of p. We define $\Delta(p)$ to be ± 1 depending on whether Df_p^n preserves or reverses orientation of E^u. It is the same for any two points on the same periodic orbit and we will refer to it as the *twist sign* of the orbit. The following result can be found in [F2] or as a special case of a result in [Fr] (also see [F1]).

(1.2) PROPOSITION. *If $f^n : M \to M$ has only isolated fixed points in the compact set Λ for every $n > 0$ then, as a formal power series,*

$$Z(f, \Lambda) = \prod_{\gamma \in PO} Z(f, \gamma),$$

where PO is the set of periodic orbits of f in Λ. Moreover, if the points of γ are transversal periodic points,

$$Z(f,\gamma) = \begin{cases} 1 - \Delta(\gamma) t^{p(\gamma)} & \text{if the Morse index of } \gamma \text{ is odd,} \\ 1/(1 - \Delta(\gamma) t^{p(\gamma)}) & \text{otherwise,} \end{cases}$$

$p(\gamma)$ is the least period of γ, and $\Delta(\gamma)$ is the twist sign of γ as defined above.

We shall also have need of a fact from [F2] about formal power series with integer coefficients. We denote by G the set of formal power series with constant term 1 and integer coefficients. It is easy to see that G is a group under multiplication since one can solve recursively for the coefficients of $\beta(t)^{-1}$ for any $\beta(t) \in G$.

(1.3) PROPOSITION. *If $\beta(t) \in G$ then $\beta(t)$ can be written uniquely as an infinite product*

$$\beta(t) = \prod_{n=1}^{\infty} (1 - \Delta_n t^n)^{m(n)},$$

where $m(n)$ is an integer ≥ 0 and Δ_n is 1 or -1.

This means that for each $\beta(t) \in G$ there are unique choices for $m(n)$, and for Δ_n if $m(n) \neq 0$, which make the above equality valid. A proof can be found in [F].

§2 PROOF OF THE MAIN THEOREM

In this section, through a series of lemmas, we develop the proof that the Lefschetz function for a point is arbitrary for a homeomorphism of a manifold of dimension greater than 2. We begin with some constructions in the plane.

(2.1) LEMMA. *Let $A[a, b]$ denote the annulus in the plane, centered at the origin, with inner radius a and outer radius b. Given $n \geq 0$ and an irrational angle θ_0, there exists a diffeomorphism*

$$f : A[a, b] \to A[a, b]$$

with the following properties:

i) The restriction of f to either boundary component is a rotation through the angle θ_0.

ii) The only periodic points of f are a periodic sink of period n and a periodic saddle of period n with twist sign $\Delta(p) = 1$.

PROOF: See Fig. 1 for a picture of such a diffeomorphism. Its explicit construction is an easy exercise. ■

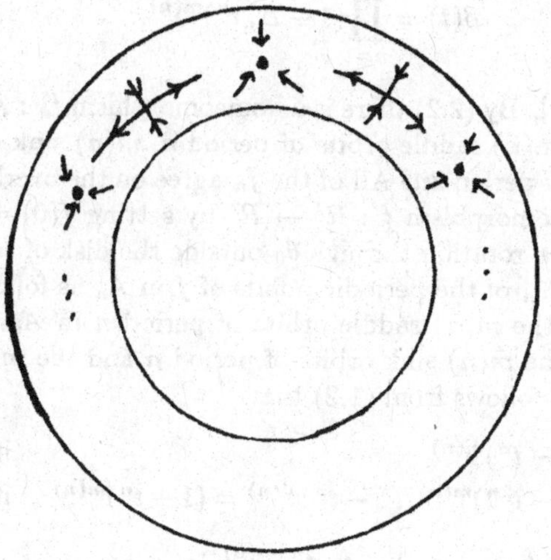

Figure 1

(2.2) **LEMMA.** *Let $A[a,b]$ be as above. Given m and $n \geq 0$ and an irrational angle θ_0, there exists a diffeomorphism*

$$f : A[a,b] \to A[a,b]$$

with the following properties:

i) The restriction of f to either boundary component is a rotation through the angle θ_0.

ii) The only periodic orbits of f are $2m$ orbits which are sinks, m of period n and m of period $2n$, and $2m$ saddles, m of period n and m of period $2n$. Moreover, the saddle orbits all have twist sign $\Delta(p) = 1$.

PROOF: Let $h = (b-a)/2m$. Using lemma (2.1) construct f_i for $1 \leq i \leq m$ on $A[a+(i-1)h, a+ih]$ with a sink and saddle of period n. Similarly on $A[(a+b)/2 + (i-1)h, (a+b)/2 + ih]$ construct g_i with a sink and saddle of period $2n$. Define f on $A[a,b]$ by

$$f(x) = \begin{cases} f_i(x) & \text{if } x \in A[a+(i-1)h, a+ih], \\ g_i(x) & \text{if } x \in A[(a+b)/2 + (i-1)h, (a+b)/2 + ih]. \end{cases}$$

■

(2.3) THEOREM. *Let $\beta(t)$ be an arbitrary element of $\{1+tZ[[t]]\}$ and let M be a manifold of dimension at least three. There exists a homeomorphism $f: M \to M$ and a point $p \in M$ such that p is an isolated fixed point of f^n for all $n > 0$ and $Z(f,p) = \beta(t)$.*

PROOF: By (1.3) $\beta(t)$ can be written as

$$\beta(t) = \prod_{n=1}^{\infty} (1 - \Delta_n t^n)^{m(n)}.$$

Let $A_n = A[1/2^{n+1}, 1/2^n]$. By (2.2) there is a homeomorphism $f_n : A_n \to A_n$ with $m(n)$ sink orbits of period n, $m(n)$ saddle orbits of period n, $m(n)$ sink orbits of period $2n$, and $m(n)$ saddle orbits of period $2n$. All of the f_n agree on the overlap of their domains so we can define a homeomorphism $f : R^2 \to R^2$ by setting $f(0) = 0, f(x) = f_n(x)$ if $x \in A_n$ and letting f be a rotation through θ_0 outside the disk of radius 1/2.

We define a subset K_n of the periodic points of f in A_n as follows. If $\Delta_n = 1$ then let K_n be the union of the $m(n)$ saddle orbits of period n in A_n. If $\Delta_n = -1$ then let K_n be the union of the $m(n)$ sink orbits of period n and the $m(n)$ saddle orbits of period $2n$ in A_n. It then follows from (1.2) that

$$Z(f, K_n) = \begin{cases} (1-t^n)^{m(n)} & \text{if } \Delta_n = 1, \\ (1-t^{2n})^{m(n)}/(1-t^n)^{m(n)} = (1+t^n)^{m(n)} & \text{if } \Delta_n = -1. \end{cases}$$

In either case we have $Z(f, K_n) = (1 - \Delta_n t^n)^{m(n)}$.

We define a compact f invariant set K in the plane by

$$K = \bigcup_{n=1}^{\infty} K_n \cup \{0\}.$$

The origin 0 is an isolated fixed point of f^n for every $n > 0$. Moreover, $L(f^n, 0) = 1$ for all n as can be seen by calculating the index on a boundary circle of A_N, for some $N > 2n$. Inside this circle 0 is the only fixed point of f^n and on the boundary f is a rotation. Thus one calculates easily (see [F1]) that $Z(f,0) = 1/(1-t)$.

It now follows from (1.2) that

$$Z(f,K) = Z(f,0) \prod_{n=1}^{\infty} Z(f, K_n) = (1-t)^{-1} \prod_{n=1}^{\infty} (1-\Delta_n t^n)^{m(n)}.$$

In other words $Z(f,K) = \beta(t)/(1-t)$.

Identifying the two sphere S^2 with $R^2 \cup \{\infty\}$ via stereographic projection, we obtain a homeomorphism of S^2 which we also denote by f. Considering S^2 as the unit sphere in R^3 we can extend f to a homeomorphism F of all of R^3 by setting

$$F(\rho, \phi, \theta) = (\rho^{1/2}, f(\phi, \theta)),$$

where (ρ, ϕ, θ) are spherical co-ordinates on R^3. This makes S^2 an attracting invariant manifold for F and F restricted to S^2 is equal to f. From this it follows easily (see [D]) that for any fixed point $p \in S^2$ of f^n, $L(f^n, p) = L(F^n, p)$. Hence $Z(f,K) = Z(F,K) = \beta(t)/(1-t)$. It is easy and useful at this point to alter F by an isotopy supported outside a ball of radius 2 so that outside a ball of radius 3 it will be the identity. This new homeomorphism will also be called F.

If we define Λ to be the set $\{sx : x \in K, s \in [0,1]\}$, i.e. the cone from K to $0 \in R^3$, then $F(\Lambda) \subset \Lambda$. The periodic points of F in Λ are the periodic points of f in K together with the fixed point $0 \in R^3$. The set Λ is a compact contractible subset of R^3. Since 0 is a repelling fixed point of F, it is an easy calculation that $L(F^n, 0) = -1$ for all $n > 0$. It follows that $Z(F, 0) = (1-t)$. From (1.2) we see that $Z(F, \Lambda) = Z(F, 0)Z(F, K) = \beta(t)$.

Finally if we let $M_0 = R^3/\Lambda$, i.e. if we collapse Λ to a point, then M_0 is homeomorphic to R^3 and F induces a homeomorphism on M_0 which we will denote F_0. To show that M_0 is homeomorphic to R^3, it suffices to show that $R^3 - \Lambda$ is homeomorphic to $R^3 - 0$. The function $h : (R^3 - \Lambda) \to (R^3 - 0)$, given by

$$h(x) = dist(x, \Lambda)\frac{x}{\|x\|}$$

is easily seen to be a homeomorphism, since it preserves rays through the origin and is monotonic on each such ray.

If p is the point to which Λ is collapsed then $L(F^n, \Lambda) = L(F_0^n, p)$ for all $n > 0$, since if r is sufficiently large, both these integers are the degree of the map $g : S_r \to S_r$, where S_r is the sphere of radius r, and $g(x) = r(x - F(x))/\|x - F(x)\|$. Hence $Z(F_0, p) = Z(F, \Lambda) = \beta(t)$.

Since outside of some compact neighborhood of p the homeomorphism F_0 is the identity, it is easy to embed this example in an arbitrary three manifold M. For higher dimensions let $F_1 : R^k \to R^k$ be a homeomorphism which has 0 as a contracting fixed point and which is the identity outside of the ball of radius 1. Then $F_0 \times F_1 : R^{k+3} \to R^{k+3}$ satisfies $Z(F_0 \times F_1, p \times 0) = \beta(t)$. As before it can easily be embedded in an example in an arbitary manifold of dimension greater than three. ∎

References

[B] M. Brown, *On the index of Iterates of Planar Homeomorphisms*, preprint.

[D] A. Dold, *Fixed point index and fixed point theorem for Euclidean neighborhood retracts*, Topology, **4** (1965), 1-8.

[F1] J. Franks, *Homology and Dynamical Systems*, CBMS Regional Conf. Ser. in Math., **49** Amer. Math. Soc., Providence, P.R., 1982.

[F2] J. Franks, *Period Doubling and the Lefschetz Formula*, Trans. Amer. Soc. **287** (1985), 275–283.

[Fr] D. Fried, *Periodic Points and Twisted Coefficients*, in *Geometric Dynamics*, Proceedings Rio de Janeiro 1981, Springer Lecture Notes in Math. **1007**, 261-293.

NIELSEN-TYPE NUMBERS FOR PERIODIC POINTS, I

Philip Heath, Renzo Piccinini and Chengye You
Memorial University of Newfoundland and Peking University
St.John's, Canada and Beijing, China

0 - Introduction - † For simplicity of exposition, all spaces considered here will be compact, path-connected ANR's and thus, admitting an index theory (cf. [2], where an axiomatic index theory is developed in a more general context).

Let $f : X \to X$ be a given self-map and n a positive integer; the fixed points of its iterates are called *periodic points;* the elements of $\Phi(f^n) = \{x \in X : f^n(x) = x\}$ are said to have *period n*. If $m \mid n$ and x is a periodic point of period m, then x is also a periodic point of period n. In what follows

$$P_n(f) = \Phi(f^n) \setminus \bigcup_{m<n} \Phi(f^m)$$

denotes the set of periodic points of *least period n*. We are interested in the numbers

$$M\Phi_n(f) := \min\{\#\ \Phi(g^n) : g \simeq f\},$$

$$MP_n(f) := \min\{\#\ P_n(g) : g \simeq f\},$$

where the symbol $\#$ denotes the cardinality of the set to which it refers. Note that

$$M\Phi_1(f) = MP_1(f) = M\Phi(f) = \min\{\#\ \Phi(g) : g \simeq f\}$$

(the number $M\Phi(f)$ is defined in [6], Chapter 1, Section 6).

---

† - This paper is in final form and no version of it will be submitted for publication elsewhere.

The actual computation of these numbers is not, in general, easy; the following example is useful in elucidating the definitions given above and motivating what follows. Define $f : S^1 \to S^1$ by $f(e^{i\theta}) = e^{3i\theta}$. For every $n \geq 1$, $\Phi(f^n) = \{ e^{2\pi i k/(3^n-1)} : k = 1,\ldots,3^n - 1 \}$. Since $\# \Phi(f^n) = 3^n - 1$ and $N(f^n) = 3^n - 1$ (see [6], page 2) then it is easy to see that $M\Phi_n(f) = 3^n - 1$.

The situation for $MP_n(f)$ in this example is more complex. There are 80 points on S^1 with period 4; however, the complex numbers $e^{2\pi i k/(3^n-1)}$, with k a multiple of 10, are also points of period ≤ 2; so, $MP_4(f) = (3^4-1) - (3^2-1) = 72$. Later on, we shall see that in our example, if $n = p^r$, where p is a prime, $MP_n(f) = 3^n - 3^{n/p}$.

One objective of this paper and its sequel is to give Nielsen type lower bounds $NP_n(f)$ and $N\Phi_n(f)$ of $MP_n(f)$ and $M\Phi_n(f)$, respectively. Some attemps have already been made to define a lower bound for $MP_n(f)$, for example, by Halpern [4] and Jiang [6]. Our definition follows Jiang's more closely than Halpern's, but agreeing with the philosophy of [5], we work only with the fundamental group rather than with the fundamental group and the covering space. Our aim is to present this simpler theory and to make some calculations of these lower bounds.

In this first paper we will consider only the number $NP_n(f)$, reserving the study of $N\Phi_n(f)$ to the sequel. We start in Section 1 by discussing the geometry, restricting our considerations to the purely topological. In some instances this gives better information. We then relate the geometry to algebraic considerations on the orbit sets of the Reidemeister operations on the fundamental group. In order to do this we need to choose a base point $x_o \in X$ and a path $\omega : x_o \to f(x_o)$. At first sight this may appear more restrictive than Jiang's approach which is base point free. However, in practice Jiang must always "choose coordinates" [6], page 59, and so, in effect he is choosing a base point. One of the supposed advantages of the covering space approach is that it is able to count possibly empty fixed point classes. As in [5] we get around this problem by assigning an index to the Reidemeister classes. In this way not only are we able to count possibly empty classes, but we are also able to easily distinguish between them. Our approach then differs from Jiang's in two ways: firstly, because we distinguish between geometric and algebraic fixed point classes and secondly, because we use only the fundamental group in our approach.

In the second section we define $NP_n(f)$, prove some of its properties, and show that a sum of such numbers is a provisional lower bound for $M\Phi_n(f)$. In the third section we prove an analogue of Dold's theorem on the divisibility of the index of periodic points, and make some calculations. As corollaries of our considerations we obtain some results on tori. The last section is an appendix in which some of the technical details on the operations with algebraic fixed point classes are proved.

We would like to thank E. Jespers and S. Wilson for useful conversations.

1 - Geometric and Algebraic Periodic Point Classes - Two periodic points $x, y \in \Phi(f^n)$ are said to be *equivalent* if there exists a path $\lambda : I \to X$ such that $\lambda(0) = x$, $\lambda(1) = y$ and λ is homotopic to $f^n(\lambda)$ rel. end points (notation: $\lambda \simeq f^n(\lambda)$). We denote by $\Phi(f^n)/\sim$ the set of equivalence classes of $\Phi(f^n)$ by this equivalence relation. The elements of $\Phi(f^n)/\sim$ are called *(geometric) periodic point classes* of f^n.

Let $\mathbf{F}^{(n)} \in \Phi(f^n)/\sim$ and $x \in \mathbf{F}^{(n)}$ be given; let $f(\mathbf{F}^{(n)})$ be the unique class defined by $f(x)$; observe that $f(\mathbf{F}^{(n)})$ is independent of the choice of x because if $\lambda : I \to X$ is a path connecting $x, y \in \Phi(f^n)$ and such that $\lambda \simeq f^n(\lambda)$, then the path $f(\lambda)$ connects $f(x)$ to $f(y)$ and $f(\lambda) \simeq f(f^n(\lambda)) = f^n(f(\lambda))$. Thus f defines a function $\Phi(f^n)/\sim \to \Phi(f^n)/\sim$ which, by abuse of notation, we also denote by f. Notice that $f^n : \Phi(f^n)/\sim \to \Phi(f^n)/\sim$ is the identity function. We take advantage of this fact to give the following:

Definition 1.1 - The f *-orbit* of the periodic point class $\mathbf{F}^{(n)}$ is the set

$$< \mathbf{F}^{(n)} > = \{ \mathbf{F}^{(n)}, f(\mathbf{F}^{(n)}), \cdots f^{l(<\mathbf{F}^{(n)}>)-1}(\mathbf{F}^{(n)}) \}$$

where $l(<\mathbf{F}^{(n)}>)$ - called the *length* of the orbit $<\mathbf{F}^{(n)}>$ - is the smallest positive integer r such that $f^r(\mathbf{F}^{(n)}) = \mathbf{F}^{(n)}$; using the division algorithm one easily checks that $l(\mathbf{F}^{(n)}) \mid n$.

For $m \mid n$, let $\mathbf{F}^{(m)} \in \Phi(f^m)/\sim$ and $x \in \mathbf{F}^{(m)}$ be given. Define $\gamma(\mathbf{F}^{(m)})$ as the unique class of $\Phi(f^n)/\sim$ determined by x. This correspondence gives a well-defined function $\gamma : \Phi(f^m)/\sim \to \Phi(f^n)/\sim$. Notice that γ need not be injective;

however, as a set $\mathbf{F}^{(m)} \subset \gamma(\mathbf{F}^{(m)})$ and thus, if $\mathbf{F}^{(n)} = \gamma(\mathbf{F}^{(m)})$, we say that $\mathbf{F}^{(m)}$ is *contained* in $\mathbf{F}^{(n)}$ and we write $\mathbf{F}^{(m)} \subset \mathbf{F}^{(n)}$. These ideas lead to the next definition:

Definition 1.2 - The *(geometric) depth* of $\mathbf{F}^{(n)}$ - denoted by $d(\mathbf{F}^{(n)})$ - is the smallest integer m for which there exists a geometric periodic point class $\mathbf{F}^{(m)} \in \Phi(f^m)/\sim$ such that $\mathbf{F}^{(m)} \subset \mathbf{F}^{(n)}$. If the depth of $\mathbf{F}^{(n)}$ is n itself, the class $\mathbf{F}^{(n)}$ is said to be *irreducible.*

It is easy to see that if $m = d(\mathbf{F}^{(n)})$ then any element of the orbit $< \mathbf{F}^{(n)} >$ has the same depth m, that is to say, depth is a property of orbits.

Proposition 1.3 - *Let $\mathbf{F}^{(n)} \in \Phi(f^n)/\sim$ be a periodic point class with orbit length l and depth d. Then, $l \mid d$ and $\mathbf{F}^{(n)}$ contains at least d/l points. In particular, an irreducible periodic point class $\mathbf{F}^{(n)}$ of length l has at least n/l periodic points.*

Proof - Let $\mathbf{F}^{(d)} \subset \mathbf{F}^{(n)}$ be irreducible; since f^d is the identity on $\Phi(f^d)/\sim$ and $f^d : \Phi(f^n)/\sim \rightarrow \Phi(f^n)/\sim$ is well defined, we have that $f^d(\mathbf{F}^{(n)}) = \mathbf{F}^{(n)}$ and so $l \mid d$ by the division algorithm. Further, for any $x \in \mathbf{F}^{(d)}$, the set

$$S = \{x, f(x), \cdots, f^{d-1}(x)\} \subset < \mathbf{F}^{(d)} > \subset < \mathbf{F}^{(n)} >$$

has distinct elements by the minimality of d.

Let $d = nl$ and consider the set

$$S_1 = \{x, f^l(x), f^{2l}(x), \cdots, f^{(n-1)l}(x)\}.$$

Clearly $S_1 \subset S$ and contains d/l distinct elements. Finally, since $f^l(\mathbf{F}^{(n)}) \subset \mathbf{F}^{(n)}$, then $S_1 \subset \mathbf{F}^{(n)}$, proving our result. □

Corollary - *If $\Phi(f) = \emptyset$ and p is a prime number with $\Phi(f^p) \neq \emptyset$ then,* $\# \Phi(f^p) \geq p$.

Proof - Let $\mathbf{F}^{(p)}$ be a non-empty periodic point class with orbit length l. Since $\Phi(f) = \emptyset$, $\mathbf{F}^{(p)}$ is irreducible and so, each element of its orbit contains at least p/l periodic points. The result follows. □

We can regard the periodic points of a self-map from an algebraic point of view (compare with [5] where this is done for the fixed point classes of a self-map). As in [5], we shall be forced to choose a base-point x_o for the space X and also, select a path ω in X connecting x_o to $f(x_o)$; it turns out that our definitions are actually independent of both the base point x_o and the path ω (see Appendix). Note that we shall fail to distinguish between a path in X and its class in the fundamental groupoid of X.

Throughout the paper we denote the fundamental group of X based at x_o by

$$\pi := \pi_1(X, x_o).$$

For $n \geq 1$, let $n(\omega): I \to X$ be the path from x_o to $f^n(x_o)$ defined by $\omega + f(\omega) + \cdots + f^{n-1}(\omega)$; then, for each integer $n \geq 1$, define the homomorphism $f^{n\omega}: \pi \to \pi$ by $f^{n\omega}(\alpha) = n(\omega) + f^n(\alpha) - n(\omega)$, for every $\alpha \in \pi$ (we use the commutative notation although, in general, π need not be abelian).

From now on, the function $f^{1\omega}$ will be denoted simply by f^{ω}; if ω is the constant path, $f^{n\omega}$ will be denoted simply by f^n (or f, if $n = 1$) when it is clear from the context if f^n is the topological or algebraic function.

We notice that a simple induction procedure shows that, for every $n \geq 1$,

(*F* 1) $\quad (f^{\omega})^n = f^{n\omega}$,

(*F* 2) $\quad (n+1)(\omega) = n(\omega) + f^n(\omega) = \omega + f(n(\omega))$,

(*F* 3) $\quad$ for every $m, r \geq 1$, $f^{m\omega} \circ f^{r\omega} = f^{(m+r)\omega}$. Notice that $(f^n)^{n(\omega)} = f^{n\omega}$.

As in the case for $n = 1$ (see [5]), the group homomorphism $f^{n\omega}$ defines an equivalence relation on π by setting $\alpha \underset{n}{\sim} \alpha'$ if there exists a $\beta \in \pi$ such that $\alpha = \beta + \alpha' - f^{n\omega}(\beta)$. Let *Coker* $(1 - f^{n\omega})$ be the quotient set of π by this equivalence relation. The *Reidemeister number of* f^n is the number $R(f^n) = \#$ *Coker* $(1 - f^{n\omega})$. In what follows, $j: \pi \to$ *Coker* $(1 - f^{n\omega})$ denotes the quotient function; moreover, the following notation will be used consistently: if $\alpha \in \pi$, then $j(\alpha)$ will be denoted by $[\alpha]^{(n)} \in$ *Coker* $(1 - f^{n\omega})$. An element $[\alpha]^{(n)} \in$ *Coker* $(1 - f^{n\omega})$ is said to be an *(algebraic) periodic point class of* f^n. Notice that the set *Coker* $(1 - f^{n\omega})$ is a based set with base point $[0]^{(n)} = j(0)$. Recall that $f: X \to X$ is *eventually commutative* if $f^{k\omega}(\pi)$ is abelian, for some k.

Let $Fix\ f^{n\omega}$ be the subgroup of all elements of π which are left fixed by $f^{n\omega}$. We wish to warn the reader to distinguish carefully between $Fix\ f^{n\omega}$ (algebraic) and $\Phi(f^n)$ (geometric).

Proposition 1.4 - *The sequence of based sets and base-point preserving functions*

$$1 \rightarrow Fix\ f^{n\omega} \rightarrow \pi \xrightarrow[1-f^{n\omega}]{} \pi \xrightarrow[j]{} Coker\ (1-f^{n\omega}) \rightarrow 1$$

is exact. The function $1-f^{n\omega}$ *is injective if, and only if,* $Fix\ f^{n\omega} = 0$. *Moreover, if* f *is eventually commutative there is a canonical abelian group structure on* $Coker\ (1-f^{n\omega})$ *under which* j *is a group homomorphism.*

Proof - Replace f by $f^{n\omega}$ in [5], Section 1. □

Lemma 1.5 - *The homomorphism* $f^{\omega}: \pi \rightarrow \pi$ *induces a function* $[f^{\omega}]: Coker\ (1-f^{n\omega}) \rightarrow Coker\ (1-f^{n\omega})$ *satisfying the following properties:*

(i) $[f^{\omega}] \cdot j = j \cdot f^{\omega}$

(ii) $([f^{\omega}])^n$ *is the identity function; in particular,* $[f^{\omega}]$ *is a bijection;*

(iii) if f *is eventually commutative,* $[f^{\omega}]$ *is a homomorphism of abelian groups.*

Proof - We first prove that $[f^{\omega}]$ is well-defined. Suppose that α, $\alpha' \in \pi$ are in the same equivalence class, i.e., $\alpha' = \beta + \alpha - f^{n\omega}(\beta)$ for some $\beta \in \pi$. Then,

$$f^{\omega}(\alpha') = f^{\omega}(\beta) + f^{\omega}(\alpha) - f^{\omega}(f^{n\omega}(\beta)) = f^{\omega}(\beta) + f^{\omega}(\alpha) - f^{n\omega}(f^{\omega}(\beta))$$

and thus, $f^{\omega}(\alpha) \underset{\pi}{\sim} f^{\omega}(\alpha')$.

Part (i) of the proof just follows from the definitions. To see (ii), replace f by $f^{n\omega}$ in [5], 1.4. For (iii), use the techniques of [5], 1.8, 1.11 and 1.13. □

The first part of (ii) in the previous lemma leads us to the following:

Definition 1.6 - The *(algebraic) orbit* of the element $[\alpha]^{(n)} \in Coker\ (1-f^{n\omega})$ is the set $< [\alpha]^{(n)} > = \{ [\alpha]^{(n)}, [f^{\omega}]([\alpha]^{(n)}), \cdots, ([f^{\omega}])^{l(<[\alpha]^{(n)}>)-1}([\alpha]^{(n)}) \}$ where $l(< [\alpha]^{(n)} >)$ - called *length* of the orbit $< [\alpha]^{(n)} >$ - is the smallest positive integer r such that $([f^{\omega}])^r([\alpha]^{(n)}) = [\alpha]^{(n)}$; as in the geometric case, $l(< [\alpha]^{(n)} >) \mid n$. Note that $[f^{\omega}]$ is the identity on $Coker\ (1-f^{\omega})$, so $l(< [\alpha]^{(1)} >) = 1$ for all $[\alpha]^{(1)}$.

Example 1.7 - Define $f : S^1 \rightarrow S^1$, as in the Introduction, by $f(e^{i\theta}) = e^{3i\theta}$. Take

$e^{i0}=1$ as base point of S^1 and choose the path ω to be constant. Then, $1-f^{n}:\mathbf{Z}\to\mathbf{Z}$ is multiplication by $1-3^n$, $Coker\,(1-f^{n})=\mathbf{Z}_{3^n-1}$ and $[f]$ is multiplication by 3. Let us denote the elements of $Coker\,(1-f^{n})$ by $[i]^{(n)}$, $i=0,1,\ldots,3^n-1$. Then $l(\langle [0]^{(4)}\rangle) = l(\langle [40]^{(4)}\rangle)=1$ and $\langle [1]^{(4)}\rangle = \{[1]^{(4)},[3]^{(4)},[9]^{(4)},[27]^{(4)}\}$; so, $l(\langle [1]^{(4)}\rangle)=4$.

For $m \mid n$, consider the following function from π to itself

$$\iota_{m,n} = \sum_{i=0}^{(n/m)-1} f^{\,im\omega} = 1 + f^{\,m\omega} + \cdots + f^{\,(n-m)\omega}.$$

The following result can be proved easily by induction.

Lemma 1.8 - *The function $\iota_{m,n}$ induces a function*

$$[\iota_{m,n}] : Coker\,(1-f^{\,m\omega}) \to Coker\,(1-f^{\,n\omega})$$

such that the following diagram is commutative:

$$\begin{array}{ccccccc}
Fix\, f^{\,m\omega} & \longrightarrow & \pi & \xrightarrow{1-f^{\,m\omega}} & \pi & \xrightarrow{j} & Coker\,(1-f^{\,m\omega}) \\
\downarrow i & & \downarrow 1 & & \downarrow \iota_{m,n} & & \downarrow [\iota_{m,n}] \\
Fix\, f^{\,n\omega} & \longrightarrow & \pi & \xrightarrow{1-f^{\,n\omega}} & \pi & \xrightarrow{j} & Coker\,(1-f^{\,n\omega})
\end{array}$$

here i is the restriction of the identity; in particular, $Fix\, f^{\,m\omega}$ is a subgroup of $Fix\, f^{\,n\omega}$. Moreover, if $k \mid m \mid n$, $[\iota_{m,n}]\cdot[\iota_{k,m}] = [\iota_{k,n}]$. Finally, if f is eventually commutative, then $[\iota_{m,n}]$ is an abelian group homomorphism. □

Definition 1.9 - An element $[\alpha]^{(n)} \in Coker\,(1-f^{\,n\omega})$ is *reducible to* m if there exists $[\beta]^{(m)} \in Coker\,(1-f^{\,m\omega})$ such that $[\iota_{m,n}]([\beta]^{(m)}) = [\alpha]^{(n)}$; we say that $[\beta]^{(m)}$ is *contained* in $[\alpha]^{(n)}$ or that $[\alpha]^{(n)}$ *contains* $[\beta]^{(m)}$. Notice that if $[\alpha]^{(n)}$ is reducible to m, then $m \mid n$. If $[\alpha]^{(n)}$ is not reducible to any $m < n$ it is said to be *irreducible*.

Definition 1.10 - The *(algebraic) depth* of an element $[\alpha]^{(n)} \in Coker\,(1-f^{\,n\omega})$ is the smallest positive integer m to which $[\alpha]^{(n)}$ is reducible. Notation: $d([\alpha]^{(n)})$; notice that $d([\alpha]^{(n)}) \mid n$ and for any n, $d([0]^{(n)}) = 1$.

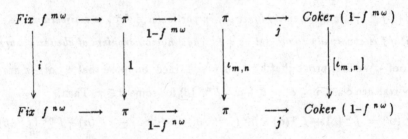

Example 1.11 - Take the function $f : S^1 \to S^1$ of Example 1.7. In the particular cases $n = 1,2$ and 4, $Coker\ (1\text{-}f^{n\omega})$ is equal to $\mathbf{Z}_2$, $\mathbf{Z}_8$ and $\mathbf{Z}_{80}$, respectively; moreover, $[\iota_{1,4}]$ is multiplication by $1+3+3^2+3^3 = 40$ and $[\iota_{2,4}]$ is multiplication by $1+3^2 = 10$. Then, $d([0]^{(4)}) = d([40]^{(4)}) = 1$, $d([10]^{(4)}) = 2$ and $d([1]^{(4)}) = 4$.

The next result shows that depth is a property of orbits (cf. the geometric case).

Lemma 1.12 - *If $[\alpha]^{(n)}$ is reducible to m, so is any element of its orbit.*

Proof - An induction argument shows that the diagram

$$\begin{array}{ccc} Coker\ (1\text{-}f^{m\omega}) & \xrightarrow{[\iota_{m,n}]} & Coker\ (1\text{-}f^{n\omega}) \\ \Big\downarrow [f^{\omega}] & & \Big\downarrow [f^{\omega}] \\ Coker\ (1\text{-}f^{m\omega}) & \xrightarrow{[\iota_{m,n}]} & Coker\ (1\text{-}f^{n\omega}) \end{array}$$

is commutative; the result follows. □

Now we can define the *depth of the orbit* $[\alpha]^{(n)}$: $d(<[\alpha]^{(n)}>) = d([\alpha]^{(n)})$.

Lemma 1.13 - *The length l of an orbit divides its depth d.*

Proof - Suppose that $[\alpha]^{(n)} \in Coker\ (1\text{-}f^{n\omega})$ and $[\beta]^{(m)} \in Coker\ (1\text{-}f^{m\omega})$ are such that $< [\alpha]^{(n)} > = [\iota_{m,n}](< [\beta]^{(m)} >)$; the commutativity of the previous diagram and the fact that $([f^{\omega}])^m$ is the identity on $Coker\ (1\text{-}f^{m\omega})$, imply that

$$([f^{\omega}])^m\ ([\alpha]^{(n)}) = ([f^{\omega}])^m\ [\iota_{m,n}]([\beta]^{(m)}) = [\iota_{m,n}]\ (([f^{\omega}])^m\ ([\beta]^{(m)})) = [\alpha]^{(n)}$$

and so, $l([\alpha]^{(n)}) \mid m$ by the division algorithm. □

In Section 3 we shall introduce a condition under which $l = d$.

Now it is time to intertwine the algebraic and geometric notions of the theory. We relate the geometric and algebraic periodic point classes by defining a function

$$\rho : \Phi(f^n)/\!\sim\ \to Coker\ (1\text{-}f^{n\omega})$$

by $\rho(\mathbf{F}^{(n)}) = j(c - f^n(c) - n(\omega))$, where c is any path in X with $c(0) = x_o$, $c(1) = x$, for any $x \in \mathbf{F}^{(n)}$ (cf. $n = 1$ in [5]).

We establish next the relationship between ρ and the functions $[f^{\omega}]$ and $[\iota_{m,n}]$.

Proposition 1.14 - *The following properties hold true:*

(i) the correspondence ρ is a well-defined injective function;

(ii) for any integers m and n such that $m \mid n$, the following diagrams commute:

$$\begin{array}{ccc} \Phi(f^{n})/\sim & \xrightarrow{\rho} & Coker\,(1-f^{n\omega}) \\ \big\downarrow f & & \big\downarrow [f^{\omega}] \\ \Phi(f^{n})/\sim & \xrightarrow[\rho]{} & Coker\,(1-f^{n\omega}) \end{array} \qquad \text{and} \qquad \begin{array}{ccc} \Phi(f^{m})/\sim & \xrightarrow{\rho} & Coker\,(1-f^{m\omega}) \\ \big\downarrow \gamma & & \big\downarrow [\iota_{m,n}] \\ \Phi(f^{n})/\sim & \xrightarrow[\rho]{} & Coker\,(1-f^{n\omega}) \end{array}$$

Proof - (i) : Replace f^{ω} with $f^{n\omega}$ in [5], 2.5.

(ii) : Take $x \in \mathbf{F}^{(n)}$ and a path $c : I \to X$ connecting x_0 to x; then,

$$f^{\omega}(c - f^{n}(c) - n(\omega)) = \omega + f(c) - f(f^{n}(c)) - f(n(\omega)) - \omega .$$

Notice that $f(x) \in \Phi(f^{n})$ and that $\omega + f(c)$ is a path in X connecting x_0 to $f(x)$; the equality

$$(\omega + f(c)) - f^{n}(\omega + f(c)) - n(\omega) = \omega + f(c) - f(f^{n}(c)) - f(n(\omega)) - \omega$$

establishes the commutativity of the first diagram. As for the second, without loss of generality, we may assume that $m = 1$. Let $\mathbf{F}^{(1)}$ and $\mathbf{F}^{(n)}$ be periodic point classes containing the same $x \in \Phi(f)$. If c is a path in X from x_0 to x, then $\rho(\mathbf{F}^{(1)}) = [c - f(c) - \omega]^{(1)}$, and

$$\begin{aligned} [\iota_{1,n}](\rho(\mathbf{F}^{(1)})) &= [\sum_{i=0}^{n-1} f^{i\omega}(c - f(c) - \omega)]^{(n)} \\ &= [c - f^{n}(c) - n(\omega)]^{(n)} = \rho(\mathbf{F}^{(n)}) . \square \end{aligned}$$

Definition 1.15 - By analogy with the case $n = 1$ in [5], the *index* of an element $[\alpha]^{(n)} \in Coker\,(1-f^{n\omega})$ is the index of $\mathbf{F}^{(n)}$ (as defined in [2], page 87) if $[\alpha]^{(n)} = \rho(\mathbf{F}^{(n)})$, for some $\mathbf{F}^{(n)}$, or is 0, otherwise. The elements $[\alpha]^{(n)} \in Coker\,(1-f^{n\omega})$ whith non-zero index are said to be *essential* (otherwise, they are called *inessential*).

Part (ii) of Proposition 1.14 and the commutativity property of the index (see, for example, [2], page 53, and take the map g there to be f^{n-1}) allows us to deduce:

Corollary 1 - *The function* $[f^{\omega}] : Coker\,(1-f^{n\omega}) \to Coker\,(1-f^{n\omega})$ *is index preserving.* $\square$

This corollary shows that essentiality is a property of orbits.

Another consequence of the previous proposition is the following:

Corollary 2 - *For every* $\mathbf{F}^{(n)} \in \Phi(f^n)/\sim$, $l(\rho(\mathbf{F}^{(n)})) = l(\mathbf{F}^{(n)})$ *and* $d(\rho(\mathbf{F}^{(n)})) \leq d(\mathbf{F}^{(n)})$. *Moreover, if all the algebraic periodic point classes are essential, the algebraic and geometric depths coincide.* □

Corollary 3 - *Let* $\mathbf{F}^{(n)} \in \Phi(f^n)/\sim$ *be given. Consider* $[\alpha]^{(n)} = \rho(\mathbf{F}^{(n)})$. *Then,* $<\mathbf{F}^{(n)}>$ *has at least* $d(<[\alpha]^{(n)}>)$ *points of period n. In particular,* $\mathbf{F}^{(n)}$ *has at least* $d(<[\alpha]^{(n)}>)/l(<[\alpha]^{(n)}>)$ *periodic points of period* n.

Proof - See Proposition 1.3 and the previous corollary. □

As an illustration of Corollary 2 we give an example, showing that the algebraic and geometric depths do not necessarily coincide.

Example 1.16 - Let $f: S^2 \to S^2$ be given by the antipodal map $f(p) = -p$; then, for any choice of x_o and ω, *Coker* $(1 - f^{2\omega})$ and *Coker* $(1 - f^{\omega})$ reduce to the trivial group, so that the algebraic class $[0]^{(2)} = \rho(\mathbf{F}^{(2)})$ - where $\mathbf{F}^{(2)}$ is the only periodic point class of f^2 - is reducible to 1 (i.e., $d([0]^{(2)}) = 1$); however, $\Phi(f) = \emptyset$ and so $\mathbf{F}^{(2)}$ is irreducible, i.e., $d(\mathbf{F}^{(2)}) = 2$.

Observe that in examples 1.7 and 1.11 all classes are essential and so, the algebraic and geometric depths do indeed coincide. We give an example in which depth is actually greater than length, illustrating Corollary 3.

Example 1.17 - Let $X = S^2 \vee S^1 \vee S^2$ with $S^2 \cap S^1 = x_o$ and $S^1 \cap S^2 = x_1 = -x_o$. Consider X as embedded in the vector space $\mathbf{R}^3$ with $x_o = (0,-1,0)$, S^1 the unit circle of the plane y , z as shown:

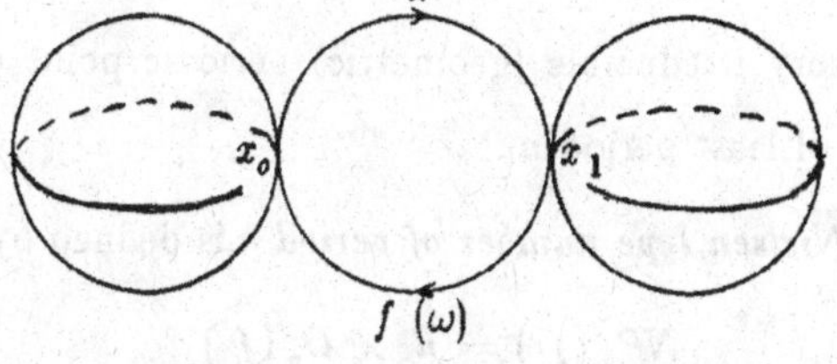

Let $f: X \to X$ be the restriction to X of the linear map which takes a vector $(x,y,z) \in \mathbf{R}^3$ into $(-x,-y,-z)$ (in particular, $f(x_o) = x_1$). Then, f^2 is the identity map, $\pi \cong \mathbf{Z}$ is generated by the path α obtained by travelling S^1 once, starting at x_o, in the counter-clockwise direction. Now select the path ω to be the arc of the circle S^1 from x_o to x_1 passing through the north pole. It is easy to see from the geometry that $(i)\quad \omega + f(\alpha) = \alpha + \omega$ and $(ii)\, \omega + f(\omega) = -\alpha$. From (i) we have $f^{\omega}(\alpha) = \omega + f(\alpha) - \omega = \alpha$ and thus, one easily checks that

$$Coker\,(1 - f^{\omega}) = Coker\,(1 - f^{2\omega}) = \pi \cong \mathbf{Z}$$

and $[\iota_{1,2}] : Coker\,(1 - f^{\omega}) \rightarrow Coker\,(1 - f^{2\omega})$ is multiplication by 2. From (ii) and the definition of ρ we have $\rho(\Phi(f^2)/\sim) = [\alpha]^{(2)}$, is non-empty and irreducible (because $[\iota_{1,2}]$ is multiplication by 2) thus $d(<[\alpha]^{(2)}>) = 2$ but $l(<[\alpha]^{(2)}>) = 1$ (because $f^{\omega} = 1$); so, the class $[\alpha]^{(2)}$ contains at least two periodic points.

In certain cases the geometry may give better results. If $\pi = 0$, $\Phi(f) = \emptyset$ but if $\Phi(f^p) \neq \emptyset$ for some prime p, we can, from the algebra, only deduce that $\#\,\Phi(f^p) \geq 1$ but from the geometry, that $\#\,\Phi(f^p) \geq p$ (see the corollary after 1.3).

The following result of Jiang ([6], page 66) is easily translated into our language.

Proposition 1.19 - *Let* $[\alpha]^{(n)} \in Coker\,(1-f^{n\omega})$ *be reducible to both m and r. If f is eventually commutative,* $[\alpha]^{(n)}$ *reduces to* (m, r). □

2 - A lower bound for the number of periodic points of least period n - In this section we define a lower bound for $MP_n(f)$, the so-called *Nielsen type number of period n.* This number will be denoted by $NP_n(f)$. Let $O_n(f)$ be the number of irreducible, essential periodic point orbits of $Coker\,(1-f^{n\omega})$. Notice that because of Proposition 1.14 (i), if $[\alpha]^{(n)} \in Coker\,(1-f^{n\omega})$ is essential and irreducible, there is a unique $\mathbf{F}^{(n)} \in \Phi(f^n)/\sim$ such that $\rho(\mathbf{F}^{(n)}) = [\alpha]^{(n)}$. It follows from Corollary 3 that the f-orbit of every irreducible (geometric) periodic point class $\mathbf{F}^{(n)}$ contains at least n periodic points of least period n.

Definition 2.1 - The *Nielsen type number of period n* is defined by the formula

$$NP_n(f) = n \times O_n(f).$$

Remarks - (i) It is clear from the definition that n divides $NP_n(f)$ (the reader should compare this result with Dold's result [3], 1.1 on the divisibility by n of the periodic point index).

(ii) Our definition is equivalent to Jiang's (see [6], Definition 4.8, Chapter III).

(iii) The number $NP_n(f)$ differs from the number $D^n(f)$ introduced by Halpern (see [4]) in that $D^n(f)$ in effect counts the irreducible, essential orbits that are of length exactly n. Thus, $D^n(f) \leq NP_n(f)$ and the inequality is strict if there is an essential orbit with $l < d$; for example, in 1.17, since $L(f^2) = 2$, $NP_2(f) = 2$ while $D^2(f) = 0$. We will introduce in Section 3 - see Remark 3.6 (ii) - conditions under which $D^n(f) = NP_n(f)$.

Proposition 2.2 - *(Jiang [6]) The following properties of $NP_n(f)$ hold true:*

(i) $NP_n(f)$ is a homotopy invariant;

(ii) $\# P_n(f) \geq MP_n(f) \geq NP_n(f)$.

Proof - (i): A homotopy $H : f \simeq g : X \to X$ induces a homotopy $H^n : f^n \simeq g^n$ in the obvious way. Let $\nu = \omega + H(x_o, -)$; then, one can prove that $n(\nu) \simeq n(\omega) + H^n(x_o, -)$ and so, H^n induces an index preserving bijection $H_{\#}{}^n : Coker(1 - f^{n\omega}) \to Coker(1 - g^{n\nu})$ for every positive integer n, given by $H_{\#}{}^n([\alpha]^{(n)}) = [\alpha]^{(n)}$ (cf. [5], 3.3) and such that the two diagrams

$$
\begin{array}{ccc}
Coker(1-f^{n\omega}) & \xrightarrow{[f^{\omega}]} & Coker(1-f^{n\omega}) \\
\downarrow H_{\#}{}^{n} & & \downarrow H_{\#}{}^{n} \\
Coker(1-g^{n\nu}) & \xrightarrow[{[g^{\nu}]}]{} & Coker(1-g^{n\nu})
\end{array}
\qquad
\begin{array}{ccc}
Coker(1-f^{m\omega}) & \xrightarrow{[\iota_{m,n}]} & Coker(1-f^{n\omega}) \\
\downarrow H_{\#}{}^{m} & & \downarrow H_{\#}{}^{n} \\
Coker(1-g^{m\nu}) & \xrightarrow[{[\iota_{m,n}]}]{} & Coker(1-g^{n\nu})
\end{array}
$$

are commutative. It follows easily that $NP_n(f) = NP_n(g)$.

(ii): Note that different irreducible f-orbits of $\Phi(f^n)/\sim$ have no common periodic points; also, use (i). □

Example 2.3 - Consider the projective space $\mathbf{R}P^3$ as the quotient of $S^3 = \{(ue^{i\phi}, ve^{i\theta}) : u^2+v^2 = 1\}$ modulo identification of antipodal points and let $f : \mathbf{R}P^3 \to \mathbf{R}P^3$ denote the map induced by $\tilde{f}(ue^{i\phi}, ve^{i\theta}) = (ue^{3i\phi}, ve^{3i\theta})$. Then, for any choice of x_o and ω, the following statements hold true: (i) $L(f^m) = 1 - 9^m$ (see [6], page 67) for every m; (ii) $\pi = J(f) = \mathbf{Z}_2$;

(iii) $f^{\omega}: \pi \to \pi$ is the identity function; (iv) $Coker\ (1-f^{m\omega}) \cong \mathbf{Z}_2$ for every m; (v) $[\iota_{m,n}]$ is multiplication by n/m, for every m dividing n. Thus, by (iii) and (iv), for every m, $Coker\ (1-f^{m\omega})$ has exactly two orbits $< [0]^{(m)} >$ and $< [1]^{(m)} >$; these are essential by (i) and (ii) and with length 1, by (iii). If $n = 2^r$, with $r > 0$, then n/m must be even for every m that divides n and hence, $< [1]^{(n)} >$ is irreducible. Thus, $NP_n(f) = 2^r$, confirming Jiang's inequality $M\Phi_n(f) \geq 2^r$ (see [6], page 68).

Although it is not our intention to discuss the lower bound $M\Phi_n(f)$ in this paper, we do have the following easy consequence of the previous proposition:

Corollary - *(Jiang [6])* $M\Phi_n(f) \geq \sum_{m|n} NP_m(f)$.

Proof - Notice that the set $\{ P_m(f) : m \mid n \}$ defines a partition of $\Phi_n(f)$. Then, $\# \Phi_n(f) = \sum \# P_m(f)$; now use part (ii) of 2.2. □

In the sequel to this paper we will give an example showing that $\sum NP_m(f)$ is not an optimal lower bound for $M\Phi_n(f)$.

3 - Some Computations - The main thrust of this section is to give a criterion under which $NP_n(f)$ can be related to the Nielsen numbers of f^m, for $m \mid n$.

We begin by relating the homotopy to the homology; more specifically, we wish to relate the group $Fix\ f^{\omega}$ to the group $Fix\ f_* = Ker\ (1 - f_*)$, where f_* is the endomorphism of $H_1(X)$ induced by the self-map $f : X \to X$. Let $\theta : \pi \to H_1(X)$ be the standard epimorphism with $Ker\ \theta = [\pi,\pi]$, the commutator subgroup of π. The restriction of θ to $Fix\ f^{\omega}$ will also be denoted by θ.

Lemma 3.1 - *If $f : X \to X$ is eventually commutative, the homomorphism $\theta : Fix\ f^{\omega} \to Fix\ f_*$ is an isomorphism.*

Proof - (i) θ is injective: given that $\beta \in Fix\ f^{\omega}$ is such that $\theta(\beta) = 0$, it follows that $\beta \in Ker\ \theta$ thus, $\beta = f^{k\omega}(\beta) = 0$, when $f^{k\omega}(\pi)$ is abelian.

(ii) θ is surjective: Let $\alpha \in Fix\ f_*$ be given; take a $\beta \in \pi$ so that $\theta(\beta) = \alpha$. Then, $\theta(\beta f^{\omega}(\beta^{-1})) = \alpha - \alpha = 0$ so, $\beta f^{\omega}(\beta^{-1}) \in [\pi,\pi]$ and therefore,

$$f^{k\omega}(\beta) f^{(k+1)\omega}(\beta^{-1}) = f^{k\omega}(\beta f^{\omega}(\beta^{-1})) = 1 .$$

It follows that $f^{k\omega}(\beta) = f^{\omega}(f^{k\omega}(\beta))$, so $f^{k\omega}(\beta) \in Fix\ f^{\omega}$ and

$$\theta(f^{k\omega}(\beta)) = f_*^k\,\theta(\beta) = f_*^k(\alpha) = \alpha . \quad \square$$

We note that this lemma also follows from [5], 1.5, 1.8 and 1.13.

Corollary - *If* $f : X \to X$ *is eventually commutative, then* $Fix\ f^{\omega} = 0$ *iff* $Fix\ f_* = 0$. $\square$

We wish to note that in the context of this corollary, if f is not eventually commutative, neither condition implies the other. We give two examples to this effect.

Example 3.2 - Take $X = S^1 \vee S^1$; then π is the free group with two generators a and b. Let $f : X \to X$ be the map which interchanges the circles; choose $x_0 = S^1 \cap S^1$ and $\omega = 0$. Then, $f^{\omega}(a) = b$ and $f^{\omega}(b) = a$. Now $H_1(X)$ is the free abelian group with generators $\alpha = \theta(a)$ and $\beta = \theta(b)$. Note that $f_*(\alpha) = \beta$ and $f_*(\beta) = (\alpha)$. It is now easy to see that $Fix\ f^{\omega} = 0$ but $Fix\ f_* \neq 0$.

Example 3.3 - Let A_n be the alternating group of degree n, with $n \geq 5$; take an abelian group B and form the product $\pi = A_n \times B$. Now $[\pi,\pi] = A_n$ and so, if $X = K(\pi,1)$, then $H_1(X) \cong B$. Next take the self-map f of X inducing the endomorphism 1×0 of π; then $Fix\ f_* = 0$ but $Fix\ f^{\omega} = A_n$.

Definition 3.4 - A map $f : X \to X$ is said to be n*-toral* if the following two conditions hold:

(i) : For every $m \mid n$ and every $[\alpha]^{(m)} \in Coker\ (1 - f^{m\omega})$, $d(<[\alpha]^{(m)}>) = l(<[\alpha]^{(m)}>)$;

(ii) : for every $m \mid n$, $[\iota_{m,n}]$ is injective.

Note that any self-map $f : X \to X$ is 1-toral. Furthermore, if f is n-toral then it is m-toral, for every $m \mid n$.

Proposition 3.5 - *A map* $f : X \to X$ *is n-toral if it is eventually commutative and* $Fix\ f^{n\omega} = 0$ *(or equivalently,* $Fix\ f_*^n = 0$ *).*

Proof - Since for f eventually commutative $Coker\ (1 - f^{n\omega}) \cong Coker\ (1 - f_*^n)$ and $Fix\ f^{n\omega} \cong Fix\ f_*^n$ we can, without loss of generality, assume π abelian. (i) : Let us denote the length (respectively, depth) of the orbit $<[\alpha]^{(m)}>$ by the letter l

(respectively, d). Notice that $l \mid m$. Also, recall from Lemma 1.13 that $l \mid d$; thus, we must just prove that $l \geq d$, i.e., that $[\alpha]^{(m)}$ is reducible to l.

Observe that $[(1-f^{l\omega})(\alpha)]^{(m)} = [\alpha]^{(m)} - f^{l\omega}([\alpha]^{(m)}) = 0$ and so, $(1-f^{l\omega})(\alpha) \in \mathrm{Im}(1-f^{m\omega})$ (see Proposition 1.4). Let $\gamma \in \pi$ be such that

$$(1-f^{l\omega})(\alpha) = (1-f^{m\omega})(\gamma) = (1-f^{l\omega})(1+f^{l\omega}+\cdots+f^{(m-l)\omega})(\gamma).$$

Since $1-f^{l\omega}$ is injective (*Fix* $f^{n\omega}=0$ and the fact that $l \mid m \mid n$ imply that *Fix* $f^{l\omega}=0$, using 1.4 an 1.8) it follows that $\alpha = (1+f^{l\omega}+\cdots+f^{(m-l)\omega})(\gamma)$ and so, $[\alpha]^{(m)} = [\iota_{l,m}]([\gamma]^{(l)})$ by the commutativity of the diagram in Lemma 1.8.

(ii) : Suppose that $[\iota_{m,n}]([\beta]^{(m)}) = 0$ and let $j(\beta) = [\beta]^{(m)}$. Then, since $[(1+f^{m\omega}+\cdots+f^{(n-m)\omega})(\beta)]^{(n)} = 0$, by 1.4 there is a $\gamma \in \pi$ such that

$$(1+f^{m\omega}+\cdots+f^{(n-m)\omega})(\beta) = (1-f^{n\omega})(\gamma).$$

Applying $1-f^{m\omega}$ at the left of each side of the above equality and using commutativity we obtain $(1-f^{n\omega})(\beta) = (1-f^{n\omega})(1-f^{m\omega})(\gamma)$. But $1-f^{n\omega}$ is injective and so, $\beta = (1-f^{m\omega})(\gamma)$, implying thereby that $[\beta]^{(m)} = 0$. □

Remark 3.6 - (i) We choose the name "n-toral" because of the following example: let $f : T^k \to T^k$ be represented by a matrix A with characteristic values $\omega_1, \cdots, \omega_k$; thus, f is n-toral if no $\omega_i{}^n = 1$, $i = 1,\ldots,k$. Note that we are not excluding roots of unity (cf. [6], Example 3, page 64 and see our Example 3.10).

(ii) Under the hypothesis of the previous proposition, the irreducible, essential orbits in *Coker* $(1-f^{n\omega})$ do have length n and so, in this case, Halpern's $D^n(f)$ is equal to $NP_n(f)$.

We now recall that the *Jiang group J(f)* of a self-map $f : X \to X$ is the set of all $\alpha \in \pi$ for which there exists a cyclic homotopy $H : f \simeq f$ such that $H(x_o, t) = \alpha$. Recall that if $J(f) = \pi$ then (i) if $L(f) \neq 0$, $N(f) = R(f)$ (see [2], Corollary VII.B.6) and (ii), f is eventually commutative: [6], II.3.7.

The following theorem is an analogue of [3], 1.1.

Theorem 3.7 - *Let $f : X \to X$ be an n-toral map such that, for every $m \mid n$, $L(f^m) \neq 0$ and $J(f) = \pi$. Let $\mathbf{P}(n) = \{p(1), \cdots, p(k)\}$ be the set of all* *distinct primes which divide n. Then,*

$$NP_n(f) = \sum_{\tau \subseteq \mathbf{P}(n)} (-1)^{\#\tau} N(f^{n:\tau})$$

where $n:\tau = n(\prod_{p\in\tau} p)^{-1}$.

Remark - We note that since the empty set is a subset of $\mathbf{P}(n)$, the sum always includes the term $N(f^n)$.

Proof - For each $i = 1, \cdots, k$, let A_i be the image of the homomorphism $[\iota_{n/p(i),n}]$. Since f is n-toral, $NP_n(f)$ is equal to the number of irreducible, essential classes of f^n; since $J(f) = \pi$ and $L(f^n) \neq 0$, $NP_n(f) = N(f^n) - \# \bigcup_{i=1}^{k} A_i$.

If $[\alpha]^{(n)} \in Coker\,(1-f^{n\omega})$ is a reducible class, then $[\alpha]^{(n)} \in A_i$ for some (not necessarily unique) i. To see this, suppose that $[\alpha]^{(n)}$ is reducible to m, that is to say, there exists a $[\beta]^{(m)} \in Coker\,(1-f^{m\omega})$ such that $[\iota_{m,n}]([\beta]^{(m)}) = [\alpha]^{(n)}$. Let $p(i)$ be a prime that divides n/m and set $[\gamma]^{(n/p(i))} = [\iota_{m,n/p(i)}]([\beta]^{(m)}) \in Coker\,(1-f^{(n/p(i))\omega})$; then $[\alpha]^{(n)} = [\iota_{n/p(i),n}]([\gamma]^{(n/p(i))}) \in A_i$.

We now prove that for any $\tau \neq \emptyset \subseteq \mathbf{P}(n)$, $\bigcap_{p(i)\in\tau} A_i = \mathrm{Im}([\iota_{m,n}])$, where $m = n:\tau$. In fact, we first notice that, for every $p(i) \in \tau$, the abelian group homomorphism $[\iota_{m,n}]$ can be written as $[\iota_{m,n}] = [\iota_{n/p(i),n}] \cdot [\iota_{m,n/p(i)}]$ and thus, $\bigcap_{p(i)\in\tau} A_i \supseteq \mathrm{Im}([\iota_{m,n}])$. On the other hand, if $[\alpha]^{(n)} \in \bigcap_{p(i)\in\tau} A_i$ then, $[\alpha]^{(n)}$ is reducible to $n/p(i)$ for each $p(i) \in \tau$ and thus, is reducible to m (see 1.19). This shows that $[\alpha]^{(n)} \in \mathrm{Im}([\iota_{m,n}])$. Now since f is n-toral and $L(f^m) \neq 0$, we have that $\#\,\mathrm{Im}([\iota_{m,n}]) = N(f^m)$. The formula described in the statement follows from an elementary result of combinatorics. □

Example 3.8 - Consider the function $f: S^1 \to S^1$ given by $f(e^{i\theta}) = e^{3i\theta}$; we claim that if $n = p^r$, p prime, $MP_n(f) = 3^n - 3^{n/p}$, as announced in the introduction.

In fact, the previous theorem implies that $NP_n(f) = N(f^n) - N(f^{n/p})$. This latter number is equal to $3^n - 3^{n/p}$ because $N(f^m) = 3^m - 1$. The equality

$MP_n(f) = NP_n(f)$ holds true since for this f

$$\# P_n(f) = \# (\Phi(f^n) \setminus \bigcup_{m<n} \Phi(f^m)) = N(f^n) - N(f^{n/p}).$$

In order to further illustrate Theorem 3.7, we need the following result giving the Nielsen numbers of some iterates of a self-map of the torus T^n.

Lemma 3.9 - *Let $f : T^n \to T^n$ be represented by an $n \times n$ matrix A with characteristic equation $\chi_A = \lambda^n + 1$. Then, if $m/(m,n)$ is odd, $N(f^m) = 2^{(m,n)}$.*

Proof - We use the fact that

$$N(f^m) = | L(f^m) | = | det(I - A^m) | = | \chi_{A^m}(1) |$$

(see [1]). Let $\omega = e^{i\pi/n}$; then the characteristic values of A are $\omega, \omega^3, \dots, \omega^{2n-1}$ and those of A^m are $\omega^m, \omega^{3m}, \dots, \omega^{(2n-1)m}$. The latter are roots of $\lambda^{n/(m,n)} + 1 = 0$, since $(\omega^{(2k-1)m})^{n/(m,n)} = (\omega^n)^{(2k-1)m/(m,n)} = (-1)^{(2k-1)m/(m,n)}$. Notice that

$\omega^{(2k-1)m} = \omega^{(2k'-1)m} \Leftrightarrow (2k-1)m \equiv (2k'-1)m \mod 2n \Leftrightarrow k \equiv k' \mod n/(m,n)$;

so, the distict numbers $\omega^m, \omega^{3m}, \dots, \omega^{((2n/(m,n))-1)m}$ exhaust the roots of $\lambda^{n/(m,n)} + 1 = 0$. Then, $\chi_{A^m}(1) = \prod_{k=1}^{n} (1 - \omega^{(2k-1)m}) =$

$$= (\prod_{k=1}^{n/(m,n)} (1 - \omega^{(2k-1)m})^{(m,n)}) = (1+1)^{(m,n)} = 2^{(m,n)}. \quad \square$$

Example 3.10 - Let $f : T^{24} \to T^{24}$ be represented by a matrix with characteristic equation $\lambda^{24} + 1 = 0$. Then, $NP_{12}(f) = N(f^{12}) - N(f^6) - N(f^4) + N(f^2)$ and thus, $NP_{12}(f) = 4020$ by the previous lemma.

4 - Appendix - The aim of this appendix is to sketch the proof of the following:

Theorem A - *The concepts of (1) reducibility, (2) depth and (3) length of orbit are independent of the base point x_0 and the path $\omega : x_0 \to f(x_0)$.*

The proof (as in [5]) can be split essentially in two parts: one, which considers a change in the base point and the other, which considers the change of the basic path; we first recall two lemmas (see [5], Sections 2 and 3).

Lemma A.1 - *Let $f : X \to X$, $x_o \in X$ and ω and μ be paths in X connecting x_o to $f(x_o)$. Then, there is an index preserving bijection $r_{\omega,\mu}$: Coker $(1 - f^{\omega}) \to$ Coker $(1 - f^{\mu})$ given by $r_{\omega,\mu}([\alpha]^{(1)}) = [\alpha + \omega - \mu]^{(1)}$.*

Lemma A.2 - *Let $f : X \to X$, $x_o \in X$ and $\omega : I \to X$ be given, with $\omega(0) = x_o$, $\omega(1) = f(x_o)$. Let $y_o \in X$ be another base point and let $u : I \to X$ be a path in X connecting x_o to y_o. Then, u_* : Coker $(1 - f^{\omega}) \to$ Coker $(1 - f^{-u+\omega+f(u)})$ defined by $u_*([\alpha]^{(1)}) = [-u + \alpha + u]^{(1)}$ is an index preserving bijection.*

The next two results will be useful in what follows.

Lemma A.3 - *Let $u : I \to X$ be a path in X connecting x_o to y_o. Then,*

$$n(-u + \omega + f(u)) = -u + n(\omega) + f^{n}(u) .$$

Since we change paths in what follows, we shall attach the name of the path volved to the notation of $[\iota_{1,n}]$ and we write, for example, $[\iota^{\omega}{}_{1,n}]$. We use Lemma A.3 show:

Proposition A.4 - *With u as above, define $\theta = -u + \omega + f(u)$. Then, the diagram*

$$\begin{array}{ccc}
\text{Coker}\,(1 - f^{\omega}) & \xrightarrow{[\iota^{\omega}{}_{1,n}]} & \text{Coker}\,(1 - (f^{n})^{n(\omega)}) \\
\Big\downarrow u_* & & \Big\downarrow u_* \\
\text{Coker}\,(1 - f^{\theta}) & \xrightarrow{[\iota^{\theta}{}_{1,n}]} & \text{Coker}\,(1 - (f^{n})^{n(\theta)})
\end{array}$$

is commutative.

Lemma A.5 - *For every path ω connecting x_o to $f(x_o)$ and every loop α of X based at x_o,* $\sum_{i=0}^{n-1} f^{i\omega}(\alpha) = \sum_{i=0}^{n-1} f^{i}(\alpha + \omega) - n(\omega)$.

Proof - $\displaystyle\sum_{i=0}^{n-1} f^{i\omega}(\alpha) = \sum_{i=0}^{n-1} (i(\omega) + f^{i}(\alpha) - i(\omega))$

$$= \sum_{i=0}^{n-1} (i(\omega) + f^{i}(\alpha + \omega) - (i+1)(\omega)) = \sum_{i=0}^{n-1} f^{i}(\alpha + \omega) - n(\omega) . \square$$

Proposition A.6 - *The diagram*

$$\begin{array}{ccc} Coker\,(1-f^{\omega}) & \xrightarrow{[\iota^{\omega}{}_{1,n}]} & Coker\,(1-(f^{n})^{n(\omega)}) \\ \Big\downarrow r_{\omega,\mu} & & \Big\downarrow r_{n(\omega),n(\mu)} \\ Coker\,(1-f^{\mu}) & \xrightarrow{[\iota^{\mu}{}_{1,n}]} & Coker\,(1-(f^{n})^{n(\mu)}) \end{array}$$

is commutative.

Proof - The following equalities follow from Lemma A.5 and properties of the functions involved:

$$[\iota^{\mu}{}_{1,n}]\,(r_{\omega,\mu})\,([\alpha]^{(1)}) = [\sum_{i=0}^{n-1} f^{i\mu}(\alpha+\omega-\mu)]^{(n)} = [\sum_{i=0}^{n-1} f^{i}(\alpha+\omega) - n(\mu)]^{(n)}$$

$$= [\sum_{i=0}^{n-1} f^{i\omega}(\alpha) + n(\omega) - n(\mu)]^{(n)} = r_{n(\omega),n(\mu)}\,[\iota^{\omega}{}_{1,n}]\,([\alpha]^{(1)})\,.$$

Parts (1) and (2) of Theorem A now follow; part (3) follows by similar arguments.

REFERENCES

[1] Brooks,R., Brown,R.F., Pak,J. and Taylor,D. - *Nielsen numbers of maps of tori,* Proc.Amer.Math.Soc., 52 (1975), 398-400.

[2] Brown,R.F. - The Lefschetz Fixed Point Theory, Scott, Foresman and Co., Glenview 1971.

[3] Dold A. - *Fixed point indices of iterated maps,* Inventiones math. (1983), 419-435.

[4] Halpern,B. - *Nielsen type numbers for periodic points* (unpublished).

[5] Heath,P. - *Product formulae for Nielsen numbers of fibre maps,* Pacific J.Math. 117 (1985), 267-289.

[6] Jiang,B. - Lectures on Nielsen Fixed Point Theory, Contemporary Mathematics # 14, American Mathematical Society, Providence 1983.

Braids and Periodic Solutions

HAI-HUA HUANG AND BO-JU JIANG

Nankai Institute of Mathematics
and
Peking University

INTRODUCTION

Matsuoka studied in [7], [8], [9] the periodic solutions to a periodic system of differential equations on the plane. Based on the braiding information of known solutions, he obtained lower bounds for the number of extra periodic solutions. Some very interesting applications to subharmonic solutions of second order differential equations were given in [8]. The core of his method is Proposition 2 in §4 of [7], which is an elaborate application of the Lefschetz fixed point theorem. This Proposition is in fact more naturally accessible from the Nielsen fixed point theory. The purpose of the present paper is to give (in §2) such an alternative approach to the foundation of Matsuoka's theory, and then, with the insight thus gained, to attack (in §3) the corresponding problem for the torus. For the convenience of the reader, some relevant material in Nielsen theory is reviewed in §1. In principle, our approach applies to surfaces of higher genus as well.

We would like to emphasize the approach rather than the best possible estimates in various cases. So, in §2 we confine ourselves to recovering one of Matsuoka's estimates, and in §3 we will not go into the possible refinements of our estimates.

1. A BRIEF ACCOUNT OF NIELSEN THEORY

(A) Nielsen number and Reidemeister trace.

Let X be a compact connected polyhedron, $f : X \to X$ be a map. The fixed point set $\mathrm{Fix} f := \{x \in X | x = f(x)\}$ splits into a disjoint union of *Nielsen fixed point classes*. Two fixed points are in the same class iff they can be joined by a path which is homotopic (relative to end-points) to its own f-image. Each fixed point class is an isolated subset of $\mathrm{Fix} f$ hence its fixed point index is defined. The number of fixed point classes with non-zero index is called the *Nielsen number* $N(f)$ of f. It is a homotopy invariant of f, so that every map homotopic to f must have at least $N(f)$ fixed points. (Cf. [6] p.19.)

Choose a base point $x_0 \in X$ and a path w from x_0 to $f(x_0)$ once and for all. (The effect of changing w will be discussed in (F).) Let $\pi = \pi_1(X, x_0)$ and let $f_\pi : \pi \to \pi$ be the composition

$$\pi_1(X, x_0) \xrightarrow{f_*} \pi_1(X, f(x_0)) \xrightarrow{w_*} \pi_1(X, x_0).$$

Two elements $\alpha, \beta \in \pi$ are said to be *f_π-conjugate* (or Reidemeister equivalent) if there is a $\gamma \in \pi$ such that $\beta = f_\pi(\gamma)\alpha\gamma^{-1}$. (This definition differs from that of [6] p.26 by an inversion.) Thus π is divided into f_π-conjugacy classes. Let π_R denote the set of f_π-conjugacy classes, and $\mathbb{Z}\pi_R$ denote the abelian group freely generated by π_R. We will use the bracket notation $\alpha \mapsto [\alpha]$ for both projections $\pi \to \pi_R$ and $\mathbb{Z}\pi \to \mathbb{Z}\pi_R$.

The second named author partially supported by a TWAS grant.

Let x be a fixed point of f. Take a path c from x_0 to x. The f_π-conjugacy class of $\langle w(f \circ c)c^{-1}\rangle \in \pi$, which is evidently independent of the choice of c, will be called the coordinate of x. Two fixed points are in the same Nielsen class iff they have the same coordinates. This f_π-conjugacy class is thus called the *coordinate* of the Nielsen class. (This also differs from the definition in [6] p.26 by an inversion.)

REMARK: The point x_0 and the path w chosen at the beginning serve as a reference frame for the coordinate. If x_0 is a fixed point and w is the constant path, the Nielsen class containing x_0 will have coordinate $[1] \in \mathbb{Z}\pi_R$.

The *Reidemeister trace* of f, denoted $L_R(f)$, is the element of $\mathbb{Z}\pi_R$ in which the coefficient of $[\alpha] \in \pi_R$ equals the index of the fixed point class with coordinate $[\alpha]$. It is also a homotopy invariant of f, indeed a more powerful one since $N(f)$ is merely the number of (non-zero) terms in $L_R(f)$. Yet it can be computed as an alternating sum of traces on the chain level, similar to that for the Lefschetz number, as follows (cf. [10], [11]).

A cellular decomposition $\{e_i^q\}$ of X lifts to a π-invariant cellular structure on the universal covering $\widetilde{X}$ of X, where π is identified with the group of covering translations of $\widetilde{X}$. Choose an arbitrary lift $\tilde{e}_i^q$ for each e_i^q. They constitute a free $\mathbb{Z}\pi$-basis for the cellular chain complex of $\widetilde{X}$. Without loss we assume f to be a cellular map, and consider its lift $\tilde{f} : \widetilde{X} \to \widetilde{X}$ corresponding to x_0 and w (cf. [6] p.25). In every dimension q, the cellular chain map $\tilde{f}_q$ gives rise to a $\mathbb{Z}\pi$-matrix $\widetilde{F}_q$ with respect to the above basis. Then we have

$$L_R(f) = \sum_q (-1)^q \left[\operatorname{tr}\widetilde{F}_q\right] \in \mathbb{Z}\pi_R.$$

(B) The abelianized version.

Everything in (A) has a homological version, obtained by abelianizing π into the homology group $H_1(X)$, which is less powerful but easier to compute. (Cf. [6] §III.2 on mod K Nielsen theory. Consider $K = [\pi, \pi]$, the commutator subgroup of π.) Thus, two fixed points are in the same *homological Nielsen class* iff they can be joined by a path which is homologous to its f-image. A homological Nielsen class is a union of (ordinary) Nielsen classes, its index equals the sum of indices of these latter classes. The *abelianized Nielsen number* $N^{\mathrm{Ab}}(f)$ is the number of homological Nielsen classes with non-zero index. Obviously $N^{\mathrm{Ab}}(f) \leq N(f)$.

The abelianization $\mathrm{Ab} : \pi \to H_1(X)^{\cdot}$ induces a projection $\mathrm{Ab} : \pi_R \to H$, where $H := \mathrm{Coker}\big(1 - f_* : H_1(X) \to H_1(X)\big)^{\cdot}$. (The superscript $\cdot$ indicates that an additive group is being regarded as a multiplicative group.) Let μ denote the projection $H_1(X)^{\cdot} \to H = \mathrm{Coker}(1 - f_*)^{\cdot}$ as well as its extension $\mathbb{Z}H_1(X)^{\cdot} \to \mathbb{Z}H$.

The coordinate of a homological Nielsen class is the element of H represented by the loop $w(f \circ c)c^{-1}$, where c is a path from x_0 to a fixed point in the class. The *abelianized Reidemeister trace* $L_H(f)$ is an element of the group-ring $\mathbb{Z}H$, the coefficient of $h \in H$ being the index of the homological Nielsen class with coordinate h. Thus $L_H(f) = L_R(f)^{\mathrm{Ab}}$ and, in the setting of cellular computation in (A),

$$L_H(f) = \sum_q (-1)^q \operatorname{tr}\widetilde{F}_{H,q} \in \mathbb{Z}H$$

where $\widetilde{F}_{H,q} := \mu \widetilde{F}_q^{\mathrm{Ab}}$ are $\mathbb{Z}H$-matrices.

(C) Surface maps.

Fadell and Husseini [4] devised a method of computing $L_R(f)$ for surface maps. Let X be a surface with boundary, and $f: X \to X$ be a map. Let a base point x_0 and a reference path w be chosen. Suppose $\{a_1, \cdots, a_n\}$ is a free basis for $\pi_1(X)$. Then f has the homotopy type of a self-map of a bouquet of n circles which can be decomposed into one 0-cell and n 1-cells corresponding to the a_i's. The trace formula in (A) leads to the following method of computing $L_R(f)$. Let

$$D(f_\pi) = \left(\frac{\partial f_\pi(a_i)}{\partial a_j}\right)$$

be the Jacobian in Fox calculus, an $n \times n$ matrix in $\mathbb{Z}\pi$. Then ([4] p.64, Theorem 2.3)

$$L_R(f) = [1] - [\mathrm{tr} D(f_\pi)] = [1] - \sum_{i=1}^{n} \left[\frac{\partial f_\pi(a_i)}{\partial a_i}\right] \in \mathbb{Z}\pi_R,$$

$$L_H(f) = 1 - \mathrm{tr} D_H(f_\pi) = 1 - \sum_{i=1}^{n} \mu\left(\left(\frac{\partial f_\pi(a_i)}{\partial a_i}\right)^{\mathrm{Ab}}\right) \in \mathbb{Z}H,$$

where $D_H(f_\pi)$ is the projection of $D(f_\pi)$ into a $\mathbb{Z}H$-matrix.

(D) Periodic points.

Periodic points of f are fixed points of iterates of f. We can study $N(f^p)$, $L_R(f^p)$, etc. for a natural number p. By definition, $(L_R(f^p))^{\mathrm{Ab}}$ should be in $\mathbb{Z}H^{(p)}$ where $H^{(p)} := \mathrm{Coker}(1 - f_*^p)$. There is a natural projection from $H^{(p)}$ onto $H = H^{(1)}$, because $1 - f_*$ is a factor of $1 - f_*^p : H_1(X) \to H_1(X)$. We will write $L_H(f^p)$ for the projection of $(L_R(f^p))^{\mathrm{Ab}}$ in $\mathbb{Z}H$, a common group-ring for all p. Now in $L_H(f^p) \in \mathbb{Z}H$ the coefficient of $h \in H$ equals the total f^p-index of the homological f^p-fixed point classes whose H-coordinate is h (i.e. for a path c from the base point x_0 to a point of the class, the loop $w(f \circ w) \cdots (f^{p-1} \circ w)(f^p \circ c)c^{-1}$ represents h).

An advantage of working with H instead of $H^{(p)}$ is that f_* induces the identity automorphism on H. This fact is crucial for the following

PROPOSITION. *Suppose h is the H-coordinate of a fixed point class of f^p containing a point x.*

(i) *The points $x, f(x), \cdots, f^{p-1}(x)$ lie in f^p-fixed point classes with the same H-coordinate h.*

(ii) *If x is a fixed point of f^q where $q|p$, then h is a p/q-th power of some element in H.*

PROOF: We turn back to additive notation for $\mathrm{Coker}(1 - f_*)$. Since it is an abelian quotient of π, [6] Remark III.4.5 tells us that [6] Lemma III.4.3 remains true with obvious changes. The proof of (ii): Lemma III.4.3(iii) says $h = \mu \circ (1 + f_*^q + \cdots + f_*^{p-q})(k)$ for some $k \in H_1(X)$. But $\mu \circ f_* = \mu : H_1(X) \to \mathrm{Coker}(1 - f_*)$, so that $h = (p/q)\mu(k)$ as we claimed. The claim (i) follows similarly from Lemma III.4.3(iv). ∎

COROLLARY. *Let K_p be the number of elements of $h \in H$ appearing (with non-zero coefficient) in $L_H(f^p)$ such that h is not a q-th power in H for any prime factor q of p. Then f has at least pK_p points of least period p.* ∎

In the language of [6] pp.69–70, pK_p is nothing but $NP_{p,K}$ with $K = \mathrm{Ker}(\pi \to H)$. It follows from (B) that

$$L_H(f^p) = \sum_q (-1)^q \mathrm{tr} \widetilde{F}^p_{H,q} \in \mathbb{Z}H.$$

(E) Zeta function.

A recent development in Nielsen theory is the introduction by Fried [5] of the *twisted Lefschetz zeta function*. It is a powerful tool both for the computation of $L_H(f^p)$ for all p and for the analysis of periodic points (e.g. proving the existence of infinitely many periodic points). By definition,

$$\tilde{\zeta}_H = \exp\left(\sum_{p=1}^{\infty} \frac{L_H(f^p)}{p} t^p\right).$$

It is a formal power series in t with coefficients in $\mathbb{Z}H$, i.e. $\tilde{\zeta}_H \in \mathbb{Z}H[[t]]$. Actually it is in the multiplicative subgroup $1 + t\mathbb{Z}H[[t]]$ of $\mathbb{Z}H[[t]]$. In terms of cellular chains, we have ([5] p.269 Theorem 2)

$$\tilde{\zeta}_H = \prod_q \det(I - t\widetilde{F}_{H,q})^{(-1)^{q+1}},$$

where I is the identity matrix. Note that our H here is the $H_{\max}$ on [5] p.265.

Specializing to the case of bounded surfaces, we get

$$\tilde{\zeta}_H = \frac{\det(I - tD_H(f))}{1-t}.$$

(F) Change of coordinates.

Suppose we change the reference path w to w' (also leading from x_0 to $f(x_0)$). Let $\omega \in \pi = \pi_1(X, x_0)$ be the element represented by the loop $w'w^{-1}$. Looking through (A)–(E), we can easily find the relation between the invariants calculated with respect to the new reference system x_0, w' versus the old x_0, w.

The homomorphism $f'_\pi = w'_* \circ f_*$ clearly differs from $f_\pi = w_* \circ f_*$ by an inner automorphism: $f'_\pi(\gamma) = \omega f_\pi(\gamma)\omega^{-1}$ for all $\gamma \in \pi$. For a fixed point class, the new coordinate is obtained from the old by multiplying ω on the left. Thus for the Reidemeister trace we might think of $L'_R(f)$ as $\omega L_R(f)$, but be cautious since the f_π- and f'_π-conjugacy relations in π are different. Under abelianization, the difference between f'_π and f_π disappears. Hence we can safely write

$$L'_H(f) = \omega L_H(f) \in \mathbb{Z}H.$$

The Nielsen numbers are of course unaffected.

For a homological f^p-fixed point class, the new H-coordinate is obtained from the old by multiplying $\omega f_\pi(\omega)\cdots f_\pi^{p-1}(\omega)$. But in H we have $\omega = f_\pi(\omega)$. So $L'_H(f^p) = \omega^p L_H(f^p) \in \mathbb{Z}H$. Therefore

$$\tilde{\zeta}'_H(t) = \tilde{\zeta}_H(\omega t) \in 1 + t\mathbb{Z}H[[t]].$$

For surfaces with boundary, L_H and $\tilde{\zeta}_H$ can be computed from f_π as in (C). So in this case we have

$$1 - \operatorname{tr}D_H(f'_\pi) = \omega\,(1 - \operatorname{tr}D_H(f_\pi)) \qquad \text{in } \mathbb{Z}H,$$

$$\frac{\det(I - tD_H(f'_\pi))}{1-t} = \frac{\det(I - \omega t D_H(f_\pi))}{1 - \omega t} \qquad \text{in } 1 + t\mathbb{Z}H[[t]].$$

2. Periodic Differential Equations on the Plane

Consider a differential equation

$$(1) \qquad \frac{dx}{dt} = f(t,x), \qquad t \in \mathbb{R}^1, x \in \mathbb{R}^2.$$

We always assume that:

(i) $f : \mathbb{R}^1 \times \mathbb{R}^2 \to \mathbb{R}^2$ is a C^1-map;
(ii) $f(t+1,x) = f(t,x)$ for all $(t,x) \in \mathbb{R}^1 \times \mathbb{R}^2$;
(iii) there exists a general solution $x = \phi(t;t_0,x_0)$ of the equation (1) defined on $t_0 \le t < \infty$ with any initial condition $(t_0,x_0) \in \mathbb{R}^1 \times \mathbb{R}^2$; and
(iv) $C = \{c_1(t), c_2(t), \cdots, c_n(t)\}$ is a set of n periodic solutions of the equation, such that $\{c_1(0), c_2(0), \cdots, c_n(0)\} = \{c_1(1), c_2(1), \cdots, c_n(1)\}$.

Our purpose is to estimate the number of periodic solutions of the equation (1).

Let $\phi_t(x) = \phi(t;0,x)$ and let $T = \phi_1$. The map $T : \mathbb{R}^2 \to \mathbb{R}^2$ is usually called the Poincaré transformation of the differential equation.

Clearly, for a natural number $p \in \mathbb{N}$, $x(t)$ is a p-periodic solution of the differential equation $\iff$ $x(0)$ is a p-periodic point of T. By "p-periodic" we always mean "of minimal integer period p". So, finding periodic solutions other than the given $\{c_i(t)\}$ is equivalent to finding periodic points of the map $T : \mathbb{R}^2 \setminus \{c_1(0), c_2(0), \cdots, c_n(0)\} \to \mathbb{R}^2 \setminus \{c_1(0), c_2(0), \cdots, c_n(0)\}$.

Let $S^2 = \mathbb{R}^2 \cup \{\infty\}$ be the 2-sphere. Extend the general solution $x = \phi(t;t_0,x_0)$ to S^2 by defining $\phi(t;t_0,\infty) = \infty$ for all $t \ge t_0$, and extend ϕ_t and T accordingly. The extended ϕ, ϕ_t and T are continuous, but in general not C^1 at ∞. For convenience we will write $c_0(t) = \infty$ for all $t \in \mathbb{R}^1$, and write $M = S^2 \setminus \{c_0(0), c_1(0), \cdots, c_n(0)\} = \mathbb{R}^2 \setminus \{c_1(0), c_2(0), \cdots, c_n(0)\}$.

After preparation in (A) and (B) on braids, we use Nielsen theory to analyze the fixed points and periodic points of $T : M \to M$ under a smoothness condition at infinity introduced in (C). This additional condition is then removed in (E).

(A) Braids as automorphisms of a free group.

We use [1] as our standard reference for braid theory. Let $F_n = \langle a_1, a_2, \cdots, a_n \rangle$ be the free group on free generators $a_1, a_2, \cdots, a_n$. By [1] p.25, Corollary 1.8.3, we can faithfully represent the n-braid group $\pi_1 B_{0,n}$ into AutF_n, the standard generators σ_i, $i = 1, 2, \cdots, n-1$, of $\pi_1 B_{0,n}$ being represented as

$$\sigma_i : \begin{cases} a_i & \mapsto a_i a_{i+1} a_i^{-1}, \\ a_{i+1} & \mapsto a_i, \\ a_j & \mapsto a_j \quad \text{if } j \neq i, i+1. \end{cases}$$

Thus for arbitrary $\sigma \in \pi_1 B_{0,n}$, $a_i\sigma$ is conjugate to $a_{(i)\nu(\sigma)}$, where ν denotes the homomorphism from $\pi_1 B_{0,n}$ to the symmetric group Σ_n on $\{1, \cdots, n\}$ defined by $\nu(\sigma_i) = (i, i+1)$ (cf. [1] p.19).

(B) The "twisted" representation.

Let $g_i = a_1 a_2 \cdots a_i$, $1 \leq i \leq n$. Then $\{g_1, g_2, \cdots, g_n\}$ is also a basis of F_n and we have (cf. [1] p.121)

$$\sigma_i : \begin{cases} g_k \mapsto g_k & \text{if } k \neq i, \\ g_i \mapsto g_{i+1} g_i^{-1} g_{i-1} & \text{if } i \neq 1, \\ g_1 \mapsto g_2 g_1^{-1} & \text{if } i = 1. \end{cases}$$

Let $\Lambda = \mathbb{Z}[a_1^{\pm 1}, \cdots, a_n^{\pm 1}]$ be the ring of integral Laurent polynomials in the variables a_i, $i = 1, 2, \cdots, n$.

Define a map $B' : \pi_1 B_{0,n} \to GL(n, \Lambda)$ by

$$B'(\sigma) = \left(\left(\frac{\partial (g_i \sigma)}{\partial g_j} \right)^{\mathrm{Ab}} \right) \quad \text{for } \sigma \in \pi_1 B_{0,n},$$

where Ab denotes the abelianization operator of $\mathbb{Z}F_n$.

The last row of these matrices is always $(0, \cdots, 0, 1)$. Hence the last row and column may be deleted to obtain another map $B : \pi_1 B_{0,n} \to GL(n-1, \Lambda)$.

In particular, the generators $\sigma_1, \sigma_2, \cdots, \sigma_{n-1}$ will be mapped as follows:

$$B(\sigma_1) = \begin{pmatrix} -a_2 & 1 & \\ 0 & 1 & \\ & & I_{n-3} \end{pmatrix}, \quad B(\sigma_j) = \begin{pmatrix} I & & & \\ & 1 & 0 & 0 & \\ & a_{j+1} & -a_{j+1} & 1 & \\ & 0 & 0 & 1 & \\ & & & & I \end{pmatrix} \begin{matrix} \\ \\ \leftarrow j\text{-th row,} \\ 1 < j < n. \\ \\ \end{matrix}$$

Suppose $\alpha, \beta \in \pi_1 B_{0,n}$ are arbitrary with $g_i\alpha = W_i(g_1, \cdots, g_n)$, $g_j\beta = V_j(g_1, \cdots, g_n)$. Then

$$\begin{aligned} B'(\alpha\beta) &= \left(\left(\frac{\partial (g_i \alpha\beta)}{\partial g_j} \right)^{\mathrm{Ab}} \right) = \left(\left(\frac{\partial W_i(V_1, V_2, \cdots, V_n)}{\partial g_j} \right)^{\mathrm{Ab}} \right) \\ &= \left(\left(\sum_{k=1}^{n} \frac{\partial W_i}{\partial g_k} \bigg|_{g_k := V_k(g_1, g_2, \cdots, g_n)} \frac{\partial V_k}{\partial g_j} \right)^{\mathrm{Ab}} \right) \\ &= \left(\sum_{k=1}^{n} \left(\left(\frac{\partial W_i}{\partial g_k} \right)^{\mathrm{Ab}} \right)^{\nu(\beta)} \left(\frac{\partial V_k}{\partial g_j} \right)^{\mathrm{Ab}} \right) \\ &= B'(\alpha)^{\nu(\beta)} B'(\beta). \end{aligned}$$

(Here the superscript $\nu(\beta)$ denotes the substitution $a_h := a_h^{\nu(\beta)}$ for all a_h. Note that $(x\beta)^{\mathrm{Ab}} = (x^{\mathrm{Ab}})^{\nu(\beta)}$ for all $x \in F_n$.) Hence we get the product formula for the map B

$$B(\alpha\beta) = B(\alpha)^{\nu(\beta)} B(\beta).$$

B is not a representation in the usual sense because of the appearance of $\nu(\beta)$ in the product formula. So we call it a twisted representation.

REMARK: By identifying $a_1, a_2, \cdots, a_n$ in $B(\sigma)$ to a single symbol a, we obtain the reduced Burau representation (cf. [1] p.121).

(C) Compactification of the space under a restriction at infinity.

We introduce an additional assumption: There is a closed disk $D \subset \mathbb{R}^2$ such that $f(x,t) = 0$ for all $t \in \mathbb{R}^1$ and $x \notin D$. This makes ϕ, ϕ_t and T differentiable all over S^2, and $\phi_t(x) = x$ on $U = \mathbb{R}^2 \setminus D$.

Let X be the compactification of M obtained from S^2 by "blowing up" each $c_i(0)$, $0 \leq i \leq n$, to a circle S_i^1 (cf. [3] p.24). Let $f: X \to X$ be the extension of $T: M \to M$. (Here we need the differentiability of T at ∞). We shall apply Nielsen fixed point theory to the surface map $f: X \to X$.

(D) Nielsen theory for f.

Choose a base point $x_0 \in U$ for both M and X. Identify both $\pi_1(M, x_0)$ and $\pi_1(X, x_0)$ with $F_n = \langle a_1, a_2, \cdots, a_n \rangle$ as indicated in Figure 1.

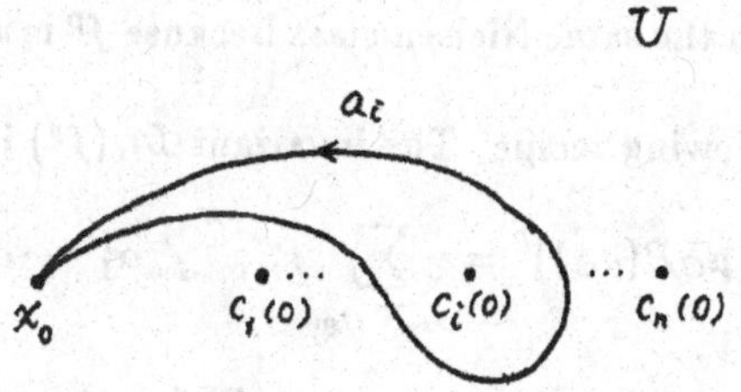

Figure 1

Let S be the geometric braid $\{(c_i(t), t) \mid 0 \leq t \leq 1,\ 1 \leq i \leq n\} \subset \mathbb{R}^2 \times I$, and let $\sigma_C \in \pi_1(B_{0,n}\mathbb{R}^2, \{c_1(0), c_2(0), \cdots, c_n(0)\})$ be the braid represented by S (cf. [1] p.6). It is not hard to see that under the above identification of F_n and the representation in (A) of $\pi_1 B_{0,n}$ into $\mathrm{Aut}F_n$, the following diagram commutes:

$$\begin{array}{ccccc} F_n & = & \pi_1(M, x_0) & \xrightarrow{i_{0*}} & \pi_1(\mathbb{R}^2 \times I \setminus S, (x_0, 0)) \\ \downarrow \sigma_C & & \downarrow T_* & & \downarrow u_* \\ F_n & = & \pi_1(M, x_0) & \xrightarrow{i_{1*}} & \pi_1(\mathbb{R}^2 \times I \setminus S, (x_0, 1)) \end{array}$$

where $i_t: x \mapsto (x, t)$ is the obvious inclusion map, u_* is the isomorphism induced by the vertical path $x_0 \times I$ joining the base points.

Since $f_*: \pi_1(X, x_0) \to \pi_1(X, x_0)$ is naturally identified with $T_*: \pi_1(M, x_0) \to \pi_1(M, x_0)$, we have:

PROPOSITION. *If we take the reference path w to be the constant path at x_0, then $f_\pi = f_* : \pi_1(X, x_0) \to \pi_1(X, x_0)$ is the same as $\sigma_C : F_n \to F_n$.* ∎

Now apply the Nielsen theory sketched in §1 to the map f of the bounded surface X. In our case $H_1(X)$ is the free abelian group generated by $a_1, \cdots, a_n$ and $\mathbb{Z}H_1(X)^{\cdot} = \Lambda$. The homomorphism $f_* : H_1(X) \to H_1(X)$ sends a_i to $a_i^{\nu_C}$ where ν_C is short-hand for $\nu(\sigma_C)$. Let the permutation ν_C be decomposed as a product of m disjoint cycles, the length of j-th cycle being ℓ_j and $\ell_1 + \ell_2 + \cdots + \ell_m = n$. The quotient $\mathrm{Coker}(1 - f_*)$ is obtained from $H_1(X)$ by identifying the a_i's in the j-th cycle to a single symbol α_j. So we have

$$\Lambda_C := \mathbb{Z}H = \mathbb{Z}[\alpha_1^{\pm 1}, \cdots, \alpha_m^{\pm 1}].$$

Let μ_C stand for the projection $H_1(X)^{\cdot} \to H$ as well as for its extension $\Lambda \to \Lambda_C$. Then we get from §1 (C) to (E) (since $D_H(f_\pi) = \mu_C B'(\sigma_C)$):

$$\begin{aligned} L_H(f) &= 1 - \mu_C(\mathrm{tr} B'(\sigma_C)) = -\mu_C \mathrm{tr} B(\sigma_C), \\ L_H(f^p) &= -\mathrm{tr}(\mu_C B(\sigma_C))^p \\ \zeta_H &= \det(I - t\mu_C B(\sigma_C)). \end{aligned}$$

These formulas provide information about periodic points of $f : X \to X$.

To relate it with periodic points of our original map $T : M \to M$, we need two observations. First, the subset $S_0^1 \cup U$ contains in a single Nielsen class of f^p, namely the one with coordinate $1 \in H$ (cf. §1(D)). Second, on each S_i^1, $1 \le i \le n$, the fixed points of f^p (if any) are in the same Nielsen class because f^p is an orientation preserving homeomorphism.

So we arrive at the following recipe: The invariant $L_H(f^p)$ is a finite sum

$$-\mathrm{tr}(\mu_C B(\sigma_C))^p = \sum_{j_1, \cdots, j_m} r_{j_1, \cdots, j_m} \alpha_1^{j_1} \cdots \alpha_m^{j_m}.$$

Count the number K_p of terms (with non-zero coefficients) such that the greatest common divisor $(p, j_1, j_2, \cdots, j_m) = 1$, excluding the term with $j_1 = j_2 = \cdots = j_m = 0$. Then by the Corollary in §1(D) the number of p-periodic points of $f : X \to X$ is at least pK_p. If furthermore let n_p be the number of indices i, $1 \le i \le n$, such that the least period of $c_i(t)$ divides p, then the number of p-periodic points of $T : M \to M$ is not smaller than $p(K_p - n_p)$.

(E) Removing the restriction at infinity.

We now consider an equation

$$\frac{dx}{dt} = f'(t, x) \tag{2}$$

satisfying the conditions (i)–(iv). The ϕ, ϕ_t, T, etc. for (2) will be denoted by ϕ', ϕ'_t, T', etc. As before we are interested in the p-periodic points of $T' : M \to M$. We shall construct an auxiliary equation (1),

$$\frac{dx}{dt} = f(t, x),$$

which also satisfies the additional assumption of (C), and compare the periodic points of T and T'.

The construction follows. Take closed disks $D_1 \subset D_2 \subset \mathbb{R}^2$ with $\{c_1(0), \cdots, c_n(0)\} \subset D_1$ and $T'^p(D_1) \subset D_2$. Let $\lambda : \mathbb{R}^2 \to [0,1]$ be a smooth function with compact support such that $\lambda = 1$ on the compact set $\bigcup_{0 \le t \le p} \phi'_t(D_2)$. Let $f(t,x) = \lambda(x) f'(t,x)$. Then the equation (1) certainly satisfies our requirements.

Obviously $\phi_t(x) = \phi'_t(x)$ for all $x \in D_2$ and $0 \le t \le p$, so that T and T' have the same p-periodic points on D_2. Now suppose x is a fixed point of T^p on $(\mathbb{R}^2 \setminus T^p(D_2)) \cup (\mathbb{R}^2 \setminus D_2) = \mathbb{R}^2 \setminus (T'^p(D_2) \cap D_2) \subset \mathbb{R}^2 \setminus T'^p(D_1)$ because of the construction $T'^p(D_1) \subset T'^p(D_2) \cap D_2$. But $T'^p(D_1)$ is a closed disc in $\mathbb{R}^2$, so by the Jordan Curve Theorem $\mathbb{R}^2 \setminus T'^p(D_1)$ is an open annulus. Hence $\mu_C[(f^p \circ h)h^{-1}] = \mu_C(a_1 a_2 \cdots a_n)^k = (\alpha_1^{\ell_1} \alpha_2^{\ell_2} \cdots \alpha_m^{\ell_m})^k$ for some $k \in \mathbb{Z}$, where h is a path from x_0 to x.

Conclusion: When we use the method of (C)–(D) to estimate the number of p-periodic points for $T : M \to M$, if we disregard the terms in $-\mathrm{tr}(\mu_C B(\sigma_C))^p$ such that $(j_1, j_2, \cdots, j_m) = k(\ell_1, \ell_2, \cdots, \ell_m)$ for some $k \in \mathbb{Z}$, then we get an estimate for the number of p-periodic points for the original map $T' : M \to M$. In other words, we arrive at the following strengthened form of Matsuoka [9] Theorem 1.

THEOREM. *Suppose equation (1) satisfies conditions (i)–(iv). Let p be a positive integer. Denote by $N(C,p)$ the number of p-periodic solutions not contained in C. Let $K(C,p)$ be the number of terms in $-\mathrm{tr}(\mu_C B(\sigma_C))^p$ such that $(p, j_1, j_2, \cdots, j_m) = 1$ and $(j_1, j_2, \cdots, j_m)$ is not a multiple of $(\ell_1, \ell_2, \cdots, \ell_m)$. Let n_p be the number of indices i, $1 \le i \le n$, such that the least period of $c_i(t)$ divides p. Then*

$$N(C,p) \ge p(K(C,p) - n_p). \qquad \blacksquare$$

3. Periodic Differential Equation on the Torus

Consider a differential equation

$$\frac{dx}{dt} = f(t,x) \tag{3}$$

on the torus T^2, where $f_t := f(t, \cdot)$ is a vector field on the torus, and f is of class C^1 on $\mathbb{R}^1 \times T^2$. We always assume that

(i) $f_{t+1} = f_t$ for all $t \in \mathbb{R}$, and
(ii) $C = \{c_1(t), c_2(t), \cdots, c_n(t)\}$ is a set of periodic solutions of the equation, such that $\{c_1(0), c_2(0), \cdots, c_n(0)\} = \{c_1(1), c_2(1), \cdots, c_n(1)\}$.

REMARK: There always exists a general solution $x = \phi(t; t_0, x_0)$ on $-\infty < t < \infty$ with any initial condition $(t_0, x_0) \in \mathbb{R}^1 \times T^2$, because T^2 is a closed manifold.

Our purpose is to estimate the number of periodic solutions of the equation (3). The treatment below parallels that of §2. Essential differences arise in (A) and (C) because we are now working on a closed surface.

(A) Braids as outer automorphisms of a free group.

Let us first recall some facts about braids on the torus. The special braids σ_i, ρ_i, τ_i used below are illustrated in Figure 2. We use the commutator notation $[a,b] = aba^{-1}b^{-1}$ in groups.

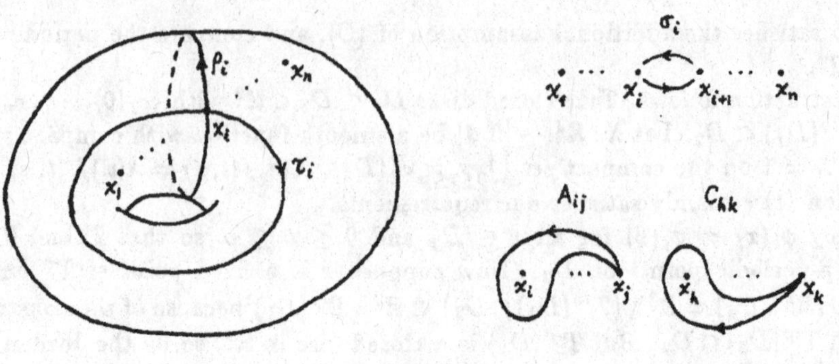

Figure 2

PROPOSITION. (Birman [2]) *The pure n-braid group $\pi_1(F_{0,n}T^2,(x_1\cdots,x_n))$ admits the following presentation:*
Generators: $\rho_1,\rho_2,\cdots,\rho_n,\tau_1,\tau_2,\cdots,\tau_n$;
Relations:

$$
\begin{aligned}
&[\rho_i,\rho_j]=[\tau_i,\tau_j]=1,\\
&[A_{jk},\tau_i]=[A_{jk},\rho_i]=1,\\
&A_{ij}=\tau_j^{-1}\rho_i\tau_j\rho_i^{-1},\qquad C_{ij}=\rho_j^{-1}\tau_i\rho_j\tau_i^{-1},\\
&C_{ij}=(\tau_i\tau_j)A_{ij}^{-1}(\tau_j^{-1}\tau_i^{-1}),\qquad A_{ij}=(\rho_i\rho_j)C_{ij}^{-1}(\rho_j^{-1}\rho_i^{-1}),\\
&C_{ij}=(A_{j-1,j}^{-1}A_{j-2,j}^{-1}\cdots A_{i+1,j}^{-1})A_{ij}^{-1}(A_{i+1,j}\cdots A_{j-1,j}),\\
&\tau_1\rho_1\tau_1^{-1}\rho_1^{-1}=A_{12}A_{13}\cdots A_{1n},
\end{aligned}
$$

where $1\le i<j<k\le n$, but $j\le n$ if k is absent.

The full n-braid group $\pi_1(B_{0,n}T^2,\{x_1,\cdots,x_n\})$ admits a presentation obtained from above by adding generators $\sigma_1,\sigma_2,\cdots,\sigma_{n-1}$ and relations:

$$
\begin{aligned}
&\sigma_i\sigma_j=\sigma_j\sigma_i,\qquad \text{if } |i-j|\ge 2,\\
&\sigma_i\sigma_{i+1}\sigma_i=\sigma_{i+1}\sigma_i\sigma_{i+1},\\
&\sigma_i^2=A_{i,i+1},\\
&[\rho_1,\sigma_i]=[\tau_1,\sigma_i]=1\qquad \text{for } i>1,\\
&\rho_{i+1}=\sigma_i\rho_i\sigma_i,\qquad \tau_{i+1}=\sigma_i^{-1}\tau_i\sigma_i^{-1}.
\end{aligned}
$$

We shall not make use of the above presentation of $\pi_1 B_{0,n}T^2$ except for obtaining a convenient set of generators:

COROLLARY. *The group $\pi_1 B_{0,n}T^2$ is generated by $\sigma_1,\sigma_2,\cdots,\sigma_{n-1},\rho_n,\tau_n$.*

PROOF: $\pi_1 B_{0,n}T^2$ is generated by $\sigma_1,\sigma_2,\cdots,\sigma_{n-1},\rho_1,\cdots,\rho_n,\tau_1,\cdots,\tau_n$. Since $\rho_i=\sigma_i^{-1}\rho_{i+1}\sigma_i^{-1}$ and $\tau_i=\sigma_i\tau_{i+1}\sigma_i$ for $1\le i<n$, they can be deleted from the list of generators. ∎

For brevity we shall write $M=T^2\setminus\{x_1,x_2,\cdots,x_n\}$. We now introduce a representation of $\pi_1 B_{0,n}T^2$ into the outer automorphism group

$$\mathrm{Out}\pi_1(M):=\mathrm{Aut}\pi_1(M)\,/\,\mathrm{Inn}\pi_1(M),$$

where $\mathrm{Inn}\pi_1(M)$ denotes the inner automorphisms of $\pi_1(M)$.

Pick a base point x_0 for M. Every braid $\sigma \in \pi_1 B_{0,n}T^2$ can be represented by strings $S = \{(y_i(t),t) \mid 0 \le t \le 1,\ 1 \le i \le n\} \subset (T^2 \setminus x_0) \times I$. Such a geometric braid S defines an automorphism S_* of $\pi_1(M,x_0)$ by the commutativity of the diagram

$$\begin{array}{ccc} \pi_1(M,x_0) & \xrightarrow[\cong]{i_{0*}} & \pi_1(T^2 \times I \setminus S,(x_0,0)) \\ \Big\downarrow{\scriptstyle S_*} & & \Big\downarrow{\scriptstyle u_*} \\ \pi_1(M,x_0) & \xrightarrow[\cong]{i_{1*}} & \pi_1(T^2 \times I \setminus S,(x_0,1)) \end{array}$$

where u_* is the isomorphism induced by the vertical path u from $(x_0,1)$ to $(x_0,0)$, and $i_t : x \mapsto (x,t)$ is the obvious inclusion at level t. Suppose $S' \subset (T^2 \setminus x_0) \times I$ is another geometric braid representing σ. By the isotopy extension theorem, there is a homeomorphism $\psi : T^2, S' \to T^2, S$ which is identity on $T^2 \times \{0,1\}$. Let $u' = \psi \circ u$. So we have a commutative diagram

$$\begin{array}{ccc} \pi_1(M,x_0) & = & \pi_1(M,x_0) \\ {\scriptstyle i_{0*}}\Big\downarrow{\scriptstyle \cong} & & {\scriptstyle \cong}\Big\downarrow{\scriptstyle i_{0*}} \\ \pi_1(T^2 \times I \setminus S',(x_0,0)) & \xrightarrow[\cong]{\psi_*} & \pi_1(T^2 \times I \setminus S,(x_0,0)) \\ {\scriptstyle u_*}\Big\downarrow & & \Big\downarrow{\scriptstyle u'_*} \\ \pi_1(T^2 \times I \setminus S',(x_0,1)) & \xrightarrow[\cong]{\psi_*} & \pi_1(T^2 \times I \setminus S,(x_0,1)) \\ {\scriptstyle i_{1*}}\Big\uparrow{\scriptstyle \cong} & & {\scriptstyle \cong}\Big\uparrow{\scriptstyle i_{1*}} \\ \pi_1(M,x_0) & = & \pi_1(M,x_0). \end{array}$$

The composition of the vertical arrows on the left is S'_* by definition. Hence the commutative diagram

$$\begin{array}{ccc} \pi_1(M,x_0) & \xrightarrow[\cong]{i_{0*}} & \pi_1(T^2 \times I \setminus S,(x_0,0)) \\ \Big\downarrow{\scriptstyle S'_*} & & \Big\downarrow{\scriptstyle u'_*} \\ \pi_1(M,x_0) & \xrightarrow[\cong]{i_{1*}} & \pi_1(T^2 \times I \setminus S,(x_0,1)). \end{array}$$

Compare this with the diagram defining S_*, we see S_* and S'_* differ by an inner automorphism induced by a loop. Thus our construction gives us a representation of $\pi_1 B_{0,n}T^2$ into $\mathrm{Out}\pi_1(M)$.

We now describe this representation explicitly. The group $F_{n+1} := \pi_1(M,x_0)$ is a free group of rank $n+1$. Define elements $a_1, a_2, \cdots, a_n, b_1, b_2$, and $g_1, g_2, \cdots, g_n \in \pi_1(M)$ as shown in Figure 3. Obviously $g_i = a_1 a_2 \cdots a_i$ for $1 \le i \le n$, and $g_n = b_1 b_2 b_1^{-1} b_2^{-1}$. Two useful bases of F_{n+1} are the standard basis $\{a_2, a_3, \cdots, a_n, b_1, b_2\}$ and the auxiliary basis $\{g_1, g_2, \cdots, g_{n-1}, b_1, b_2\}$.

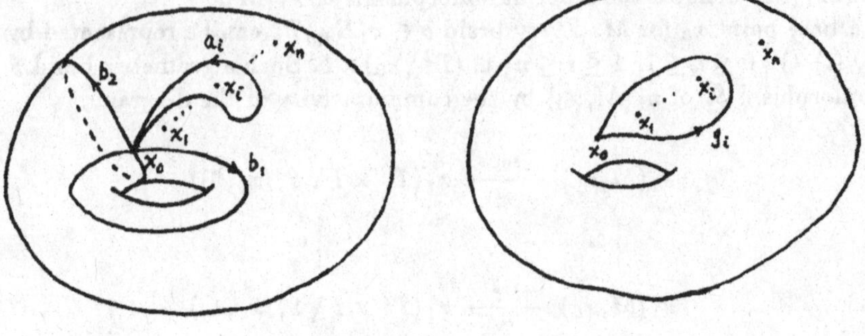

Figure 3

Define the automorphisms $\sigma_i, \rho_n, \tau_n : F_{n+1} \to F_{n+1}$ to be those determined by the corresponding geometric braids indicated in Figure 2. By geometric inspection we can write them down in terms of both the standard and auxiliary bases:

$$\sigma_j : \underset{j<n-1}{} \begin{cases} a_i \mapsto a_i & \text{if } i \neq j, j+1, \\ a_j \mapsto a_j a_{j+1} a_j^{-1}, \\ a_{j+1} \mapsto a_j, \\ b_1 \mapsto b_1, \\ b_2 \mapsto b_2; \end{cases} \qquad \begin{cases} g_i \mapsto g_i & \text{if } i \neq j, \\ g_j \mapsto g_{j+1} g_j^{-1} g_{j-1} & \text{if } j \neq 1, \\ g_1 \mapsto g_2 g_1^{-1} & \text{if } j = 1, \\ b_1 \mapsto b_1, \\ b_2 \mapsto b_2; \end{cases}$$

$$\sigma_{n-1} : \begin{cases} a_i \mapsto a_i & \text{if } i < n-1, \\ a_{n-1} \mapsto a_{n-1} a_n a_{n-1}^{-1}, \\ a_n \mapsto a_{n-1}, \\ b_1 \mapsto b_1, \\ b_2 \mapsto b_2; \end{cases} \qquad \begin{cases} g_i \mapsto g_i & \text{if } i < n-1, \\ g_{n-1} \mapsto [b_1, b_2] g_{n-1}^{-1} g_{n-2}, \\ b_1 \mapsto b_1, \\ b_2 \mapsto b_2; \end{cases}$$

$$\rho_n : \begin{cases} a_i \mapsto a_i & \text{if } i < n, \\ a_n \mapsto b_2 [b_1, b_2] a_n [b_2, b_1] b_2^{-1}, \\ b_1 \mapsto [b_1, b_2] a_n^{-1} [b_2, b_1] b_1, \\ b_2 \mapsto b_2; \end{cases} \qquad \begin{cases} g_i \mapsto g_i, \\ b_1 \mapsto g_{n-1} b_2 b_1 b_2^{-1}, \\ b_2 \mapsto b_2; \end{cases}$$

$$\tau_n : \begin{cases} a_i \mapsto a_i & \text{if } i < n, \\ a_n \mapsto a_n [b_2 b_1 b_2^{-1}, a_n], \\ b_1 \mapsto b_1, \\ b_2 \mapsto a_n b_2; \end{cases} \qquad \begin{cases} g_i \mapsto g_i, \\ b_1 \mapsto b_1, \\ b_2 \mapsto g_{n-1}^{-1} b_1 b_2 b_1^{-1}. \end{cases}$$

REMARK: The above action of these generators of $\pi_1 B_{0,n} T^2$ on F_{n+1} is meant to give a representation of $\pi_1 B_{0,n} T^2$ into $\mathrm{Out} F_{n+1}$, not a representation into $\mathrm{Aut} F_{n+1}$. In fact, in $\pi_1 B_{0,n} T^2$ there is a relation

$$\tau_n^{-1} \rho_n^{-1} \tau_n \rho_n = \sigma_{n-1} \sigma_{n-2} \cdots \sigma_2 \sigma_1^2 \sigma_2 \cdots \sigma_{n-2} \sigma_{n-1},$$

but according to the above formulas

$$\tau_n^{-1}\rho_n^{-1}\tau_n\rho_n(a_2) \neq \sigma_{n-1}\sigma_{n-2}\cdots\sigma_2\sigma_1^2\sigma_2\cdots\sigma_{n-2}\sigma_{n-1}(a_2).$$

This is in contrast to §2 (A) where the base point near infinity would not interfere with the braiding of the strings, hence the representation is into $\mathrm{Aut}F_n$.

(B) The "twisted representation".

Let $\Lambda = \mathbb{Z}H_1(M)^{\cdot}$. It can be regarded as $\mathbb{Z}[a_2^{\pm1},\cdots,a_n^{\pm1},b_1^{\pm1},b_2^{\pm1}]$, the ring of integral Laurent polynomials in the variables a_i, $1 < i \leq n$, and b_j, $1 \leq j \leq 2$. Alternatively, it is the quotient of $\mathbb{Z}[a_1^{\pm1},\cdots,a_n^{\pm1},b_1^{\pm1},b_2^{\pm1}]$ modulo the relation $a_1a_2\cdots a_n = 1$.

Define a map $B : \mathrm{Aut}F_{n+1} \to GL(n+1,\Lambda)$ by

$$B(\varphi) = \begin{pmatrix} \dfrac{\partial(g_i\varphi)}{\partial g_j} & \dfrac{\partial(g_i\varphi)}{\partial b_{j'}} \\ \dfrac{\partial(b_{i'}\varphi)}{\partial g_j} & \dfrac{\partial(b_{i'}\varphi)}{\partial b_{j'}} \end{pmatrix}^{\mathrm{Ab}}_{1\leq i,j\leq n-1,\,1\leq i',j'\leq 2}$$

for any $\varphi \in \mathrm{Aut}F_{n+1}$, where Ab denotes the abelianization operator of $\mathbb{Z}F_{n+1}$.

Please note that the Fox derivatives are taken with respect to the auxiliary basis $\{g_1,\cdots,g_{n-1},b_1,b_2\}$, while the elements of Λ should be written in terms of the standard basis $\{a_2,\cdots,a_n,b_1,b_2\}$. Since $(a_1a_2\cdots a_n)^{\mathrm{Ab}} = (g_n)^{\mathrm{Ab}} = (b_1b_2b_1^{-1}b_2^{-1})^{\mathrm{Ab}} = 1$, we have $(g_i)^{\mathrm{Ab}} = a_{i+1}^{-1}\cdots a_n^{-1}$, $(b_j)^{\mathrm{Ab}} = b_j$.

B is not a homomorphism. To discuss the product formula for B, we must understand the (right) action of $\mathrm{Aut}F_{n+1}$ on Λ. An automorphism of $F_{n+1} = \pi_1(M)$ naturally induces automorphisms on $H_1(M)$ and $\Lambda = \mathbb{Z}H_1(M)^{\cdot}$. The notation we use is indicated in the commutative diagram

$$\begin{array}{ccccc} F_{n+1} & \xrightarrow{\mathrm{Ab}} & H_1(M)^{\cdot} & \longrightarrow & \Lambda \\ \downarrow{\scriptstyle\varphi} & & \downarrow{\scriptstyle\nu(\varphi)} & & \downarrow{\scriptstyle\nu(\varphi)} \\ F_{n+1} & \xrightarrow{\mathrm{Ab}} & H_1(M)^{\cdot} & \longrightarrow & \Lambda. \end{array}$$

The effect of this action on Λ is in the form of substitutions. To obtain the result $\lambda^{\nu(\varphi)}$ of the operator $\nu(\varphi)$ on an element $\lambda \in \Lambda$, we write λ as a Laurent polynomial in the variables $\{a_1,a_2,\cdots,a_n,b_1,b_2\}$, then for each monomial $a_1^{i_1}\cdots a_n^{i_n}b_1^{j_1}b_2^{j_2}$ substitute the monomial $(a_1^{\nu(\varphi)})^{i_1}\cdots(a_n^{\nu(\varphi)})^{i_n}(b_1^{\nu(\varphi)})^{j_1}(b_2^{\nu(\varphi)})^{j_2}$.

This kind of substitutions is what we need for a product formula for B. By exactly the same argument as that in §2 (B), we see

$$B(\alpha\beta) = B(\alpha)^{\nu(\beta)}B(\beta),$$

where the superscript $\nu(\beta)$ means applying the substitution $\nu(\beta)$ to every entry of the matrix $B(\alpha)$.

We are concerned with the B-images of the automorphisms of F_{n+1} that arise from braids in $\pi_1 B_{0,n}T^2$. In particular, under B the automorphisms $\sigma_1, \cdots, \sigma_{n-1}, \rho_n, \tau_n : F_{n+1} \to F_{n+1}$ specified in (A) will become:

$$B(\sigma_j) = \begin{pmatrix} I & & & \\ & 1 & 0 & 0 & \\ & a_{j+1} & -a_{j+1} & 1 & \\ & 0 & 0 & 1 & \\ & & & & I \end{pmatrix} \leftarrow j\text{-th row}, \ 1 < j < n-1,$$

$$B(\sigma_1) = \begin{pmatrix} -a_2 & 1 & \\ 0 & 1 & \\ & & I \end{pmatrix}, \qquad B(\sigma_{n-1}) = \begin{pmatrix} I & & & & \\ & 1 & 0 & & \\ & a_n & -a_n & 1-b_2 & b_1 - 1 \\ & & & 1 & 0 \\ & & & 0 & 1 \end{pmatrix},$$

$$B(\rho_n) = \begin{pmatrix} I & & & \\ & 1 & & \\ & 1 & a_n^{-1} b_2 & a_n^{-1}(1-b_1) \\ & 0 & 0 & 1 \end{pmatrix}, \quad B(\tau_n) = \begin{pmatrix} I & & & \\ & 1 & & \\ & 0 & 1 & 0 \\ & -a_n & a_n(1-b_2) & a_n b_1 \end{pmatrix}.$$

Their action on Λ is

$$\nu(\sigma_j) : \begin{cases} a_i & \mapsto a_i \quad \text{if } i \neq j, j+1, \\ a_j & \mapsto a_{j+1}, \\ a_{j+1} & \mapsto a_j, \\ b_k & \mapsto b_k; \end{cases}$$

$$\nu(\rho_n) : \begin{cases} a_i \mapsto a_i, \\ b_1 \mapsto a_n^{-1} b_1, \\ b_2 \mapsto b_2; \end{cases} \qquad \nu(\tau_n) : \begin{cases} a_i \mapsto a_i, \\ b_1 \mapsto b_1, \\ b_2 \mapsto a_n b_2. \end{cases}$$

These expressions and the product formula enables one to calculate $B(\varphi)$ for all φ that is written as a product of $\{\sigma_1, \cdots, \sigma_{n-1}, \rho_n, \tau_n\}$ and their inverses.

(C) Compactification of the space.

As in §2, we write $\phi_t = \phi(t;\cdot)$. $T = \phi_1$ is the Poincaré map for equation (3). Let X be the compactification of $M = T^2 \setminus \{c_1(0), \cdots, c_n(0)\}$ obtained from T^2 by "blowing up" each $c_i(0)$, $1 \le i \le n$, to a circle S_i^1. Since $T : T^2 \to T^2$ is a C^1 map, $T : M \to M$ can always be extended to a map $f : X \to X$. Note that our situation here is much simpler than that in §2 where the smoothness at infinity is not granted.

(D) Nielsen theory for f.

Let $x_i = c_i(0)$ for $1 \le i \le n$. The given periodic solutions $C = \{c_1(t), c_2(t), \cdots, c_n(t)\}$ provides a geometric braid $S = \{(c_i(t), t) \mid 0 \le t \le 1, \ 1 \le i \le n\} \subset T^2 \times I$ that determines a braid $\sigma_C \in \pi_1 B_{0,n}T^2$. Write σ_C as a word in the generators $\sigma_1, \cdots, \sigma_{n-1}, \rho_n, \tau_n$ of $\pi_1 B_{0,n}T^2$. Using this word and the automorphisms σ_i, ρ_n, τ_n defined in (B) we obtain an automorphism $\sigma_C : F_{n+1} \to F_{n+1}$. This automorphism may depend on the word used, but we always have

PROPOSITION. *There exists a path w in X from x_0 to $f(x_0)$ such that $f_\pi = w_* \circ f_* = \sigma_C : F_{n+1} \to F_{n+1}$.*

PROOF: We examine $T_* : \pi_1(M, x_0) \to \pi_1(M, f(x_0))$ because it is the same as f_*. Without loss we may assume $S \subset (T^2 \setminus x_0) \times I$ (perturb x_0 if necessary). Consider the diagram

$$\begin{array}{ccc}
\pi_1(M, x_0) & \xrightarrow[\cong]{T_* = \phi_{1*}} & \pi_1(M, f(x_0)) \\
{\scriptstyle i_{0*}} \downarrow \cong & & \cong \downarrow {\scriptstyle i_{1*}} \\
\pi_1(T^2 \times I \setminus S, (x_0, 0)) & \xleftarrow[\cong]{v_*} & \pi_1(T^2 \times I \setminus S, (f(x_0), 1)) \\
\Vert & & \downarrow {\scriptstyle (i_1 \circ w)_*} \\
\pi_1(T^2 \times I \setminus S, (x_0, 0)) & \xrightarrow[\cong]{u'_*} & \pi_1(T^2 \times I \setminus S, (x_0, 1)) \\
{\scriptstyle i_{0*}} \uparrow \cong & & \cong \uparrow {\scriptstyle i_{1*}} \\
\pi_1(M, x_0) & \xrightarrow[\cong]{\sigma_C} & \pi_1(M, x_0).
\end{array}$$

The path v in $T^2 \times I \setminus S$ is given by $v(t) = (\phi_t(x_0), t)$, hence the first square commutes. The third square commutes with a suitable path u' in $T^2 \times I \setminus S$ since σ_C and S_* differ by an inner automorphism, as explained in (A). The product path $u'v$ can be deformed in $T^2 \times I \setminus S$ to a path in $M \times 1$, of the form $i_1 \circ w$ for some path w in M, making the second square also commute. This w is the required path. ∎

Apply the Nielsen theory to the map f of the bounded surface X. Now $H_1(X)^{\cdot}$ is the abelian group generated by $a_1, \cdots, a_n, b_1, b_2$ with a relation $a_1 \cdots a_n = 1$, and $\mathbb{Z}H_1(X)^{\cdot} = \Lambda$. The homomorphism $f_* : H_1(X) \to H_1(X)$ is in the form of the substitution $\nu_C := \nu(\sigma_C)$. $H = \mathrm{Coker}(1 - f_*)^{\cdot}$ is a quotient of $H_1(X)^{\cdot}$ obtained by identifying each a_i with $a_i^{\nu\sigma}$ and each b_k with $b_k^{\nu\sigma}$. The effect of identifying each a_i with $a_i^{\nu\sigma}$ is similar to that in §2: Suppose the permutation on $\{1, 2, \cdots, n\}$ induced by the braid C decomposes as a product of m disjoint cycles, the length of the j-th cycle being ℓ_j. Then the a_i's in the j-th cycle will be identified into a single symbol α_j. The relation $a_1 \cdots a_n = 1$ then becomes $\alpha_1^{\ell_1} \cdots \alpha_m^{\ell_m} = 1$. The j-th cycle corresponds to a ℓ_j-periodic solution in C which represents a homology class $b_1^{t_j} b_2^{s_j} \in H_1(T^2)^{\cdot}$. Then $b_1^{\nu\sigma} = b_1 \alpha_1^{-s_1} \cdots \alpha_m^{-s_m}$ and $b_2^{\nu\sigma} = b_2 \alpha_1^{t_1} \cdots \alpha_m^{t_m}$. Thus H has generators $\alpha_1, \cdots, \alpha_m, b_1, b_2$ and relations

$$\alpha_1^{\ell_1} \cdots \alpha_m^{\ell_m} = \alpha_1^{s_1} \cdots \alpha_m^{s_m} = \alpha_1^{t_1} \cdots \alpha_m^{t_m} = 1,$$

and Λ_C equals $\mathbb{Z}[\alpha_1^{\pm 1}, \cdots, \alpha_m^{\pm 1}, b_1^{\pm 1}, b_2^{\pm 1}]$ modulo the ideal generated by $1 - \alpha_1^{\ell_1} \cdots \alpha_m^{\ell_m}$, $1 - \alpha_1^{s_1} \cdots \alpha_m^{s_m}$ and $1 - \alpha_1^{t_1} \cdots \alpha_m^{t_m}$. This is an important difference from §2: generally speaking Λ_C is no longer a Laurent polynomial ring, because $\mathrm{Coker}(1 - f_*)$ may have torsion.

Let μ_C stand for the projection $H_1(M)^{\cdot} \to H$ as well as for its extension $\Lambda \to \Lambda_C$.

Then just as in §2 we get from §1 (C) to (E) (since $D_H(f_\pi) = \mu_C B(\sigma_C)$):

$$L_H(f) = 1 - \mu_C\big(\mathrm{tr}B(\sigma_C)\big) = 1 - \mu_C \mathrm{tr}B(\sigma_C),$$
$$L_H(f^p) = 1 - \mathrm{tr}\,\big(\mu_C B(\sigma_C)\big)^p,$$
$$\zeta_H = \frac{\det(I - t\mu_C B(\sigma_C))}{1-t}.$$

The map $f : X \to X$ may have more periodic points than our original map $T : M \to M$, namely those on the boundary $\bigcup_{i=1}^n S_i^1$. However, we know that on each S_i^1, the fixed points of f^p (if any) are in the same Nielsen class because f^p is an orientation preserving homeomorphism.

We summarize our result as

THEOREM. *Suppose equation (3) satisfies the conditions (i), (ii). Let p be an integer. Denote by $N(C,p)$ the number of p-periodic solutions to (3) not contained in C. Suppose*

$$1 - \mathrm{tr}\big(\mu_C B(\sigma_C)\big)^p = \sum_{\substack{j_1,\cdots,j_m\\ k_1,k_2}} r_{j_1,\cdots,j_m,k_1,k_2}\alpha_1^{j_1}\cdots\alpha_m^{j_m} b_1^{k_1} b_2^{k_2} \in \mathbb{Z}H.$$

Let K_p be the number of terms in this finite sum such that, when $p > 1$, in the additive group $\mathrm{Coker}(1 - f_*)$ the element $j_1\alpha_1 + \cdots + j_m\alpha_m + k_1 b_1 + k_2 b_2$ is not divisible by any prime factor of p. Let n_p be the number of indices i, $1 \le i \le n$, such that the least period of the solution $c_i(t)$ divides p. Then

$$N(C,p) \ge p(K_p - n_p). \quad \blacksquare$$

Let us come back to the question of torsion in $\mathrm{Coker}(1 - f_*)$. If there are elements of finite order in H, the ring $\mathbb{Z}H$ will have zero divisors. This algebraic complication can be avoided by means of a homomorphism from H to a torsion-free abelian group H' and computing the image $L_{H'}(f^p) \in \mathbb{Z}H'$ of $L_H(f^p)$.

As an example, consider the following interesting case. Suppose the braid σ_C is written as a word in $\sigma_1, \cdots, \sigma_{n-1}$ and $\rho_1, \cdots, \rho_n, \tau_1, \cdots, \tau_n$. Suppose the total exponent of all the ρ's and the total exponent of all the τ's are both 0. This implies $s_1 + \cdots + s_m = t_1 + \cdots + t_m = 0$. Suppose further that one periodic solution in C, say the one corresponding to α_1, is null-homologous on T^2 (i.e. $s_1 = t_1 = 0$). Let H' be the abelian group freely generated by a, b_1, b_2. A homomorphism $\mu'_C : H_1(X) \to H'$ is defined by assigning $\mu'_C(a_i) = a^{\ell_1 - n}$ if $\mu(a_i) = \alpha_1$, $\mu'_C(a_i) = a^{\ell_1}$ if $\mu(a_i) \ne \alpha_1$, and $\mu'_C(b_k) = b_k$ for $k = 1,2$. Evidently μ'_C factors through H. It extends to $\mu'_C : \Lambda \to \mathbb{Z}H' = \mathbb{Z}[a^{\pm1}, b_1^{\pm1}, b_2^{\pm1}]$. Now $L_{H'}(f^p) = 1 - \mathrm{tr}(\mu'_C B(\sigma_C))^p$ is a finite sum $\sum_{j,k_1,k_2} r_{j,k_1,k_2} a^j b_1^{k_1} b_2^{k_2}$. Let K'_p be the number of terms in this sum such that the greatest common divisor $(p, j, k_1, k_2) = 1$. Then

$$N(C,p) \ge p(K'_p - n_p).$$

EXAMPLE: Suppose $p(t)$ and $q(t)$ are two periodic solutions to the equation (3) which satisfies the conditions (i), (ii) and $\mathrm{period}(p(t)) = 1$, $\mathrm{period}(q(t)) = 2$. Suppose the braid σ_C determined by $C = \{p(t), q(t), q(t+1)\}$ is $\rho_2^2\rho_3^{-2}\sigma_2$. Then

$$\mu'_C B(\sigma_C) = \begin{pmatrix} 1 & 0 & 0 & 0 \\ * & -a^{-3}b_2^2 & -a^{-4}(b_2^3 - b_2^2) & * \\ * & a^{-2}b_2\frac{1-a^4b_2^{-4}}{1-ab_2^{-1}} & b_2^{-2} + a^{-3}(b_2^2 - b_2)\frac{1-a^4b_2^{-4}}{1-ab_2^{-1}} & * \\ 0 & 0 & 0 & 1 \end{pmatrix},$$

$$L_{H'}(f) = 1 - \mathrm{tr}(\mu'_C B(\sigma_C)) = -1 - a^{-1} + a^{-2} - b_2^{-1} + a^{-1}b_2^{-1} - a^{-2}b_2 + a^{-3}b_2,$$

$$\tilde{\zeta}_{H'} = (1-t)\left(1 - (1-a^{-1})b_2(a^{-2} + a^{-1}b_2^{-1} + b_2^{-2})t - a^{-3}t^2\right).$$

From $L_{H'}(f)$ it follows that there are at least six more 1-periodic solutions other than $p(t)$. It can be deduced from the theory in [5] and the formula for $\tilde{\zeta}_{H'}$ that there are infinitely many periodic solutions.

REFERENCES

[1] Birman, J.S., "Braids, Links, and Mapping Class Groups," Ann. Math. Studies vol. 82, Princeton Univ. Press, Princeton, 1974.

[2] Birman, J.S., *On braid groups*, Comm. Pure Appl. Math. **22** (1969), 41–72.

[3] Bowen, R., *Entropy and the fundamental group*, in "Structure of Attractors in Dynamical Systems," N. G. Markley et al. (eds.), Lecture Notes in Math. vol. 668, Springer, Berlin, Heidelberg, New York, 1978, pp. 21–29.

[4] Fadell, É., Husseini, S., *The Nielsen number on surfaces*, in "Topological Methods in Nonlinear Functional Analysis," Contemp. Math. vol.21, AMS, Providence, 1983, pp. 59–98.

[5] Fried, D., *Periodic points and twisted coefficients*, in "Geometric Dynamics," J. Palis Jr. (ed.), Lecture Notes in Math. vol. 1007, Springer, Berlin, Heidelberg, New York, 1983, pp. 261–293.

[6] Jiang, B., "Lectures on Nielsen Fixed Point Theory," Contemp. Math. vol.14, AMS, Providence, 1983.

[7] Matsuoka, T., *The number and linking of periodic solutions of periodic systems*, Invent. Math. **70** (1983), 319–340.

[8] Matsuoka, T., *Waveform in dynamical systems of ordinary differential equations*, Japan. J. Appl. Math. **1** (1984), 417–434.

[9] Matsuoka, T., *The number and linking of periodic solutions of non-dissipative systems*, J. Diff. Eqs. **76** (1988), 190–201.

[10] Reidemeister, K., *Automorphismen von Homotopiekettenringen*, Math. Ann. **112** (1936), 586–593.

[11] Wecken, F., *Fixpunktklassen, II*, Math. Ann. **118** (1942), 216–234.

1980 *Mathematics subject classifications*: 58F22, 34C25

Current addresses:
Department of Mathematics, Zhongshan University, Guangzhou 510275, China
Department of Mathematics, Peking University, Beijing 100871, China

COINCIDENCES FOR GRASSMANNIAN AND ASSOCIATED SPACES

S. Y. Husseini
Department of Mathematics
University of Wisconsin
Madison, WI 53706 USA

Introduction

Let $G_{n,m}$ be the Grassmannian manifold of complex m-planes in n-space C^n, and denote by $\sigma : G_{n,m} \to G_{n,n-m}$ the usual duality map which sends an m-plane to its orthogonal complement. Consider the following question: <u>Given a set of s maps</u> $f_1, \cdots, f_s$ <u>of</u> $G_{n,m}$ <u>to</u> C^n, <u>can one find an</u> m-<u>plane</u> $<P>$ <u>of</u> $G_{n,m}$ <u>such that</u> $f_1(<P>), \cdots, f_s(<P>) \in \sigma<P>$? (Clearly, this question can be also formulated as a fixed-point question.)

In the real case, it was shown in [2] that the answer is positive if $s \leq n-m$, and it is not hard to see that it is negative otherwise. (See [1], §4 or §1 below.) In [2] the question is considered for the bundle of m-flags of the universal vector bundle over $G_{n,m}$. This latter space is another homogeneous space associated to the Stiefel manifold $O_{n,m}$ of orthonormal m-frames in C^n.

Here we study the question for quotient spaces $O_{n,m}/G$, where G is a closed subgroup of U_m. The answer depends on the rank of G: For G of maximal rank, the answer is positive if, and only if, $s \leq (n-m)$. The situation for subgroups of positive but not maximal rank is more complicated. First note that if the answer is negative for $O_{n,m}/U_m = G_{n,m}$ then it is also negative for $O_{n,m}/G$: if there are maps $f_1, \cdots, f_s$: $G_{n,m} \to C^n$ such that for any $<P>$ in $G_{n,m}$, there is an f_i such that

This work was supported in part by NSF Grant No. DMS-8722295.

$f_i(<P>) \notin \sigma<P>$, then on composing these maps with the natural projection $O_{n,m}/G \to O_{n,m}/U_m$ we obtain a family of maps on $O_{n,m}/G$ with the same property. Thus if $s>(n-m)$ then the answer is negative. On the other hand, we show that if $s < \frac{n-m}{m} + \frac{1}{m}$ then the answer is positive. We obtain this estimate by passing to the easier problem when rank $G = 1$. It would be interesting to determine the sharpest estimate for s, and what happens in the range $\frac{n-m}{m} + \frac{1}{m} \leq s \leq (n-m)$.

The point of view here is similar to that of [2, §4]. We show that the coincidence-like property formulated in the question is equivalent to a section property as well as a Borsuk-Ulam property which we then solve.

§1. Statement of Results

Regard the Stiefel manifold $O_{n,m}$ of orthonormal m-frames in C^n as a space of m-tuples of n-dimensional column vectors and let U_m act on $O_{n,m}$ by post-multiplication. Then $G_{n,m}$, the Grassmannian of complex m-planes in C^n can be naturally identified with $O_{n,m}/U_m$. Let G be a closed subgroup of U_m.

The Coincidence-like Property C_s (1.1): *The manifold* $O_{n,m}/G$ *is said to satisfy Property* C_s *if, and only if, given a set of maps* $f_1, \cdots, f_s$: $G_{n,m}/G \to C^n$, *then one can find a* $[P] \in O_{n,m}/G$ *such that* $f_1([P]), \cdots, f_s([P])$ *lie in* $\sigma(<P>)$, *where* $<P>$ *is the image of* $[P]$ *under the natural projection* $O_{n,m}/G \to G_{n,m}$.

Intuitively speaking, $<P>$ is the m-plane spanned by the m-frame P, and the assertion that $f_1([P]), \cdots, f_s([P]) \in \sigma(<P>)$ is just the statement that $[P]$ is orthogonal to the span of the vectors $f_1([P]), \cdots, f_s([P])$.

Theorem (1.2). *Suppose that* $G \subset U_m$ *is a closed subgroup of maximal rank. Then* $O_{n,m}/G$ *satisfies the Coincidence Property* C_s *if, and only if,* $s \leq (n-m)$.

The proof of this theorem is similar to that of Theorem A′ of [2, §4]. It will be briefly indicated in §2 below.

Note that when $G = T_m$ the standard maximal torus, then

$O_{n,m}/T_m$ can be described as $\{([P], l_1,\cdots,l_m) \mid l_i$ line through O in $<P>$, $l_i \perp l_j\}$, i.e. the bundle of m-flags over $G_{n,m}$. Thus the theorem in this case is the complex analogue of Theorem A* of [2, §5].

The situation for subgroups G not of maximal rank is more complicated.

Theorem (1.3). *Suppose that* rank $G \geq 1$ *and let* $T \subset G$ *be the maximal torus. Assume that* $(C^m)^T = \{0\}$. *Then* $O_{n,m}/G$ *satisfies the Coincidence Property* C_s *if* $s < \frac{n-m}{m} + \frac{1}{m}$, *but fails to do so for* $s > (n-m)$.

§2. *Proof of Theorem* (1.1)

The proof is analogous to the proof in the real case and depends on the equivalence of Property C_s to two other properties: the section property, and the Borsuk-Ulam property [2, §4].

To see that, consider the bundle

(2.1) $$p_{n,m} : O_{n,m} \times_G (C^m)^{\oplus s} \to O_{n,m}/G$$

where G acts on $(C^m)^{\oplus s}$ diagonally. Note first that the total space can be described as $\{([P], v_1,\cdots,v_s) \mid [P] \in O_{n,m}/G, v_i \in <P>\}$. Hence the bundle (2.1) is the s-fold direct sum of the pull-back of the universal bundle over $G_{n,m}$. Now given a set of maps $f_1,\cdots,f_s : O_{n,m}/G \to C^n$, one obtains a section of (2.1) by assigning to $[P]$ the tuple $([P], pr_{<P>}v_1,\cdots,pr_{<P>}v_s) \in O_{n,m} \times_G (C^m)^{\oplus s}$, where $pr_{<P>} : C^n \to <P>$ is the orthogonal projection onto $<P>$ and $v_i = f_i(<P>)$. Conversely, a section of (2.1) gives rise to a family of s maps $O_{n,m}/G \to C^n$ in a natural fashion. Now it is easy to see that the Coincidence Property C_s is equivalent to the following section property.

The Section Property Sec_s (2.2): *Every section of the bundle* $O_{n,m} \times_G (C^m)^s \to O_{n,m}/G$ *has a zero.*

Finally, for formal reasons, Property (2.2) is equivalent to the following Borsuk-Ulam Property.

The Borsuk-Ulam Property $B-U_s$: *Every* G-*map* $O_{n,m} \to (C^m)^{\oplus s}$ *has a zero.*

To prove Theorem (1.2) assume that $O_{n,m}/G$ satisfies Property C_s. Then Property Sec_s also holds. Now if, to the contrary, $s > (n-m)$, then the bundle $O_{n,m} \times_{U_m} (\mathbb{C}^m)^s \to O_{n,m}/U_m \cong G_{n,m}$ would admit a non-zero section as $\dim_{\mathbb{R}} G_{n,m} = 2m(n-m)$, and this section would give rise to a non-zero section of $O_{n,m} \times_G (\mathbb{C}^m)^s \to O_{n,m}/G$. This would contradict the Section Property (2.2). Hence $s \leq (n-m)$ as required.

To prove the other implication, assume that $s \leq (n-m)$. Let T_m be the standard maximal torus of diagonal unitary matrices whose entries are $e^{2\pi i t_1}, \cdots, e^{2\pi i t_m}$. Regard the variables $t_1, \cdots, t_m$ as elements of $H^2(BT_m;\mathbb{Z})$ in the usual fashion. Then $H^*(BT_m;\mathbb{Z})$ is isomorphic to the algebra $\mathbb{Z}[t_1, \cdots, t_m]$ of polynomials in $t_1, \cdots, t_m$ with integral coefficients. Denote by $S(\mathbb{C}^m)^{\oplus s}$ the unit sphere in $(\mathbb{C}^m)^{\oplus s}$.

The following two propositions on the T_m-equivariant cohomology are proved in the same fashion as the corresponding propositions in the real case (see [1, §3] and [2, §5]).

<u>Proposition</u> (2.4): $H^*_{T_m}(S(\mathbb{C}^m)^{\oplus s}; \mathbb{Z})$ <u>is isomorphic to</u> $\mathbb{Z}[t_1, \cdots, t_m]/(t_1^s \cdots t_m^s)$ <u>as an</u> $H^*(BT_m;\mathbb{Z})$-<u>module</u>.

<u>Proposition</u> (2.5): $H^*_{T_m}(O_{m,n};\mathbb{Z})$ <u>is isomorphic to</u> $\mathbb{Z}[t_1, \cdots t_m]/\mathfrak{P}$ <u>as an</u> $H^*(BT_m;\mathbb{Z})$-<u>module, where</u> $\mathfrak{P}$ <u>is the ideal generated by</u> $p_1, \cdots, p_m$ <u>where</u>

$$p_i = t_i^{n-m+i} + a_{i,1} t_i^{n-m+i-1} + \cdots + a_{i,n-m+i}$$

<u>with each</u> $a_{i,j}$ <u>being a polynomial in</u> $t_1, \cdots, t_{i-1}$ <u>and</u> $\deg a_{i,j} t_i^{n-m+i-j} = 2(n-m+i)$.

Now assume that $s \leq (n-m)$. If, to the contrary, the coincidence Property C_s did not hold for $O_{n,m}/G$, then the Borsuk-Ulam Property (2.5) would not hold either and hence there would exist a T_m-map $O_{n,m} \to S(\mathbb{C}^m)^{\oplus s}$. This would imply that $\operatorname{ind}^{T_m}\left(S(\mathbb{C}^m)^{\oplus s}\right) \subset \operatorname{ind}^{T_m}(O_{n,m})$, where $\operatorname{ind}^{T_m}(X)$ is the annihilator of $H^*_{T_m}(X;\mathbb{Z})$ regarded as module over $H^*(BT_m;\mathbb{Z})$, [1, §2]. But according to Propositions (2.4) and (2.5) this is impossible if $s \leq (n-m)$. This finishes the proof of the theorem.

§3. Proof of Theorem (1.3)

The proof, like that of the previous theorem, depends on the equivalence of the three properties C_s, Sec_s and B-U_s.

Note that if $s > (n-m)$, then according to Theorem (1.2), there is a U_m-map $O_{n,m} \xrightarrow{f} S(C^m)^{\oplus s}$. Such a map f is certainly G-equivariant. This shows that $O_{n,m}/G$ does not satisfy the Coincidence Property C_s if $s > (n-m)$.

To prove the other assertion of Theorem (1.3) we need to examine how the maximal torus T of G acts on $O_{n,m}$ and on C^m. So write

$$C^m \doteq \sum_{\lambda} V^\lambda \otimes C_\lambda$$

as a complex T-module, where λ ranges over $\mathrm{Hom}(T,S^1)$, C_λ being the irreducible complex T-module defined by λ and $V_\lambda = \mathrm{Hom}_T(C_\lambda, C^m)$. This decomposition of C^m provides us with a basis for C^m, and if we are represent $O_{n,m}$ as a space of m-tuples of mutually orthogonal n-dimensional vectors, then the action of T on $O_{n,m}$ becomes just the multiplication of the columns by suitable scalars. The condition that $(C^m)^T = \{0\}$ implies that $\lambda \neq 0$ whenever $\dim_C V^\lambda \neq 0$. Put $\Lambda = \left\{\lambda \in \mathrm{Hom}(T,S^1) \mid V^\lambda \neq \{0\}\right\}$. Then Λ is a finite set, and we can find a subgroup $T' \subset T$ such that $\lambda | T' \neq 0$, for all $\lambda \in \Lambda$. This implies that $(C^m)^{T'} = \{0\}$. Also choose T' so that the restriction of the polynomial p_1 of Proposition (2.5) to $H^*(BT';\mathbb{Z})$ is non-trivial. Proposition (2.5) of §2 implies immediately the following.

Proposition (3.1): $H^*_{T'}(O_{n,m};Q)$ is isomorphic to $Q[t]/(t^{n-m+1}) \otimes E(V)$ as a $Q[t] \doteq H'(BT';Q)$-module, where $E(V)$ is the exterior algebra of the graded vector space V generated by the elements $x_{(n-m)+2}, \ldots, x_n$ with $\deg x_i = 2i-1$.

Proof: First using the morphism of equivariant cohomology induced by the homomorphism $T' \to T_m$ we note that Proposition (2.5) implies that $t^{(n-m)+1} = 0$. Next we consider the spectral sequence with rational coefficients of the map $ET' \times_{T'} O_{n,m} \to BT'$. Since the action of T' has only finite isotropy groups, we see immediately using the Begle-Vietoris

Theorem that $H^*_{T'}(O_{n,m};Q)$ is finite-dimensional. A simple calculation shows that

$$E_\infty^{*,*} \cong E_{2(n-m)+2}^{*,*} \cong Q[t]/(t^{n-m+1}) \otimes E(V),$$

where V is graded vector space generated by elements $x_{(n-m)+2}, \ldots, x_m$, with $\deg x_j = 2j - 1$. This proves the proposition.

To finish the proof of the theorem, suppose that $s < \frac{n}{m} + \frac{1}{m} - 1$. If $O_{n,m}/G$ did not satisfy the Coincidence Property C_s, then it would not satisfy the Borsuk-Ulam Property B-U_s either, and consequently there would exist a G-equivariant map $O_{n,m} \xrightarrow{f} S(C^m)^{\oplus s}$. The map f would also be T'-invariant, since T' is a subgroup of G. This would imply that $\mathrm{ind}^{T'}\left(S(C^m)^{\oplus s}\right)$, which is isomorphic to (t^{sm}), is included in $\mathrm{ind}^{T'}(O_{n,m})$, which is isomorphic to (t^{n-m+1}). This is impossible if $s < \frac{n-m}{m} + \frac{1}{m}$. This finishes the proof of the theorem.

<u>References</u>

1. Fadell, E. and Husseini, S., An ideal-valued cohomological index theorem with applications to Borsuk-Ulam and Bourgin-Yang theorems, to appear in <u>Dynamical Systems and Ergodic Theory</u>.

2. Husseini, S., Lasry, J-M. and Magill, M.J.P., Existence of equilibrium with incomplete markets, to appear.

University of Wisconsin – Madison

CONGRUENCES FOR FIXED POINT INDICES OF EQUIVARIANT MAPS AND ITERATED MAPS

Dedicated to Professor Akio Hattori on his sixtieth birthday

Katsuhiro Komiya

Department of Mathematics, Yamaguchi University
Yamaguchi 753, Japan

Introduction

Let G be a finite group and $C(G)$ the set of conjugacy classes of subgroups of G. Let $f : X \to X$ be a self G-map of a G-ENR X such that the fixed point set $\mathrm{Fix}(f)$ is compact. For any $(K) \in C(G)$, $f^{(K)} : X^{(K)} \to X^{(K)}$ denotes the restricted map of f to $X^{(K)}$, where $X^{(K)}$ is the subspace of those points whose isotropy groups contain a subgroup conjugate to K. In the author's previous paper [6] he obtained congruences for the fixed point indices $\iota(f^{(K)})$ of $f^{(K)}$:

$$\sum_{(H) \leq (K)} \mu((H),(K)) \ \iota(f^{(K)}) \equiv 0 \quad \mathrm{mod}\ |G/H| \qquad (0\text{-}1)$$

for any $(H) \in C(G)$, where μ is the Möbius function of the partially ordered set $C(G)$ (see Aigner [1; Chap. IV]).

In this note we will reformulate these congruences in terms of a short exact sequence

$$0 \longrightarrow G\text{-FIX} \xrightarrow{\ j\ } \prod_{(H) \in C(G)} Z \xrightarrow{\ \tilde{\pi}_\mu\ } \prod_{(H) \in C(G)} Z/|G/H|Z \longrightarrow 0.$$

Here G-FIX is the ring defined by Dold [2],[4]. The elements of

---

This paper is in final form and no version of it will be submitted for publication elsewhere.

G-FIX are equivalence classes $[f]$ of G-maps $f : Y \to X$ with $\mathrm{Fix}(f)$ compact, where X is a G-ENR, Y a G-invariant open subset of X. j is the homomorphism defined by $j([f]) = (\iota(f^{(H)}))_{(H) \in C(G)}$, i.e., the (H)-th component of $j([f])$ is $\iota(f^{(H)})$. $\tilde{\pi}_\mu$ is defined by

$$\tilde{\pi}_\mu(x) = \Big(\sum_{(H) \leq (K)} \mu((H),(K))\, x_{(K)} \Big)_{(H) \in C(G)}$$

for $x = (x_{(H)})_{(H) \in C(G)} \in \prod_{(H) \in C(G)} Z$.

If α is an automorphism of $\prod_{(H) \in C(G)} Z/|G/H|Z$, then

$$0 \longrightarrow \text{G-FIX} \xrightarrow{\;j\;} \Pi\, Z \xrightarrow{\;\alpha\tilde{\pi}_\mu\;} \Pi\, Z/|G/H|Z \longrightarrow 0$$

is also exact, and we will obtain in section 3 new congruences for fixed point indices from this exact sequence. Conversely it will be shown that certain kind of congruences for fixed point indices are all obtained in this manner. In section 4 we will consider the case in which G is a cyclic group and apply the results to iterated maps.

This note comes from a private communication with Professor J. Thévenaz. The author is grateful to him for his kind comments.

1. Short exact sequences

For $(H), (K) \in C(G)$ define $(H) \leq (K)$ if and only if H is conjugate to a subgroup of K. Then $C(G)$ becomes a partially ordered set with this order $\leq$, and has the Möbius function μ. Let $C(G) = \{(H_1), (H_2), \ldots, (H_k)\}$ be numbered in such a way that $i \leq j$ if $(H_i) \leq (H_j)$. By abuse of notation, μ denotes also the square matrix which has $\mu((H_i), (H_j))$ as (i,j)-component. Then the matrix μ induces an endomorphism of $\prod_{(H) \in C(G)} Z$. If we denote the endomorphism by π_μ, then

$$\pi_\mu(x) = \Big(\sum_{(H) \leq (K)} \mu((H),(K))\, x_{(K)} \Big)_{(H) \in C(G)}$$

for $x = (x_{(H)})_{(H) \in C(G)} \in \prod_{(H) \in C(G)} Z$. Since the matrix μ is a triangular matrix whose diagonal entries are all 1, π_μ is an automorphism. Let $p : \Pi\, Z \to \Pi\, Z/|G/H|Z$ be the componentwise canonical projection. Then we obtain

<u>Theorem 1</u>. The sequence

$$0 \longrightarrow \text{G-FIX} \xrightarrow{\ j\ } \Pi\, Z \xrightarrow{\ p\pi_\mu\ } \Pi\, Z/|G/H|Z \longrightarrow 0$$

is exact.

<u>Proof</u>. G-FIX is freely generated, as an additive group, by $\{[i_{(H)}] \mid (H) \in C(G)\}$ where $i_{(H)} : G/H \to G/H$ is the identity map (see Dold [2],[4]). This shows the homomorphism j is injective. The congruences (0-1) imply $\text{Im } j \subset \text{Ker } p\pi_\mu$, and $p\pi_\mu$ is epic since π_μ is an automorphism.

It then remains to show $\text{Ker } p\pi_\mu \subset \text{Im } j$. If $x = (x_{(H)})$ is in $\text{Ker } p\pi_\mu$, there are integers $y_{(H)}$ such that

$$\sum_{(H)\leq(K)} \mu((H),(K))\, x_{(K)} = y_{(H)}\, |G/H|$$

for any $(H) \in C(G)$. Applying the Möbius inversion formula (Aigner [1; 4.18]) to these equalities, we obtain

$$x_{(H)} = \sum_{(H)\leq(K)} y_{(K)}\, |G/K|$$

for any $(H) \in C(G)$. This implies

$$j\Big(\sum_{(H)\in C(G)} y_{(H)}\, [i_{(H)}] \Big) = x$$

Thus we see $\text{Ker } p\pi_\mu \subset \text{Im } j$. □

2. <u>Endomorphisms of</u> $\Pi\, Z/|G/H|Z$

A function $\alpha : C(G) \times C(G) \to Z$ gives a square matrix α with $\alpha((H_i), (H_j))$ as (i,j)-component, and this matrix induces an endomorphism π_α of $\Pi\, Z$. By straightforward observation we obtain

<u>Lemma 2</u>. Assume that a function $\alpha : C(G) \times C(G) \to Z$ satisfies

(2-1) $\alpha((H),(K)) = 0$ if $(H) \nleq (K)$,

(2-2) $\alpha((H),(K)) \equiv 0 \mod |K'/H|$ if $(H) \leq (K)$ and $H \subset K' = gKg^{-1}$ for some $g \in G$.

Then π_α induces an endomorphism $\bar{\pi}_\alpha$ of $\Pi\, Z/|G/H|Z$.

Assume further that

(2-3) $\alpha((H),(H))$ and $|G/H|$ are prime to each other for all (H).

Then $\bar{\pi}_\alpha$ is an automorphism.

Examples of α. If $(H) \nleq (K)$, then we must take $\alpha((H),(K))$ to be zero. If $(H) \leq (K)$, then we may take, for example,

(2-4) $\alpha((H),(K)) = 1$ for $(H) = (K)$, and $\alpha((H),(K)) = 0$ for $(H) < (K)$,

(2-5) $\alpha((H),(K)) = |K'/H|$, where $H \subset K' = gKg^{-1}$ for some $g \in G$,

(2-6) $\alpha((H),(K)) = |NK'/H|$, where NK' is the normalizer of K',

or

(2-7) $\alpha((H),(K)) = |G/H|$.

If α is as in (2-4) or (2-5), then α satisfies the conditions (2-1), (2-2) and (2-3), while if α is as in (2-6) or (2-7), then α satisfies only the conditions (2-1) and (2-2).

3. Congruences

Theorem 3. If a function $\alpha : C(G) \times C(G) \to Z$ satisfies (2-1), (2-2) and (2-3), then there is a short exact sequence

$$0 \longrightarrow G\text{-FIX} \xrightarrow{\ j\ } \Pi\, Z \xrightarrow{\ p\pi_{\alpha\mu}\ } \Pi\, Z/|G/H|Z \longrightarrow 0,$$

where $\pi_{\alpha\mu}$ is the automorphism of $\Pi\, Z$ induced from the product $\alpha\mu$ of matrices α and μ, i.e., $\pi_{\alpha\mu} = \pi_\alpha \pi_\mu$.

Proof. We see $p\pi_{\alpha\mu} = \bar{\pi}_\alpha(p\pi_\mu)$, and $\bar{\pi}_\alpha$ is an automorphism. Thus we see the exactness of the above sequence from Theorem 1. □

Note. Here we should note that the conditions (2-1), (2-2) ensure $\operatorname{Im} j \subset \operatorname{Ker} p\pi_{\alpha\mu}$ and the condition (2-3) ensures $\operatorname{Ker} p\pi_{\alpha\mu} \subset \operatorname{Im} j$.

Let $C(G) \times_{\leq} C(G)$ denote the subset of $C(G) \times C(G)$ which consists of pairs $((H),(K))$ with $(H) \leq (K)$.

Theorem 4. (i) Assume that a function $\alpha : C(G) \times_{\leq} C(G) \to Z$ satisfies (2-2), and define a function $\xi : C(G) \times_{\leq} C(G) \to Z$ by

$$\xi((H),(K)) = \sum_{(H)\leq(L)\leq(K)} \alpha((H),(L))\,\mu((L),(K)). \qquad (2\text{-}8)$$

Then we obtain congruences

$$\sum_{(H)\leq(K)} \xi((H),(K))\ \iota(f^{(K)}) \equiv 0 \mod |G/H| \tag{2-9}$$

for all $[f] \in$ G-FIX and all $(H) \in C(G)$.

(ii) Conversely, if a function $\xi : C(G) \times_{\leq} C(G) \to Z$ satisfies the congruences (2-9) for all $[f] \in$ G-FIX and all $(H) \in C(G)$, then ξ is given by (2-8) for some function $\alpha : C(G) \times_{\leq} C(G) \to Z$ satisfying (2-2).

Proof. (i) The assertion holds from the Note following Theorem 3.

(ii) Applying the congruences (2-9) to the identity map $i_{(K)}$: $G/K \to G/K$, we obtain

$$\sum_{(H)\leq(L)\leq(K)} \xi((H),(L))\ |G/K| \equiv 0 \mod |G/H|.$$

If we put

$$\alpha((H),(K)) = \sum_{(H)\leq(L)\leq(K)} \xi((H),(L)), \tag{2-10}$$

then this is a multiple of $|K'/H|$, where $H \subset K' = gKg^{-1}$ for some $g \in G$. By the Möbius inversion formula, we obtain from (2-10),

$$\xi((H),(K)) = \sum_{(H)\leq(L)\leq(K)} \alpha((H),(L))\ \mu((L),(K)). \quad \square$$

Note. In (2-8), if we take α as in (2-5), then $\xi((H),(K))$ coincides with the Euler function $\varphi((H),(K))$ in Komiya [6].

4. Examples

Example 1. Consider the case $G = Z_n$, the cyclic group of order n. In this case the congruences (2-9) reduce to

$$\sum_{h|k|n} \xi(h,k)\ \iota(f^{Z_k}) \equiv 0 \mod n/h$$

for all $[f] \in Z_n$-FIX and all $h \mid n$ (divisors h of n). Here the sum is taken over all k which divide n and are divisible by h, and

$$\xi(h,k) = \sum_{h|\ell|k} \ell/h\ a_{h,\ell}\ \mu(k/\ell),$$

where $a_{h,\ell}$ are arbitrary integers and μ is the Möbius function of number theory. If $a_{h,h} = 1$ and $a_{h,\ell} = 0$ $(h \neq \ell)$, then $\xi(h,k) = \mu(k/h)$. If $a_{h,\ell} = 1$ for all $h \mid \ell$, then $\xi(h,k) = \varphi(k/h)$ where φ is the Euler function of number theory.

Example 2. If we apply the results for Z_n-maps to iterated maps as done in Komiya [6], then we obtain congruences for iterated maps. If $f : X \to X$ is a self map of an ENR X with $\mathrm{Fix}(f^n)$ compact, then we obtain congruences

$$\sum_{h|k|n} \xi(h,k)\ \iota(f^{n/k}) \equiv 0 \mod n/h$$

for all $h \mid n$, where $\xi(h,k)$ is as in Example 1. When $h = 1$, we write $\xi(k)$ for $\xi(h,k)$. Then

$$\xi(k) = \sum_{\ell|k} \ell\, a_\ell\, \mu(k/\ell) \qquad (a_\ell \in Z),$$

and

$$\sum_{k|n} \xi(k)\ \iota(f^{n/k}) \equiv 0 \mod n,$$

while we see

$$\sum_{k|n} \xi(k)\ \iota(f^{n/k})$$

$$= \sum_{k|n} \Big(\sum_{\ell|k} \ell\, a_\ell\, \mu(k/\ell) \Big)\ \iota(f^{n/k})$$

$$= \sum_{\ell|n} \mu(n/\ell) \sum_{k|\ell} k\, a_k\ \iota(f^{\ell/k}).$$

Thus Dold [3; (1.3)] says that there is a map $g : Y \to Y$ of an ENR Y with $\mathrm{Fix}(g^\ell)$ compact (for any ℓ) such that

$$\iota(g^\ell) = \sum_{k|\ell} k\, a_k\ \iota(f^{\ell/k}).$$

5. Remarks

(1) Here we should give a remark on the case of G a compact Lie group. If we consider only a finite set $\mathcal{F}$ of conjugacy classes of closed subgroups of G and define $G\text{-FIX}(\mathcal{F}) = \{[f] \in G\text{-FIX} \mid f : X \to X,\ \mathrm{Iso}(X) \subset \mathcal{F}\}$, then we have the sequence

$$G\text{-FIX}(\mathcal{F}) \xrightarrow{\ j\ } \prod_{(H)\in\mathcal{F}} Z \xrightarrow{\ p\pi_{\alpha\mu}\ } \prod_{(H)\in\mathcal{F}} Z/\chi(G/H)Z,$$

where μ is the Möbius function of $\mathcal{F}$. j is, in general, not injective, but we have $\mathrm{Im}\, j = \mathrm{Ker}\, p\pi_{\alpha\mu}$ and we also have congruences (2-9) for $[f] \in$ G-FIX($\mathcal{F}$) with $\chi(G/H)$ replacing $|G/H|$.

(2) Let $A(G)$ be the Burnside ring of G. There is an isomorphism $\varepsilon : A(G) \to$ G-FIX defined by $\varepsilon([S]) = [id_S]$ for G-sets S (see Dold [2],[4], Ulrich [7]). Komiya [5] obtained some other type of congruences for the Burnside ring. So we may obtain more congruences for G-FIX through the isomorphism ε.

References

[1] M. Aigner: Combinatorial theory, Springer-Verlag, Berlin, Heidelberg, New York, 1979.

[2] A. Dold: Fixed point theory and homotopy theory, Comtemporary Math. 12 (1982), 105-115.

[3] A. Dold: Fixed point indices of iterated maps, Invent. math. 74 (1983), 419-435.

[4] A. Dold: Combinatorial and geometric fixed point theory, Riv. Mat. Univ. Parma (4) 10 (1984), 23-32.

[5] K. Komiya: Congruences for the Burnside ring, Preprint, 1987.

[6] K. Komiya: Fixed point indices of equivariant maps and Möbius inversion, Invent. math. 91 (1988), 129-135.

[7] H. Ulrich: Der äquivariante Fixpunktindex vertikaler G-Abbildungen, Thesis, Univ. Heidelberg, 1983.

NOTE: A revised version of [7] has been published as Lecture Notes in Mathematics 1343, H. Ulrich "Fixed Point Theory of Parametrized Equivariant Maps", Springer-Verlag, Berlin, Heidelberg, New York, 1988.

Clifford Asymptotics and the Local Lefschetz Index

John D. Lafferty [1], Yu Yanlin, and Zhang Weiping

In this note we indicate a direct proof of the Lefschetz fixed point formulas of Atiyah, Bott, Segal, and Singer for isometries.

Consider the standard setup of the Dirac operator D acting on sections of the spin bundle associated with a compact and smooth spin manifold M. Thinking of $D^2 = \Delta$ as a perturbed Laplacian via the Lichnerowicz formula, we are led to consider the parametrix $H_N(t,x,y)$ of Δ given by

$$H_N(t,x,y) = \frac{\exp\left(\frac{-\rho^2(x,y)}{4t}\right)}{(4\pi t)^n}\left(\sum_{i=0}^{N} t^i U^{(i)}(y,x)\right) \tag{1}$$

as a parametrized family of endomorphisms defined in a neighborhood of the diagonal in $M \times M$. Here $2N > 2n = \dim(M)$, $\rho(x,y)$ is the Riemannian distance and $U^{(i)}(y,x) : \pi^{-1}(y) \to \pi^{-1}(x)$ are endomorphisms with $U^{(0)}(x,x) = Id$ and

$$\left(\frac{\partial}{\partial t} + \Delta\right) H_N(t,\cdot,y)v = -\frac{\exp\left(\frac{-\rho^2}{4t}\right)}{(4\pi t)^n} t^N \Delta U^{(N)}(y,\cdot)v. \tag{2}$$

It was precisely this construction that was used by Patodi [5] in the case of the de Rham complex as a means of matching the local asymptotics, as t approaches zero, to evaluate the Euler-Poincaré characteristic in terms of the Chern polynomial. In [6], this program was carried through for the spin complex using the above parametrix and a detailed analysis of the associated Clifford asymptotics.

The first observation to make in such an approach is that a quick asymptotic match gives the order of the term contributing to the final answer. For the spin complex, this is of course the observation that only the term of top order $2n$ in the generators $\{e_i\}$ of the Clifford algebra $C_{2n} = \mathrm{End}(S_+ \oplus S_-)$ contributes to the evaluation of the index of D. In terms of the parametrix, this reads

$$\int_M Tr_s\, U^{(i)}(x,x)\, dx = 0 \tag{3}$$

for $i < n$, and

$$\mathrm{index}(D) = \left(\frac{1}{4\pi}\right)^n \int_M Tr_s U^{(n)}(x,x)\, dx. \tag{4}$$

The second observation to make is that choosing the geodesic moving frame greatly simplifies the analysis. Thus, working locally in normal coordinates y_i at $x \in M$, let E_{norm} be an orthonormal frame at x which is moved parallelly along geodesics through x, yielding a local frame field. One then identifies x with zero and proceeds with the analysis by taking local Taylor expansions of operators with respect to this frame field and, essentially, matching terms on both sides of equation (1). Toward this end, the following construction is extremely useful. Let multi-indices $\alpha, \beta \in \mathbf{Z}^{2n}$ and $\gamma \in (\mathbf{Z}_2)^{2n}$ be given and define

$$\chi(y^\alpha D_y^\beta e^\gamma) = |\beta| - |\alpha| + |\gamma| \tag{5}$$

when $y^\alpha = y_1^{\alpha_1} \cdots y_{2n}^{\alpha_{2n}}$, $D_y^\beta = (\partial/\partial y_1)^{\beta_1} \cdots (\partial/\partial y_{2n})^{\beta_{2n}}$ and $e^\gamma = e_1^{\gamma_1} \cdots e_{2n}^{\gamma_{2n}}$. Thus, for example, the Lichnerowicz formula then takes the form

$$D^2 = -\sum_i \frac{\partial^2}{\partial y_i^2} + \frac{1}{4} \sum_{i,j,\alpha,\beta} R_{ij\alpha\beta} y_i \frac{\partial}{\partial y_j} e_\alpha e_\beta +$$
$$+ \frac{1}{64} \sum_{i,j,k} \sum_{\alpha_1\alpha_2\alpha_3\alpha_4} y_i y_j R_{ik\alpha_1\alpha_2} R_{kj\alpha_3\alpha_4} e_{\alpha_1} e_{\alpha_2} e_{\alpha_3} e_{\alpha_4} + O(\chi < 2).$$

The point, of course, is that if ϕ (a local section of some bundle) has a zero of order m at x then there is no contribution from the supertrace $Tr_s(y^\alpha D_y^\beta e^\gamma \phi)$ at x in case $\chi(y^\alpha D_y^\beta e^\gamma) - m < 2n$. The above thus provides an efficient scheme for throwing away terms.

Now, let T act as an isometry on M, and assume that the action of dT lifts to an action of $Spin(M)$ which commutes with the $Spin(2n)$ action, thereby inducing a map on cohomology. Then a map T^* is induced as a linear operator on sections of the spin bundle $E = Spin(M) \times_{Spin(2n)} S$. As a heat problem, the evaluation of the Lefschetz number

$$L(T) = Tr\, T^*|_{ker\, D_+} - Tr\, T^*|_{ker\, D_-} \tag{6}$$

localizes on the fixed point set $F = \{x |\ Tx = x\}$, which consists of the disjoint union of a finite number of even-dimensional totally geodesic submanifolds $F_1, F_2, \ldots F_r$. There is thus no harm in assuming that $r = 1$. Let ν be the normal bundle of F and $\nu(\epsilon) = \{v \in \nu |\ \|v\| < \epsilon\}$ for $\epsilon > 0$. The bundle ν is invariant under dT and $dT|_\nu$ is nondegenerate.

We denote by $P_t^\pm(x,y) : E_\pm|_y \to E_\pm|_x$ the fundamental solutions for the heat operators $\partial/\partial t + \Delta_\pm$. The standard heat equation argument yields

$$L(T) = \int_M (Tr\, T^* P_t^+(Tx, x) - Tr\, T^* P_t^-(Tx, x))\, dx, \quad t > 0 \tag{7}$$

where dx is the Riemannian volume element. Denote the integrand by

$$\mathcal{L}(t, x) = Tr\, T^* P_t^+(Tx, x) - Tr\, T^* P_t^-(Tx, x). \tag{8}$$

Then straightforward pseudodifferential operator and parametrix estimates allow us to write

$$L(T) = \int_F L_{\mathrm{loc}}(T)(\xi)\, d\xi \tag{9}$$

where the local Lefschetz number, defined by the limit

$$L_{\mathrm{loc}}(T)(\xi) = lim_{t\to 0}\int_{\nu_\xi(\epsilon)} \mathcal{L}(t, exp\, v)\, dv \ , \tag{10}$$

exists and is independent of ϵ.

Patodi's parametrix strategy for evaluating the local index is equally appropriate here. However, a more delicate treatment of the Clifford asymptotics is now required near the fixed-point submanifold.

We have already observed that in normal coordinates, and with respect to the geodesic moving frame E_{norm}, the expression of the parametrix as a parametrized family of endomorphisms is particularly tractable. The next important observation is that in what we call "orthogonal" coordinates near F the action of our isometry T has a particularly nice form. Such coordinates are expressed in terms of geodesics in F and transversals normal to F; and, as such, they trivialize the normal bundle. In particular, the action of dT is constant along fibers of ν in this trivialization. The key to evaluating the local Lefschetz index lies in relating the geodesic moving frame E_{norm} and the moving frame E_{orthog} obtained from orthogonal coordinates (by moving parallelly along geodesics in F and then along geodesics normal to F) in terms of an infinitesimal holonomy.

Let us now fix $\xi \in F$ and work locally near ξ. Then in our orthogonal coordinates the map dT acts as the identity in directions tangential to F and as, say, $e^{-\Theta(x')}$ in the normal fiber over $x' \in F$, where $\Theta(x') \in so(2n-2n')$ and $2n' = dim(F)$. Of course, we may arrange things so that

$$\Theta(\xi) = \begin{bmatrix} 0 & \theta_1 & & & \\ -\theta_1 & 0 & & & \\ & & \ddots & & \\ & & & 0 & \theta_{n-n'} \\ & & & -\theta_{n-n'} & 0 \end{bmatrix} . \tag{11}$$

It turns out that the infinitesimal holonomy relating the frames E_{norm} and E_{orthog}, expressed as a Lie algebra-valued map $\Phi : U \to so(2n)$ defined in a neighborhood U of ξ by $E_{norm}(x) = E_{orthog}(x)e^{\Phi(x)}$ has the property that for $x = (x', v)$ (in terms of the local trivialization of ν)

$$\Phi_{ij}(x) = -\frac{1}{2}\sum_{\alpha,\beta=1}^{2n-2n'} (ve^{-\Theta(x')})_\alpha v_\beta R_{\alpha+2n',\, \beta+2n',\, i,\, j}(x') + o(|v|^2) \tag{12}$$

where the curvature R is computed with respect to the frame E_{orthog}.

Now, to investigate the Clifford asymptotics, it is best to scale the metric in the normal directions, setting $v = \sqrt{t}w$. It is also helpful to consider a modified χ operator, setting

$$\bar{\chi}\left(\phi(t)e_{i_1}\cdots e_{i_s}\right) = s - sup\{k \in \mathbf{Z} \mid \lim_{t\to 0+} \frac{|\phi(t)|}{t^{k/2}} < \infty\} \tag{13}$$

for a monoid $\phi(t)e_{i_1}\cdots e_{i_s}$ with real-valued ϕ.

The notation is simplified by letting $\bar{P}_t(x), \bar{T}^*(x) \in \mathrm{Hom}(S_\pm, S_\pm)$ be defined locally through the equivalence relations

$$P_t(Tx, x)[(\sigma(x), v)] = [(\sigma(Tx), \bar{P}_t(x)v] \tag{14}$$

and

$$T^*[(\sigma(Tx), u)] = [(\sigma(x), \bar{T}^*(x)u]. \tag{15}$$

so that

$$\mathcal{L}(t,x) = Tr\, \bar{T}^*(x)\bar{P}_t(x)|_{S_+} - Tr\, \bar{T}^*(x)\bar{P}_t(x)|_{S_-} = Tr_s \bar{T}^* \bar{P}(x)\,. \tag{16}$$

It may then be shown that

$$\bar{T}^*(x) = (-1)^{n-n'}\Big(\prod_{\alpha=1}^{n-n'} \sin\frac{\theta_\alpha}{2} \exp\{-t/4 \sum_{\alpha,\beta=1}^{2n-2n'} b_\alpha b_\beta (e^{-\Theta(\xi)}A^\perp)_{\alpha\beta}\}\Big)e_{2n'+1}\cdots e_{2n} + O(\bar{\chi} < 2(n-n'))\,. \tag{17}$$

where $A^\perp$ is the $(2n-2n') \times (2n-2n')$ matrix whose (α,β) element is given by

$$(A^\perp)_{\alpha\beta} = -\frac{1}{2}\sum_{i,j=1}^{2n'} R_{\alpha+2n',\,\beta+2n',\,i,\,j}(\xi)e_ie_j. \tag{18}$$

($A^\top$ is defined in the obvious analogous fashion, replacing $\alpha + 2n'$ and $\beta + 2n'$ by indices ranging between 1 and $2n'$.) Finally, let $\tilde{A}$ be the $2n \times 2n$ matrix given by

$$\tilde{A}_{ij} = -\frac{1}{2}\sum_{k,l=1}^{2n} \tilde{R}_{ijkl}e_ke_l, \tag{19}$$

where $\tilde{R}_{ijkl}$ are the components of the Riemannian curvature tensor now computed with respect to the frame field E_{norm}, and set

$$\tilde{A}^k(y) = \sum_{i,j=1}^{2n} y_iy_j(\tilde{A}^k)_{ij}, \quad k = 1, 2, \ldots \tag{20}$$

Then there is a operator $P(t; z_1, z_2, \ldots; w_1, w_2, \ldots)$ which is a power series in t with coefficients polynomials in z_i and w_i such that

$$\bar{P}_t(x) = \frac{\exp\frac{-\rho(x,Tx)^2}{4t}}{(4\pi t)^n} P(t; Tr\tilde{A}^2, \ldots, Tr\tilde{A}^{2k}, \ldots, Tr\tilde{A}^{2n}; \tilde{A}^2(y), \ldots, \tilde{A}^{2k}(y), \ldots, \tilde{A}^{2n}(y)) + \sum_{m\geq 0} t^m O(\bar{\chi} < 2m)\,. \tag{21}$$

Furthermore, in diagonal form we have, by solving harmonic oscillator-type equations,

$$\begin{aligned} P(t; ((-1)^k 2(u_1^{2k} + \cdots + u_n^{2k})); ((-1)^k \sum_{\alpha=1}^{n} (v_{2\alpha-1}^2 + v_{2\alpha}^2)u_\alpha^{2k})) = \\ = (4\pi t)^n e^{\|v\|^2/4t} \prod_{\alpha=1}^{n} \left(\frac{iu_\alpha}{8\pi \sinh\frac{iu_\alpha t}{2}} \exp\left(\frac{-iu_\alpha}{8}(v_{2\alpha-1}^2 + v_{2\alpha}^2)\coth\frac{iu_\alpha t}{2}\right)\right). \end{aligned} \tag{22}$$

These then, are all of the pieces necessary to a final calculation of the local Lefschetz index. It remains simply to note that to compute the supertrace it suffices to compute the coefficient of $e_1 \cdots e_{2n}$ and that since $e_i e_j = -e_j e_i + O(\bar{\chi} < 1)$, if we formally replace e_i by ω_i, where $\omega = (\omega_1, \ldots, \omega_{2n})$ is the frame dual to E_{norm}, and then substitute the associated forms $\Omega^\top$ and $\Omega^\perp$ for $A^\top$ and $A^\perp$, where

$$\begin{aligned} \Omega^\top &= -\frac{1}{2}\sum_{k,l=1}^{2n} R_{ijkl}\omega_k \wedge \omega_l \qquad 1 \le i,j \le 2n' \\ \Omega^\perp &= -\frac{1}{2}\sum_{k,l=1}^{2n} R_{ijkl}\omega_k \wedge \omega_l \qquad 2n'+1 \le i,j \le 2n' \end{aligned} \tag{23}$$

then computing the supertrace is equivalent to computing the form of the top order $2n'$ on F, if we multiply by $(\frac{2}{\sqrt{-1}})^n$, which is the so-called Berezin-Patodi constant. (To explain the appearance of this term, simply note that $Tr|_{S_\pm}(e_1 \cdots e_{2n}) = \frac{2^{n-1}}{\sqrt{-1}^n}$.)

Letting $\Omega = \begin{bmatrix} \Omega^\top & 0 \\ 0 & \Omega^\perp \end{bmatrix}$ be given formally as

$$\Omega^\top = \begin{bmatrix} 0 & u_1 & & & \\ -u_1 & 0 & & & \\ & & \ddots & & \\ & & & 0 & u_{n'} \\ & & & -u_{n'} & 0 \end{bmatrix} \quad \Omega^\perp = \begin{bmatrix} 0 & v_1 & & & \\ -v_1 & 0 & & & \\ & & \ddots & & \\ & & & 0 & v_{n-n'} \\ & & & -v_{n-n'} & 0 \end{bmatrix} \tag{24}$$

where u_i and v_i are indeterminates, a straightforward (however tedious) calculation gives at last that as a $2n'$ form on F,

$$L_{\mathrm{loc}}(T) = \prod_{\alpha=1}^{n'} \frac{u_\alpha/4\pi}{\sinh u_\alpha/4\pi} \left(\prod_{\beta=1}^{n-n'} 2\sinh(\frac{v_\beta}{4\pi} + \frac{\sqrt{-1}\theta_\beta}{2}) \right)^{-1} \tag{25}$$

whence the main result

Theorem. *The Lefschetz number $L(T)$ of the isometry T acting on the spin manifold M is expressed by*

$$L(T) = \sum_i \int_{F_i} [L_{\mathrm{loc}}(T)]_i \tag{26}$$

where in the notation used above

$$\begin{aligned} [L_{\mathrm{loc}}(T)]_i &= \sqrt{\det \frac{\Omega^\top/4\pi}{\sin \Omega^\top/4\pi}} \mathrm{Pf}\left(2\sin(\Omega^\perp/4\pi + \sqrt{-1}\frac{\Theta}{2})\right)^{-1} \\ &= \hat{A}(TF_i)\left[\mathrm{Pf}\left(2\sin(\Omega/4\pi + \sqrt{-1}\Theta/2)\right)(\nu(F_i))\right]^{-1}. \end{aligned} \tag{27}$$

Clearly the introduction of a twisting bundle results in only minor changes necessary in the above approach. For details of the above direct and purely elementary geometrical program, as well as comments on related approaches to the local index theory, we refer to [4].

Acknowledgement

The first author would like to express his gratitude to Professor S. S. Chern for his generous invitation to visit the Nankai Institute of Mathematics.

References

[1] M. F. Atiyah and R. Bott, "A Lefschetz fixed point formula for elliptic complexes", I, Ann. of Math. **86** (1967), 374- 407; II, **88** (1968), 451-491.

[2] M. F. Atiyah and G. B. Segal, "The index of elliptic operators, II", Ann. of Math **87** (1968), 531-545.

[3] E. Getzler, "A short proof of the local Atiyah-Singer index theorem," Topology, **25**, No. 1 (1986), 111-117.

[4] J. D. Lafferty, Y. L. Yu, and W. P. Zhang, "A direct geometric proof of the Lefschetz fixed point formulas," Nankai Institute of Mathematics preprint.

[5] V. K. Patodi, "Curvature and the eigenforms of the Laplace operator," J. Diff. Geom. **5** (1971) 233-249.

[6] Y. L. Yu, "Local index theorem for Dirac operator", Acta. Math. Sinica, New Series, **3**, No. 2 (1987), 152-169.

[7] W. P. Zhang, "The local Atiyah-Singer index theorem for families of Dirac operators," to appear, Springer Lecture Notes in Mathematics.

Nankai Institute of Mathematics,
Tianjin, People's Republic of China

---

[1] Present address: I.B.M. Thomas J. Watson Research Center, Yorktown Heights, NY 10598 USA.

On solutions of frame mappings into manifolds

CHAO-QUN LI
Peking University

1. Introduction.

Knaster [3] proposed in 1947 the

PROBLEM: Given a map $f : S^n \to \mathbb{R}^m$ from the standard n-sphere S^n into the Euclidean m-space $\mathbb{R}^m$ and $n-m+2$ distinct points $y_1, \ldots, y_{n+2-m} \in S^n$, does there exist a rotation $r \in SO(n+1)$ such that

$$f(ry_1) = f(ry_2) = \cdots = f(ry_{n+2-m}) \quad ?$$

Special cases of this problem have been studied by many mathematicians in the 1950's and 1960's. For example it was proved by Bourgin [1] that, if

(1) n is even and $p > 2$ is a prime with $m < 2\left[\dfrac{n}{p-1}\right]$, where $[\,]$ denotes the integer part, and

(2) the points $y_1, \ldots, y_p \in S^n$ are mutually orthogonal,

then for any map $f : S^n \to \mathbb{R}^m$, there is a rotation $r \in SO(n+1)$ such that

$$f(ry_1) = \cdots = f(ry_p).$$

In this note we show that Bourgin's result is still true if we replace the Euclidean m-space $\mathbb{R}^m$ by any m-dimensional manifold M^m. On the other hand, in Knaster's problem $\mathbb{R}^m$ can not be replaced by an arbitrary m-manifold, as shown by the classical Hopf map $S^{2k+1} \to \mathbb{C}P^k$.

I am grateful to Prof. Boju Jiang and Dr. Haibao Duan for their help.

2. Preliminaries.

From now on we make the following conventions: p is always an odd prime number, all the cohomologies are understood to be over the coefficient field $\mathbb{Z}_p$, and M^m is an m-manifold.

A $\mathbb{Z}_p$-system is a pair (X,t) where X is a topological space and $t : X \to X$ is a fixed point free $\mathbb{Z}_p$-action.

Let $E_{\mathbb{Z}_p} \to B_{\mathbb{Z}_p}$ (namely $S^\infty \to L^\infty$) be the universal $\mathbb{Z}_p$-bundle. Let (X,t) be a $\mathbb{Z}_p$-system. Then there is a map $h : X/t \to B_{\mathbb{Z}_p}$ which classifies the $\mathbb{Z}_p$-bundle $X \to X/t$.

Since $H^*(\mathbb{Z}_p) = H^*(B_{\mathbb{Z}_p}) = \Lambda(a) \otimes \mathbb{Z}_p(b)$—the tensor product of the exterior algebra $\Lambda(a)$ generated by $a \in H^1(B_{\mathbb{Z}_p})$ and the polynomial algebra $\mathbb{Z}_p(b)$ generated by $b \in H^2(B_{\mathbb{Z}_p})$, we call $h^*(a) \in H^1(X/t)$ and $h^*(b) \in H^2(X/t)$ the $\mathbb{Z}_p$-characteristic classes of (X,t), and, by an abuse of notation, simply write them as $a, b \in H^*(X/t)$ respectively. It is a standard fact that

---

Partially supported by a TWAS grant.

LEMMA 1. *If $g : (X,t) \to (Y,t)$ is an equivariant map, then $g^* : H^*(Y/t) \to H^*(x/t)$ maps the $\mathbb{Z}_p$-characteristic classes of (Y,t) into that of (X,t).*

EXAMPLE: The real Stiefel manifold $V_{n+1,p}$ of p-frames in $\mathbb{R}^{n+1}$ consists of all mutually orthogonal ordered p-tuples $(u_1,\ldots,u_p)$ in S^n. Let $t : V_{n+1,p} \to V_{n+1,p}$ be the map

$$t(u_1,\ldots,u_p) = (u_2,\ldots,u_p,u_1),$$

Then $(V_{n+1,p},t)$ is a $\mathbb{Z}_p$-system.

From the discussion in [1] we have

LEMMA 2. *$b^{2Lq-1} \neq 0$ in $H^*(V_{n+1,p}/t)$ where $L = \frac{1}{2}(p-1)$, and q is the smallest integer such that $\frac{1}{2}(n-p+1) < Lq$. Actually $2Lq-1 = (p-1)\left[\frac{n}{p-1}\right] - 1$, where [ ] denotes the integer part.*

PROOF: When n is even, by [1] we have

$$b^{2Lq-1} \neq 0.$$

When n is odd, we consider the equivariant inclusion map $i : (V_{n+1,p},t) \to (V_{n+2,p},t)$. By Lemma 1 we have

$$i^*(b^k) = b^k \in H^*(V_{n+1,p}/t).$$

If q is the smallest integer such that $\frac{1}{2}(n+1-p) < Lq$, q_1 is the smallest integer such that $\frac{1}{2}(n+2-p) < Lq_1$, then $q = q_1$. Thus we have $b^{2Lq-1} \neq 0$ for all n. ∎

Let $\pi : E_{\mathbb{Z}_p} \times_{\mathbb{Z}_p} V_{n+1,p} \to V_{n+1,p}/t$ be the map induced by the natural projection $\pi : E_{\mathbb{Z}_p} \times V_{n+1,p} \to V_{n+1,p}$. Then it follows from Alexander duality that

LEMMA 3. *For a closed subspace $B \subset V_{n+1,p}/t$, there is an isomorphism*

$$H^i(E_{\mathbb{Z}_p} \times_{\mathbb{Z}_p} V_{n+1,p}, E_{n+1,p} \times_{\mathbb{Z}_p} V_{n+1,p} - \pi^{-1}(B)) \cong H_{N-i}(B),$$

where $N = \dim V_{n+1,p} = \frac{1}{2}p(2n-p+1)$.

PROOF: Notice that for any subspace $V \subset V_{n+1,p}/t$, $\pi|_{\pi^{-1}(V)} : \pi^{-1}(V) \to V$ is a homotopy equivalence. ∎

Let M^m be an m-dimensional manifold. Let $(M)^p = M \times \cdots \times M$ (p copies of M), and $\Delta = \{(x_1,\ldots,x_p) | x_1 = \cdots = x_p\}$ be the diagonal of $(M)^p$. Let $t : (M)^p \to (M)^p$ be the map

$$t(x_1,\ldots,x_p) = (x_2,\ldots,x_p,x_1).$$

It generates a $\mathbb{Z}_p$-action on $(M)^p$. Let $E_{\mathbb{Z}_p} \times_{\mathbb{Z}_p} (M)^p$ be the twisted product of $E_{\mathbb{Z}_p}$ and $(M)^p$. In [4] Steenrod yielded that $H^*(E_{\mathbb{Z}_p} \times_{\mathbb{Z}_p} (M)^p)$ as an $H^*(\mathbb{Z}_p)$-algebra may be canonically identified with $H^*(\mathbb{Z}_p) \otimes D + N$ where D denotes the diagonal elements of $H^*((M)^p)$ and N denotes the image of the norm homomorphism $1 + t + \cdots + t^{p-1}$. Under this identification, Haefliger [2] gave the following explicit formula for the class

$\Delta_p(M) \in H^*(E_{\mathbb{Z}_p} \times_{\mathbb{Z}_p} (M)^p)$ which is dual to $B_{\mathbb{Z}_p} \times \Delta$ in $E_{\mathbb{Z}_p} \times_{\mathbb{Z}_p} (M)^p$ (we call it the mod p Haefliger class of M)

$$\Delta_p(M) = \lambda \sum_{j=0}^{[m/2p]} (-1)^j b^{(p-1)(m/2-jp)} \otimes (U^j_{(p)})^p + \delta_p(M),$$

where

$$\lambda = \begin{cases} (-1)^{m/2}, & \text{if } m \text{ even} \\ (-1)^{(m-1)/2} \left(\frac{p-1}{2}\right)!, & \text{if } m \text{ odd}, \end{cases}$$

b is the generator of the polynomial algebra in $H^*(\mathbb{Z}_p) = \Lambda(a) \otimes \mathbb{Z}_p(b)$, $U^j_{(p)} \in H^*(M)$ is the Wu class defined in [2] and $\delta_p(M)$ is the element in $H^*((M)^p)$ dual to the diagonal Δ.

Given a map $f : S^n \to M^m$, let $F : V_{n+1,p} \to (M)^p$ be the map $(u_1, \ldots, u_p) \mapsto (f(u_1), \ldots, f(u_p))$. Let $f_{\mathbb{Z}_p} = 1 \times_{\mathbb{Z}_p} F : E_{\mathbb{Z}_p} \times_{\mathbb{Z}_p} V_{n+1,p} \to E_{\mathbb{Z}_p} \times_{\mathbb{Z}_p} (M)^p$ be the map induced by F, and set $A(f,p) = F^{-1}(\Delta)$. Then $A(f,p)$ consists of all mutually orthogonal p-tuples such that $f(u_1) = \cdots = f(u_p)$.

Write $B = A(f,p)/t$. We have

$$\pi^{-1}(B) = f^{-1}_{\mathbb{Z}_p}(E_{\mathbb{Z}_p} \times_{\mathbb{Z}_p} \Delta) = f^{-1}_{\mathbb{Z}_p}(B_{\mathbb{Z}_p} \times \Delta).$$

LEMMA 4.

$$f^*_{\mathbb{Z}_p}(\Delta_p(M)) = \lambda b^{\frac{1}{2}(p-1)m} \in H^*(E_{\mathbb{Z}_p} \times_{\mathbb{Z}_p} V_{n+1,p}), \qquad \text{when } m < n.$$

PROOF: Since $f^*_{\mathbb{Z}_p} = (1 \times_{\mathbb{Z}_p} F)^*$ is $H^*(\mathbb{Z}_p)$-linear,

$$(1 \times_{\mathbb{Z}_p} F)^*(\Delta_p(M)) = \lambda \sum_{j=0}^{[m/2p]} (-1)^j b^{(p-1)(m/2-jp)} \otimes (f^*(U^j_{(p)}))^p + F^*(\delta_p(M)).$$

A simple but crucial observation is that, since $m < n$, we have $f^* = f_* = 0$ except in dimension 0. Thus $f^*(U^j_{(p)}) = 0$ if $j \neq 0$. Since $U^0_{(p)}$ is the unit cocycle in $H^0(M)$, $f^*(U^0_{(p)})$ is the unit cocycle in $H^0(S^n)$. Hence the above formula reduces to

$$(1 \times_{\mathbb{Z}_p} F)^*(\Delta_p(M)) = \lambda b^{\frac{1}{2}m(p-1)}(1 \otimes \cdots \otimes 1) + F^*(\delta_p(M)).$$

For each nonzero element $y_1 \otimes \cdots \otimes y_p$ of

$$H_{m(p-1)}((S^n)^p) = \sum_{i_1+\cdots+i_p=m(p-1)} H_{i_1}(S^n) \otimes \cdots \otimes H_{i_p}(S^n),$$

we have

$$\langle F^*(\delta_p(M)), y_1 \otimes \cdots \otimes y_p \rangle = \langle \delta_p(M), f_*(y_1) \otimes \cdots \otimes f_*(y_p) \rangle = 0,$$

because $f_* = 0$ in all positive dimensions. Thus we conclude that $F^*(\delta_p(M)) = 0$. It yields the desired result. ∎

LEMMA 5. *Let i denote the inclusion of $E_{\mathbb{Z}_p} \times_{\mathbb{Z}_p} ((M)^p - \Delta)$ into $E_{\mathbb{Z}_p} \times_{\mathbb{Z}_p} (M)^p$. Then $i^*(\Delta_p(M)) = 0$.*

PROOF: This follows immediately from [2], Theorem 5.2. ∎

3. The main result.

In order to present the generalization of Bourgin's result declared at the beginning, it suffices to show that

THEOREM 1. *Let $f : S^n \to M^m$ be a map and $p > 2$ a prime number. If $m < 2\left[\frac{n}{p-1}\right]$, then $A(f,p) \neq 0$. Moreover $\dim A(f,p) \geq \frac{1}{2}p(2n-p+1) - (p-1)m$.*

Notice that $A(f,p) \neq 0$ implies that, for any mutually orthogonal $u_1, \ldots, u_p \in S^n$, there is a rotation $r \in SO(n+1)$ such that $f(ru_1) = \cdots = f(ru_p)$.

PROOF: Consider the following commutative diagram

$$\begin{array}{ccccc}
H^*(X_1) & \xleftarrow{i_2^*} & H^*(X_2) & \xleftarrow{j_2^*} & H^*(X_2, X_1) \\
f_{\mathbb{Z}_p}^* \uparrow & & f_{\mathbb{Z}_p}^* \uparrow & & \uparrow f_{\mathbb{Z}_p}^* \\
H^*(Y_1) & \xleftarrow{i_1^*} & H^*(Y_2) & \xleftarrow{j_1^*} & H^*(Y_2, Y_1),
\end{array}$$

where

$$\begin{aligned}
X_1 &= E_{\mathbb{Z}_p} \times_{\mathbb{Z}_p} V_{n+1,p} - f_{\mathbb{Z}_p}^{-1}(B_{\mathbb{Z}_p} \times \Delta), \\
X_2 &= E_{\mathbb{Z}_p} \times_{\mathbb{Z}_p} V_{n+1,p}, \\
Y_1 &= E_{\mathbb{Z}_p} \times_{\mathbb{Z}_p} (M)^p - B_{\mathbb{Z}_p} \times \Delta, \\
Y_2 &= E_{\mathbb{Z}_p} \times_{\mathbb{Z}_p} (M)^p,
\end{aligned}$$

the horizontal sequences are the cohomology exact sequences of the respective pairs, $f_{\mathbb{Z}_p}$ is the same as that defined before in Lemma 4. By Lemma 5, we have $\Delta_p(M) \in \operatorname{Im} j_1^*$ and $i_2^* f_{\mathbb{Z}_p}^*(\Delta_p(M)) = 0$ in $H^*(X_1)$. On the other hand, it follows from Lemma 2, Lemma 4 and the hypothesis on m that $f_{\mathbb{Z}_p}^*(\Delta_p(M)) \neq 0$ in $H^*(X_2)$. Hence we conclude by Lemma 3 that

$$H_{\frac{1}{2}p(2n-p+1)-(p-1)m}(A(f,p)/t) \cong H^{(p-1)m}(X_2, X_1) \neq 0.$$

This implies our assertion. ∎

We end the paper with a few comments on the case $p = 2$. In this case the mod 2 Haefliger class $\Delta_2(M)$ also works, see [2]. Recall that $H^*(B_{\mathbb{Z}_2}, \mathbb{Z}_2)$ is an exterior algebra $\Lambda(c)$ generated by $c \in H^1(B_{\mathbb{Z}_2}, \mathbb{Z}_2)$. Since

(1) for any map $f : S^n \to M^m$ we have $f_{\mathbb{Z}_2}^*(\Delta_2(M)) = c^m$,
(2) if $n > m$ then $c^m \neq 0 \in H^m(V_{n+1,2}/t)$,

we can prove the following theorem using the same idea as that in the proof of Theorem 1.

THEOREM 2. *If M^m is a manifold with $m < n$, then for every map $f : S^n \to M^m$, $A(f,2) \neq 0$. Moreover $\dim A(f,2) \geq 2n - 1 - m$.*

References

[1] D.G.Bourgin, *Multiplicity of solution in frame mapping*, Ill. J. of Math. **9** (1966), 169–177.
[2] A.Haefliger, *Points multiples d'une application et produit cyclique reduit*, Amer. J. of Math. **83** (1961), 57–70.
[3] B.Knaster, Colloquium Math. **1** (1947), 30–31.
[4] N.E.Steenrod and D.B.Epstein, "Cohomology Operations," Annals of Math. Studies vol. 50, Princeton University Press, 1962.

Current Address: Department of Computer Science, Beijing Computer Institute, Beijing 100044, P.R.China

THE NUMBER OF PERIODIC ORBITS OF SMOOTH MAPS

Takashi Matsuoka

Department of Mathematics, Naruto University of Education
Naruto, Tokushima 772, Japan

1. Introduction

In this paper, we give two results on the number of periodic orbits of some smooth maps on compact manifolds. The first result is concerned with the problem of whether the number of periodic orbits of a given least period is even or odd. This problem has been studied by several authors[5],[6],[8]. We generalize their results, and give a condition on the period and on the topological property of the map under which the number is even.

In the second result, we generalize some results of Franks [3], [4] on the existence of infinitely many periodic orbits.

Both of these results are derived from the equalities between the fixed point index and the cardinalities of some sets of periodic orbits obtained by Dold[2].

More details will appear in Ergodic Theory and Dynamical Systems.

2. Even number of periodic orbits

Let M be a compact smooth manifold, and $f : M \to M$ a C^1 map.

Definition 1. A positive integer k is an L_2-period of f if

$$L_2(f^{i+k}) = L_2(f^i) \quad \text{for any } i \geq 1,$$

where $L_2(f^i) \in \mathbb{Z}/2\mathbb{Z}$ denotes the mod 2 reduction of the Lefschetz number $L(f^i) \in \mathbb{Z}$ of f^i. We denote the minimal L_2-period of f by

$\alpha(f)$.

The following proposition clearly implies that $\alpha(f)$ always exists and is an odd number:

Proposition. Let A be a square matrix with entries in $\mathbb{Z}/2\mathbb{Z}$. Then there exists an odd number k such that $\operatorname{tr} A^{i+k} = \operatorname{tr} A^i$ for any $i \geq 1$.

Proof. Since $\{ A^i \mid i \geq 1\}$ is a finite set, there exist positive integers I and ℓ such that $A^I = A^{I+\ell}$. Then for any $i \geq I$, $A^i = A^I A^{i-I} = A^{i+\ell}$. Let i be a positive integer. Then, $iq \geq I$ for some power q of 2. Since

$$\operatorname{tr} B^2 = \operatorname{tr} B \qquad (1)$$

for any square matrix B with entries in $\mathbb{Z}/2\mathbb{Z}$ ([1, Proposition 5],[7, Lemma]), we have

$$\operatorname{tr} A^i = \operatorname{tr} A^{iq} = \operatorname{tr} A^{iq+\ell q} = \operatorname{tr} A^{(i+\ell)q} = \operatorname{tr} A^{i+\ell}.$$

Hence ℓ is a period for $\{\operatorname{tr} A^i\}_{i \geq 1}$. Decompose ℓ as kr, where k is odd and r is a power of 2. Then, by (1), for any i, $\operatorname{tr} A^i = \operatorname{tr} A^{ir} = \operatorname{tr} A^{ir+\ell} = \operatorname{tr} A^{i+k}$. Thus k is also a period, and the proof is completed.

Definition 2. Let n be a positive integer. A fixed point x of f^n is *regular* if the derivative $Df^n(x)$ of f^n at x does not have 1 as an eigenvalue.

If a point $x \in M$ is a fixed point of f^n and is not a fixed point of f^m for any m with $0 < m < n$, then the set $\{ f^i(x) \mid i \geq 0 \}$ is called an n-*periodic orbit* of f. Then we have:

Theorem 1. Let $f : M \to M$ be a C^1 map on a compact manifold, and n an odd number with $n \geq 3$. Suppose f^n has only regular fixed points. Suppose also n is divisible by $\alpha(f)^2$ or by a prime number p which is congruent to 2^i modulo $\alpha(f)$ for some $i \geq 0$. Then f has an even number of n-periodic orbits.

Let $b(M)$ be the maximum of the Betti numbers of M. It is easy to see that if $b(M) = 1$ (resp. 2) then $\alpha(f) = 1$ (resp. 1 or 3). Hence, by Theorem 1, we have:

Corollary. Let M be a compact manifold, and n an odd number. Assume either that $b(M) = 1$, $n \geq 3$, or that $b(M) = 2$, $n \geq 5$. Let $f : M \to M$ be a C^1 map such that every fixed point of f^n is regular. Then f has an even number of n-periodic orbits.

In the case where M is a disk and f is a diffeomorphism, this corollary has been obtained by Levinson[5], Massera[6] under some additional assumptions. Theorem 1 follows from Shiraiwa[8, Theorem 3], when the maps f^i, $i \geq 1$, have the same Lefschetz number.

3. The existence of infinitely many periodic orbits

In this section, we shall generalize some theorems of Franks in [3],[4]. Here we assume the following conditions:

i) M is a compact smooth manifold and $f : M \to M$ is a C^1 map.
ii) f^n has only regular fixed points for any $n \geq 1$.
iii) All non-zero eigenvalues of $f_* : H_*(M ; \mathbb{Q}) \to H_*(M ; \mathbb{Q})$ are q-th roots of unity for some $q > 0$.

We first define homotopy invariant integers $\gamma(f,m)$, which play a central role in the generalization of the Franks's results. For a positive integer r and $i \geq 0$, let $\gamma(i,r)$ be the number of eigenvalues λ of $f_{*i} : H_i(M ; \mathbb{Q}) \to H_i(M ; \mathbb{Q})$ which are r-th primitive roots of unity (counting multiplicity) and set

$$\gamma_r(f) = \sum_{i \geq 0} (-1)^i \gamma(i,r)/\phi(r),$$

where $\phi(r)$ denotes the Euler function, i.e., $\phi(r)$ is the number of integers s such that $1 \leq s \leq r$ and that s and r are relatively prime. For a positive integer m, let

$$\gamma(f,m) = \sum_r \mu(r)\gamma_{rm}(f),$$

where the sum is taken over all odd numbers r and $\mu(r)$ is the Moebius function, i.e., $\mu(1) = 1$, $\mu(r) = (-1)^k$ if r is a product of k distinct primes, and $\mu(r) = 0$ otherwise. For example, if $q = 3$, then $\gamma(f,1) = \gamma_1(f) - \gamma_3(f)$, $\gamma(f,3) = \gamma_3(f)$, and $\gamma(f,m) = 0$ for odd $m \geq 5$.

Given an n-periodic orbit ξ of f, let $a_+(\xi)$ (resp. $a_-(\xi)$) be the number of real eigenvalues λ of $Df^n(x)$ with $\lambda > 1$ (resp. $\lambda < -1$) (counting multiplicity), where $x \in \xi$. (Note that $a_+(\xi)$ and

$a_-(\xi)$ do not depend on the choice of x.) Note that $a_+(\xi)$ is even or odd according to whether the fixed point index $I(f^n,x)$ of f^n at x is equal to 1 or -1, because x is a regular fixed point of f^n by the standing hypotheses. Note also that $a_-(\xi)$ is even or odd according to whether $I(f^{2n},x) = I(f^n,x)$ or $I(f^{2n},x) = - I(f^n,x)$.

We divide the set $Per(n)$ of all n-periodic orbits into four subsets by these numbers $a_+(\xi)$, $a_-(\xi)$, and set

$$N_{EE}(n) = \#\{ \xi \in Per(n) \mid a_+(\xi),\ a_-(\xi) \text{ are even } \},$$
$$N_{EO}(n) = \#\{ \xi \in Per(n) \mid a_+(\xi) \text{ is even, } a_-(\xi) \text{ is odd } \},$$
$$N_{OE}(n) = \#\{ \xi \in Per(n) \mid a_+(\xi) \text{ is odd, } a_-(\xi) \text{ is even } \},$$
$$N_{OO}(n) = \#\{ \xi \in Per(n) \mid a_+(\xi),\ a_-(\xi) \text{ are odd } \},$$

where # denotes the cardinality. Then we have:

<u>Theorem</u> 2. Let m be an odd number and let $i(m)$ denote the smallest $i \geq 0$ such that $2^i m$ does not divide q. Then for any $i \geq i(m) - 1$, we have

$$N_{EO}(2^i m) - N_{OO}(2^i m) = \gamma(f,m) + \sum_{j=0}^{i} (N_{OE}(2^j m) - N_{EE}(2^j m)).$$

We call an n-periodic orbit ξ an <u>inverting</u> n-periodic orbit if $I(f^{2n},x) = - I(f^n,x)$, where $x \in \xi$ (i.e., if $a_-(\xi)$ is odd). Then Theorem 2 immediately implies the following corollaries concerning the existence of infinitely many inverting periodic orbits:

<u>Corollary</u> 1. Let m be an odd number. Suppose that

$$\gamma(f,m) > \sum_{j=0}^{\infty} N_{EE}(2^j m) \quad \text{or} \quad \gamma(f,m) < - \sum_{j=0}^{\infty} N_{OE}(2^j m).$$

Then for any $i \geq i(m) - 1$, f has an inverting $2^i m$-periodic orbit.

<u>Corollary</u> 2. Let m be an odd number satisfying $\gamma(f,m) = 0$. Suppose either

i) $N_{EE}(2^i m) = N_{OO}(2^i m) = 0$ for any $i \geq 0$, or
ii) $N_{EO}(2^i m) = N_{OE}(2^i m) = 0$ for any $i \geq 0$.

Then we have

$$N_{EO}(2^i m) + N_{OO}(2^i m) = \sum_{j=0}^{i} (N_{EE}(2^j m) + N_{OE}(2^j m))$$

for any $i \geq i(m) - 1$. In particular, if f has a $2^{i_0}m$-periodic orbit for some $i_0 \geq i(m) - 1$, then for each $k \geq i_0$, f has an inverting $2^k m$-periodic orbit.

Corollary 1 has been proved by Franks [3] in the case where M is an interval, a disk, a circle, and a sphere. Corollary 2 has been proved also by Franks [4, Theorem C] in the case of $i(m) = 0$. (Note that if $i(m) = 0$, then $\gamma(f,m)$ is always zero.)

4. Proofs of the theorems

We begin the proofs by recalling the result of Dold[2]. Let M be a smooth manifold and $f : M \to M$ a C^1 map. Let n be a positive integer. Suppose the set $\mathrm{Fix}(f^n)$ of fixed points of f^n is compact. Then $\mathrm{Fix}(f^m)$ is also compact for any divisor m of n, and hence the fixed point index $I(f^m)$ of f^m is defined. Define $J_n(f) \in \mathbb{Q}$ by

$$J_n(f) = \frac{1}{n} \{ \sum_{\tau \subset P(n)} (-1)^{\#\tau} I(f^{n:\tau}) \},$$

where $P(n)$ is the set of all primes which divide n, the sum extends over all subsets τ of $P(n)$, and $n:\tau = n(\prod_{p \in \tau} p)^{-1}$ = n divided by all $p \in \tau$. Let

$$N_E(n) = N_{EE}(n) + N_{EO}(n), \quad N_O(n) = N_{OE}(n) + N_{OO}(n).$$

The following result is due to Dold[2, Proposition (6.2)].(Precisely speaking, his result is stronger than the below; in [2], the domain of f is allowed to be an open set of M.)

Theorem 3(Dold). Suppose that f^n has only finitely many fixed points and that every fixed point of f^n is regular . Then

$$\begin{aligned} J_n(f) &= N_E(n) - N_O(n) && \text{if } n \text{ is odd,} \\ &= N_E(n) - N_O(n) - N_{EO}(n/2) + N_{OO}(n/2) && \text{if } n \text{ is even.} \end{aligned}$$

Proof of Theorem 1. By Theorem 3, the number of n-periodic orbits of f is equal to $2\,N_O(n) + J_n(f)$. Hence, it is sufficient for the proof to show that the mod 2 reduction $J_{n,2}(f)$ of $J_n(f)$ is equal to zero. Let $\alpha = \alpha(f)$. Suppose first $\alpha^2 | n$. Then for any $\tau \subset P(n)$, $n:\tau$ is a multiple of α and $L_2(f^{n:\tau}) = L_2(f^{\alpha})$. Therefore

$$J_{n,2}(f) = \frac{1}{n} \{ \sum_{\tau \subset P(n)} (-1)^{\#\tau} \} L_2(f^{\alpha}) = 0.$$

Next, suppose n is divisible by a prime p with $p \equiv 2^i \pmod{\alpha}$

for some i. Decompose n as $n = p^j r$, where $j, r \geq 1$ and r is not divisible by p. Then

$$p^j J_n(f) = J_r(f^{p^j}) - J_r(f^{p^{j-1}}). \qquad (2)$$

Since α is a period for the sequence $\{J_{r,2}(f^i)\}_{i \geq 1}$, we have by (1):

$$J_{r,2}(f^{p^s}) = J_{r,2}(f^{2^{is}}) = J_{r,2}(f) \quad \text{for } s \geq 0.$$

Therefore by (2), $J_{n,2}(f) = 0$.

<u>Proof</u> <u>of</u> <u>Theorem</u> 2. Using Theorem 3, we can prove the following equalities by induction on i:

$$N_{EO}(2^i m) - N_{OO}(2^i m) = \sum_{j=0}^{i} \{J_{2^j m}(f) + N_{OE}(2^j m) - N_{EE}(2^j m)\}. \qquad (3)$$

Hence, it is sufficient to show that if $i \geq i(m) - 1$, then $\sum_{j=0}^{i} J_{2^j m}(f) = \gamma(f,m)$. For $r \geq 1$, let A_r be the set of all r-th primitive roots of unity. Let

$$\psi(r, j) = \sum_{\lambda \in A_r} \lambda^j, \text{ for } j \geq 1,$$

$$M_r(n) = \frac{1}{n}\{\sum_{\tau \subset P(n)} (-1)^{\#\tau} \psi(r, n:\tau)\}.$$

Then clearly we have:

$$J_n(f) = \sum_{r|q} \gamma_r(f) M_r(n). \qquad (4)$$

The following lemma can be proved by induction on the number of primes dividing n:

<u>Lemma</u>. For $n, r \geq 1$, $M_r(n) = \mu(r/n)$, where $\mu(r/n)$ means 0 if r is not divisible by n.

Now let $i \geq i(m) - 1$. By (4) and Lemma, since $\sum_{j=0}^{i} \mu(r/2^j m) = 0$ for r even, we have

$$\sum_{j=0}^{i} J_{2^j m}(f) = \sum_{r|q} \gamma_r(f) \sum_{j=0}^{i} \mu(r/2^j m) = \gamma(f,m).$$

Thus Theorem 2 is proved.

References

[1] F.E. Browder. The Lefschetz fixed point theorem and asymptotic fixed point theorems. Lecture Notes in Math. 446. Springer-Verlag:Berlin, 1975, 96-122.

[2] A. Dold. Fixed point indices of iterated maps. Invent. math. 74 (1983), 419-435.

[3] J. Franks. Some smooth maps with infinitely many hyperbolic periodic points. Trans. Amer. Math. Soc. 226 (1977), 175-179.

[4] J. Franks. Period doubling and the Lefschetz formula. Trans. Amer. Math. Soc. 287 (1985), 275-283.

[5] N. Levinson. Transformation theory of non-linear differential equations of the second order. Ann. of Math. 45 (1944),723-737, Corrections, ibid. 49 (1948), 738.

[6] J.L. Massera. The number of subharmonic solutions of non-linear differential equations of the second order. Ann. of Math. 50 (1949),118-126.

[7] H.O. Peitgen. On the Lefschetz number for iterates of continuous mappings. Proc. Amer. Math. Soc. 54 (1976),441-444.

[8] K. Shiraiwa. A generalization of the Levinson-Massera's equalities. Nagoya Math. J. 67 (1977), 121-138.

PARAMETRIZED BORSUK-ULAM THEOREMS AND CHARACTERISTIC POLYNOMIALS

Minoru Nakaoka
Okayama University of Science and Osaka University

Introduction

Let $\pi : X \longrightarrow B$ and $\rho : Y \longrightarrow B$ be vector bundles with fibre-preserving linear Z_2-action, and let $f: S \longrightarrow Y$ be a fibre-preserving equivariant map of the sphere bundle S $(\subset X)$ to Y. What can we say about the set

$$\{x \in S \mid f(x) = 0\} .$$

This problem is a parametrized version of the Borsuk-Ulam type problem. Dold contributes in [4] to this problem by showing a theorem (2.2 in [4]) which relates to Stiefel-Whitney classes. He formulates also the cohomology generalization of the theorem only when the action on S is free. In this paper we give the cohomology version of the theorem in presence of fixed points. We treat not only Z_2-action but also Z_p-action (p odd prime) and S^1-action in parallel. We also generalize the parametrized Borsuk-Ulam theorem due to Jaworowski [6].

Throughout this paper, cohomology is understood to be Čech theory. Unless specified, the cohomology takes coefficients from the field Z_p of integers mod p in case of Z_p-action, and the field Q of rational numbers in case of S^1-action.

This work was prompted by the manuscript of Dold [4] sent kindly the author.

§1 Preliminaries

A topological space X is called a <u>mod p cohomology n-sphere</u> if $H^*(X; Z_p) \cong H^*(S^n; Z_p)$, and is called an <u>integral cohomology n-sphere</u> if $H^*(X; Z) \cong H^*(S^n; Z)$. They are denoted by $X \sim_p S^n$ and $X \sim S^n$ respectively.

<u>Proposition 1.</u> (i) Let $G = Z_p$ (p prime), and let X be a finite dimensional paracompact G-space such that $X \sim_p S^n$. Then $X^G \sim_p S^d$ for some d $(-1 \leqq d \leqq n)$, and $n - d$ is even if p is odd.

(ii) Let $G = S^1$, and let X be an orientable (topological) G-manifold such that $X \sim S^n$ and the G-action is locally smooth. Then $X^G \sim S^d$ for some d $(-1 \leq d \leq n)$, and $n - d$ is even.

Proof. (i) is proved in Theorem III, 7.11 of [1]. (ii) follows from III. 10.2, III. 10.13, IV. 10.5 of [1] and the fact $\dim X/G \leq \dim X$ ([8]).

Let $G = Z_p$ (p prime) or S^1, and let X be a paracompact (Hausdorff) G-space. As usual, put

$$X^G = \{x \in X | gx = x \text{ for any } g \in G\} .$$

Consider the universal G-bundle $E_G \longrightarrow B_G$, and put

$$H_G^*(X,X^G) = H^*(E_G \underset{G}{\times} X, E_G \underset{G}{\times} X^G),$$

$$H_G^*(X) = H^*(E_G \underset{G}{\times} X) ,$$

$$H_G^*(X^G) = H^*(E_G \underset{G}{\times} X^G) = H^*(B_G \times X^G) .$$

For $w \in H_G^*(pt)$, let $w(X)$ denote the image of w under the homomorphism $H_G^*(pt) \longrightarrow H_G^*(X)$.

It is well known that if $G = Z_2$ then

$$H_G^*(pt) \cong Z_2[u] \text{ with } \deg u = 1 ,$$

if $G = S^1$ then

$$H_G^*(pt) \cong Q[u] \text{ with } \deg u = 2 ,$$

and if $G = Z_p$ (p odd prime) then

$$H_G^*(pt) \cong Z_p[u,v]/(v^2) \text{ with } \deg u = 2, \deg v = 1 ,$$

and u is the image of v under the Bockstein homomorphism.

Unifying the notation we write

$$u_q = u^q \qquad \text{if } G = Z_2 ,$$

$$u_{2q} = u^q,\ u_{2q+1} = u^q v \quad \text{if } G = Z_p \text{ (p odd)},$$

$$u_{2q} = u^q,\ u_{2q+1} = 0 \quad \text{if } G = S^1 .$$

Proposition 2. In the situation of Proposition 1, define $\beta \in H_G^{d+1}(X, X^G)$ by $\beta = 1$ if $d = -1$ and by $\beta = \delta^*(1 \times b)$ if $d \geq 0$, where b is a generator of the reduced cohomology group

$\tilde{H}^d(X^G)$. Then a basis of $H_G^*(X, X^G)$ is given by

$$\{u_q(X)\beta;\ 0 \leqq q \leqq n - d - 1\}$$

if $G = Z_p$, and by

$$\{u_{2q}(X)\beta;\ 0 \leqq q \leqq (n - d)/2 - 1\}$$

if $G = S^1$. It holds also that $u_{n-d}(X)\beta = 0$.

Proof. The results for $d = -1$ are immediate from the Smith-Gysin exact sequence (III. (7.8), (10.5) in [1]). The result for $d \geqq 0$ is a generalization of Proposition (1.5) in [5], and is proved as follows along the lines employed there.

Consider the orbit space X/G and its subset X^G. It holds that

$$H^*(X/G, X^G) \cong H_G^*(X, X^G)$$

(VII. Proposition 1.1 in [1]) and

$$\operatorname{rk} H^q(X/G, X^G) \leqq \sum_{r \geqq q} \operatorname{rk} H^r(X)$$

(III. Theorems 7.9 and 10.9 in [1]). Therefore we have

(1.1) $\quad H_G^q(X, X^G) = 0$ for q large.

Since

$$\sum_q \operatorname{rk} H^q(X) = 2 = \sum_q \operatorname{rk} H^q(X^G),$$

Leray-Hirsch theorem can be applied for the fibre bundle $E_G \times_G X \longrightarrow B_G$ (VII. Theorem 1.6 in [1]), and we have an isomorphism

(1.2) $\quad \phi : H_G^*(pt) \otimes H^*(X) \cong H_G^*(X)$

given by $\phi(u_q \otimes x) = u_q(X)\theta(x)$ $(q \geqq 0)$, where $\theta : H^*(X) \longrightarrow H_G^*(X)$ is a cohomology extension of the fibre.

In the exact sequence

$$\cdots \longrightarrow H_G^q(X, X^G) \xrightarrow{j^*} H_G^q(X) \xrightarrow{i^*} H_G^q(X^G) \longrightarrow \cdots$$

we see that i^* is injective.

In fact, suppose $i^*(x) = 0$ for $x \in H_G^q(X)$. Then we have $x = j^*(y)$ for some $y \in H_G^q(X, X^G)$. It follows from (1.1) that $u(X)^r y = 0$ and hence $u(X)^r x = 0$ for some r. This implies $x = 0$ by (1.2).

We have now an exact sequence

$$0 \longrightarrow H_G^q(X) \xrightarrow{i^*} H_G^q(X^G) \xrightarrow{\delta^*} H_G^{q+1}(X, X^G) \longrightarrow 0 .$$

Let a denote a generator of $H^n(X)$. Then it follows that

$$i^*(\phi(u_q \otimes 1)) = u_q \times 1 \quad (q \geq 0),$$

$$i^*(\phi(1 \otimes a)) = \varepsilon_1(u_{n-d} \times b) + \varepsilon_2(u_n \times 1)$$

with coefficients $\varepsilon_1, \varepsilon_2$. Since $\varepsilon_1 \neq 0$ is easily seen, we have

$$u_{n-d} \times b \in \operatorname{Im} i^* .$$

While, if $G = Z_p$ then

$$u_q \times b \notin \operatorname{Im} i^* \quad (0 \leq q \leq n - d - 1)$$

and if $G = S^1$ then

$$u_{2q} \times b \notin \operatorname{Im} i^* \quad (0 \leq q \leq (n-d)/2 - 1) .$$

Thus the above exact sequence yields the desired result.

Remark. Let X be a smooth G-manifold. Take a closed invariant tubular neighborhood N of X^G in X. Then we have the canonical isomorphisms

$$H_G^*(X - X^G) \cong H_G^*(X - \operatorname{int} N),$$

$$H_G^*(X, X^G) \cong H_G^*(X, N) \cong H_G^*(X - \operatorname{int} N, N - \operatorname{int} N) .$$

Therefore the cup product

$$H_G^*(X - \operatorname{int} N) \otimes H_G^*(X - \operatorname{int} N, N - \operatorname{int} N) \longrightarrow H_G^*(X - \operatorname{int} N, N - \operatorname{int} N)$$

yields a cup product

$$\smile : H_G^*(X - X^G) \otimes H_G^*(X, X^G) \longrightarrow H_G^*(X, X^G) .$$

It is easily seen that

$$w(X - X^G) \smile \alpha = w(X)\alpha$$

for $w \in H_G^*(pt)$ and $\alpha \in H_G^*(X, X^G)$. This fact and Proposition 2 show that if $G = Z_p$ and X is a smooth G-manifold which is a mod p cohomology sphere, or if $G = S^1$ and X is an orientable smooth G-manifold which is an integral cohomology sphere, then there is an isomorphism

$$\smile \beta : H_G^*(X - X^G) \cong H_G^*(X, X^G) .$$

§2 Characteristic polynomials

Let X and K be paracompact G-spaces, and let $\pi : X \longrightarrow B$ be a continuous map of X to a paracompact space B. Then $\xi = (X, \pi, B, K)$ is called a fibre bundle with G-action if the following condition is satisfied: For each point $b \in B$, $\pi^{-1}(b)$ is an invariant subspace of X, and there exist a neighborhood U of b and an equivariant homeomorphism $\psi : U \times K \approx \pi^{-1}(U)$ such that $\pi\psi(u, a) = u$ $(u \in U, a \in K)$, where $U \times K$ is regarded as a G-space by $g(u, a) = (u, ga)$, $g \in G$.

Let $\xi = (X, \pi, B, K)$ be a fibre bundle with G-action. If $G = Z_p$ (p prime) and if K is finite dimensional and $K \sim_p S^m$, then ξ is called a mod p cohomology m-sphere bundle with Z_p-action. If $G = S^1$ and if K is an orientable manifold such that $K \sim S^m$ and the action of S^1 on K is locally smooth, then ξ is called an integral cohomology m-sphere bundle with locally smooth S^1-action.

Let $\xi = (X, \pi, B, K)$ be a mod p cohomology m-sphere bundle with G-action $(G = Z_p)$, or an integral cohomology m-sphere bundle with locally smooth G-action $(G = S^1)$. We consider first the case $K^G = \phi$, and following Dold [4] we shall define the characteristic polynomial C_ξ of ξ.

By applying Leray-Hirsch theorem to the fibre bundle $\bar{\pi} : E_G \times_G X \longrightarrow B$ with fibre $E_G \times_G K$, we obtain an isomorphism

$$H^*(B) \otimes H_G^*(K) \cong H_G^*(X)$$

given by $\alpha \otimes u_q(K) \longmapsto \alpha u_q(X)$ $(q = 0,1,\cdots,m)$, where $H_G^*(X)$ is regarded as an $H^*(B)$-module via $\bar{\pi}^* : H^*(B) \longrightarrow H_G^*(X)$. Therefore there exist unique elements $c_q(\xi) \in H^q(B)$ $(q = 1,2,\cdots,m+1)$ such that

$$-u_{m+1}(X) = \sum_{q=1}^{m+1} c_q(\xi) u_{m+1-q}(X)$$

with $c_q(\xi) = 0$ for q odd if $G = S^1$. We define now C_ξ as follows.

For $G = Z_2$, $C_\xi \in H^*(B)[t]$ is a polynomial

$$C_\xi(t) = t^{m+1} + c_1(\xi)t^m + c_2(\xi)t^{m-1} + \cdots + c_{m+1}(\xi)$$

with $\deg t = 1$; For $G = Z_p$ (p odd), $C_\xi \in H^*(B)[t,s]/(s^2)$ is a polynomial

$$C_\xi(t,s) = t^{(m+1)/2} + c_1(\xi)st^{(m-1)/2} + c_2(\xi)t^{(m-1)/2}$$
$$+ \cdots + c_m(\xi)s + c_{m+1}(\xi)$$

with $\deg t = 2$, $\deg s = 1$; For $G = S^1$, $C_\xi \in H^*(B)[t]$ is a polynomial

$$C_\xi(t) = t^{(m+1)/2} + c_2(\xi)t^{(m-1)/2} + \cdots + c_{m-1}(\xi)t + c_{m+1}(\xi)$$

with $\deg t = 2$.

We shall next consider the case $K^G \neq \phi$. We have

$$K^G \sim_p S^d \quad \text{for } G = Z_p ,$$
$$K^G \sim S^d \quad \text{for } G = S^1$$

with $d \geqq 0$. Assume that the local system $\{H^d(\pi^{-1}(b)^G)\}_{b \in B}$ over B is trivial. (Of course this assumption is redundant for $G = Z_2$.) Then the <u>characteristic polynomial</u> C_ξ of ξ is defined as follows.

Take a generator b of $\tilde{H}^d(K^G)$. By the assumption, there exists $\bar{b} \in H^d(X^G)$ whose image in $H^d(K^G)$ is b. Consider a commutative diagram

$$\begin{array}{ccc} H^d_G(X^G) & \xrightarrow{\delta^*} & H^{d+1}_G(X, X^G) \\ \downarrow i^* & & \downarrow i^* \\ H^d_G(K^G) & \xrightarrow{\delta^*} & H^{d+1}_G(K, K^G) \end{array} ,$$

and put

$$\beta = \delta^*(1 \times b), \ \bar{\beta} = \delta^*(1 \times \bar{b}) .$$

We have $i^*(\bar{\beta}) = \beta$. Therefore, in virtue of Proposition 2, we can apply Leray-Hirsch theorem to the fibre-bundle pair $\bar{\pi}: (E_G \times_G X, E_G \times_G X^G) \longrightarrow B$ with fibre pair $(E_G \times_G K, E_G \times_G K^G)$, and we have an isomorphism

$$H^*(B) \otimes H^*_G(K, K^G) \cong H^*_G(X, X^G)$$

given by $\alpha \otimes u_q(K)\beta \longrightarrow \alpha u_q(X)\overline{\beta}$ $(q = 0,1,\cdots,m-d-1)$, where $H_G^*(X, X^G)$ is regarded as an $H^*(B)$-module via $\overline{\pi}^* : H^*(B) \longrightarrow H_G^*(X)$. Therefore there exist unique elements $c_q(\xi) \in H^q(B)$ $(q = 1,2,\cdots,m-d)$ such that

$$-u_{m-d}(X)\overline{\beta} = \sum_{q=1}^{m-d} c_q(\xi) u_{m-d-q}(X)\overline{\beta}$$

with $c_q(\xi) = 0$ for q odd if $G = S^1$. As before, we define C_ξ by

$$C_\xi(t) = t^{m-d} + c_1(\xi)t^{m-d-1} + c_2(\xi)t^{m-d-2} + \cdots + c_{m-d}(\xi),$$

$$C_\xi(t,s) = t^{(m-d)/2} + c_1(\xi)st^{(m-d)/2-1} + c_2(\xi)t^{(m-d)/2-1} + \cdots + c_{m-d-1}(\xi)s + c_{m-d}(\xi),$$

$$C_\xi(t) = t^{(m-d)/2} + c_2(\xi)t^{(m-d)/2-1} + \cdots + c_{m-d-2}(\xi)t + c_{m-d}(\xi)$$

according as $G = Z_2$, Z_p (p odd) or S^1.

Let N be a closed invariant subset of X, and let P denote an element $P(t) \in H^*(B)[t]$ or $P(t,s) \in H^*(B)[t,s]/(s^2)$. Regard $H_G^*(N)$ as an $H^*(B)$-module via $(\overline{\pi}|N)^* : H^*(B) \longrightarrow H_G^*(N)$. Then, by substituting $u(N)$, $v(N)$ for the indeterminates t,s, we have an element $P(u(N))$ or $P(u(N), v(N))$ of $H_G^*(N)$. We denote this element by $P|N$.

Let $\overline{\beta}$ denote 1 in the case $X^G = \phi$. Then we have $(C_\xi|X)\overline{\beta} = 0$, and the correspondence $P \longmapsto (P|X)\overline{\beta}$ yields an isomorphism

(2.1) $\quad H^*(B)[t]/(C_\xi) \cong H_G^*(X, X^G)$

for $G = Z_2$ or S^1, and an isomorphism

(2.2) $\quad H^*(B)[t,s]/(s^2, C_\xi) \cong H_G^*(X, X^G)$

for $G = Z_p$ (p odd).

<u>Proposition 3</u>. For $\xi = (X, \pi, B, K)$ as above, assume furthermore that K is a compact smooth G-manifold, and put $\xi' = (X - X^G, \pi|X - X^G, B, K - K^G)$. Then we have

$$C_\xi = C_{\xi'} .$$

<u>Proof</u>. By making use of the duality theorem ([3], p. 292), it can be proved that

$$K - K^G \sim_p S^{m-d-1} \quad \text{for} \quad G = Z_p ,$$

$$K - K^G \sim S^{m-d-1} \quad \text{for} \quad G = S^1 .$$

Therefore it holds that

$$-u_{m-d}(X - X^G) = \sum_{q=1}^{m-d} c_q(\xi')u_{m-d-q}(X - X^G) .$$

Similarly to Remark in § 1 , a cup product

$$\smile \; : H_G^*(X - X^G) \otimes H_G^*(X, X^G) \longrightarrow H_G^*(X, X^G)$$

can be defined, and $u_q(X - X^G) \smile \overline{\beta} = u_q(X)\overline{\beta}$ holds. Therefore we have

$$-u_{m-d}(X)\overline{\beta} = \sum_{q=1}^{m-d} c_q(\xi')u_{m-d-q}(X)\overline{\beta} ,$$

which shows $c_q(\xi) = c_q(\xi')$ and hence $C_\xi = C_{\xi'}$.

<u>Proposition 4</u>. Let $\xi_i = (X_i, \pi, B, K_i)$ $(i = 1,2)$ be mod p cohomology m-sphere bundles with Z_p-action such that $K_i^G \sim_p S^d$, and assume that the local system $\{H^d(\pi_2^{-1}(b)^G)\}_{b\in B}$ is trivial. Let $f : X_1 \longrightarrow X_2$ be an equivariant map such that $\pi_2 \circ f = \pi_1$ and $f^* : H^d(\pi_2^{-1}(b)) \longrightarrow H^d(\pi_1^{-1}(b))$ $(b \in B)$ is surjective. Then we have $C_{\xi_1} = C_{\xi_2}$. Similar for integral cohomology sphere bundles with local smooth S^1-action.

<u>Proof</u>. Consider the element $\overline{\beta}_i \in H_G^{d+1}(X_i, X_i^G)$ for the bundle ξ_i, and let $\overline{\beta}_i$ denote 1 in the case $d = -1$. Then it follows that $\overline{\beta}_2 = f^*(\overline{\beta}_1)$. Therefore,

$$-u_{m-d}(X_2)\overline{\beta}_2 = \sum_{q=1}^{m-d} c_q(\xi_2)u_{m-d-q}(X_2)\overline{\beta}_2$$

implies

$$-u_{m-d}(X_1)\overline{\beta}_1 = \sum_{q=1}^{m-d} c_q(\xi_2)u_{m-d-q}(X_1)\overline{\beta}_1 \quad .$$

Therefore we have $c_q(\xi_1) = c_q(\xi_2)$ and hence $C_{\xi_1} = C_{\xi_2}$.

§3 <u>Generalization of Dold theorem</u>

The following theorem is a generalization of the results in [4].

<u>Theorem 1</u>. Let Y be a G-space, and let Y_0 be a closed subset of Y^G. Assume the following (i) or (ii).

(i) $G = Z_p$ (p prime); $\xi = (X, \pi, B, K)$ and $\eta = (Y - Y_0, \rho, B, L)$ are mod p cohomology sphere bundles with Z_p-action such that $K \sim_p S^m$, $L \sim_p S^n$ and $L^G \sim_p S^d$; the local system $\{H^d(\rho^{-1}(b)^G)\}_{b\ B}$ over B is trivial.

(ii) $G = S^1$; $\xi = (X, \pi, B, K)$ and $\eta = (Y - Y_0, \rho, B, L)$ are integral cohomology sphere bundles with locally smooth S^1-action such that $K \sim S^m$, $L \sim S^n$ and $L^G \sim S^d$; the local system $\{H^d(\rho^{-1}(b)^G)\}_{b \in B}$ over B is trivial.

Let $f : X \longrightarrow Y$ be an equivariant map such that $f : X - f^{-1}(Y_0) \longrightarrow Y - Y_0$ is a fibre preserving map (i.e. $\rho f = \pi$), $f(X^G) \subset Y^G - Y_0$, and $f|\pi^{-1}(b)^G$ $(b \in B)$ induces a surjection $H^d(L^G) \longrightarrow H^d(K^G)$. Then, if $P \in H^*(B)[t]$ (or $P \in H^*(B)[t,s]/(s^2)$) satisfies $P|f^{-1}(Y_0) = 0$, we have

$$PC_\eta = C_\xi Q$$

for some $Q \in H^*(B)[t]$ (or $Q \in H^*(B)[t, s]/(s^2)$).

<u>Remark</u>. Theorem 1.3 in [4] is a special case for $G = Z_2$, $Y_0 = Y^G$, and Theorem 2.2 in [4] is a special case for $G = Z_2$, $X = S(E \oplus F)$, $Y = E' \oplus F$ and $Y_0 = \{0\}$, where E, E', F are real vector bundles, $S(E \oplus F)$ is the sphere bundle of $E \oplus F$, and X, Y are regarded as Z_2-spaces by $(a,b) \longmapsto (-a,b)$.

<u>Proof</u>. Since $P|f^{-1}(Y_0) = 0$ and $f^{-1}(Y_0) \subset X - X^G$, by tautness of Čech cohomology we see that there exists a closed invariant neighborhood N of $f^{-1}(Y_0)$ in $X - X^G$ such that $P|N = 0$.

Case 1: $Y_0 = Y^G$

In this case $X^G = \phi$. Since $f^*: H_G^*(Y - Y^G) \longrightarrow H_G^*(X - f^{-1}(Y^G))$ sends $u_q(Y - Y^G)$ to $u_q(X - f^{-1}(Y^G))$, we have

$$C_\eta|X - f^{-1}(Y^G) = f^*(C_\eta|Y - Y^G) = 0 .$$

Therefore, in a commutative diagram

$$\begin{array}{ccc} H_G^*(X, N) \otimes H_G^*(X, Y - f^{-1}(Y^G)) & \xrightarrow{\smile} & H_G^*(X, X) \\ \downarrow j^* \otimes j^* & & \downarrow j^* \\ H_G^*(X) \otimes H_G^*(X) & \xrightarrow{\smile} & H_G^*(X) \end{array},$$

it follows that $P|X$ and $C_\eta|X$ are in $\operatorname{Im} j^*$, and hence $PC_\eta|X = 0$.

This shows the desired result by (2.1) and (2.2)

Case 2: $Y_0 \neq Y^G$

Let $\bar{\beta}_\xi \in H_G^{d+1}(X, X^G)$ and $\bar{\beta}_\eta \in H_G^{d+1}(Y - Y_0, Y^G - Y_0)$ denote the element $\bar{\beta}$ for ξ and η respectively. From a commutative diagram

$$\begin{array}{ccccc} H_G^d(L^G) & \longleftarrow & H_G^d(Y^G - Y_0) & \xrightarrow{\delta^*} & H_G^{d+1}(Y - Y_0, Y^G - Y_0) \\ & & & & \downarrow f^* \\ \downarrow f^* & & \downarrow f^* & & H_G^{d+1}(X - f^{-1}(Y_0), X^G) \\ & & & & \uparrow i^* \\ H_G^d(K^G) & \longleftarrow & H_G^d(X^G) & \xrightarrow{\delta^*} & H_G^{d+1}(X, X^G) \end{array}$$

it follows that

$$f^*(\bar{\beta}_\eta) = i^*(\bar{\beta}_\xi) .$$

Since $f^*: H_G^*(Y - Y_0) \longrightarrow H_G^*(X - f^{-1}(Y_0))$ sends $u_q(Y - Y_0)$ to $u_q(X - f^{-1}(Y_0))$, we see

$$C_\eta | X - f^{-1}(Y_0) = f^*(C_\eta | Y - Y_0) .$$

Therefore it holds that

$$\begin{aligned} i^*((C_\eta|X)\bar{\beta}_\xi) &= (C_\eta|X - f^{-1}(Y_0))(f^*\bar{\beta}_\eta) \\ &= f^*((C_\eta|Y - Y_0)\bar{\beta}_\eta) = 0 . \end{aligned}$$

Consider a commutative diagram

$$\begin{array}{ccc} H_G^*(X, N) \otimes H_G^*(X, X - f^{-1}(Y_0)) & \xrightarrow{\smile} & H_G^*(X, X) \\ \downarrow j^* \otimes j^* & & \downarrow j^* \\ H_G^*(X) \otimes H_G^*(X, X^G) & \xrightarrow{\smile} & H_G^*(X, X^G) \end{array}$$

It follows that $P|X$ and $(C_\eta|X)\bar{\beta}_\xi$ are in $\operatorname{Im} j^*$, and hence $(PC_\eta|X)\bar{\beta}_\xi = 0$. This shows the desired result by (2.1) and (2.2).

<u>Corollary</u>. Under the assumption of Theorem 1, if $m > n$ then for any i there are injections

$$\bigoplus_{0 \leq q < m-n} H^{i+q}(B) \longrightarrow H_G^{i+m-n-1}(f^{-1}(Y_0)) \quad \text{for} \quad G = Z_p ,$$

$$\bigoplus_{0 \leq q < (m-n)/2} H^{i+2q}(B) \longrightarrow H_G^{i+m-n-2}(f^{-1}(Y_0)) \quad \text{for} \quad G = S^1$$

given by

$$(\alpha_0,\alpha_1,\cdots,\alpha_{m-n-1}) \longmapsto \sum_{0\leq q<m-n} \alpha_{m-n-1-q}u_q(f^{-1}(Y_0)) ,$$

$$(\alpha_0,\alpha_1,\cdots,\alpha_{(m-n)/2-1}) \longmapsto \sum_{0\leq q<(m-n)/2} \alpha_{(m-n)/2-1-q}u(f^{-1}(Y_0))^q$$

respectively.

Proof. Define a polynomial P by

$$P(t) = \sum_{0\leq q<m-n} \alpha_{m-n-1-q}t^q \quad \text{for } G = Z_2 ,$$

$$P(t,s) = \sum_{0\leq q<(m-n)/2} (\alpha_{m-n-1-2q}t^q + \alpha_{m-n-1-2q-1}t^q s) \quad \text{for } G = Z_p \ (p \text{ odd}),$$

$$P(t) = \sum_{0\leq q<(m-n)/2} \alpha_{(m-n)/2-1-q}t^q \quad \text{for } G = S^1 .$$

Then it suffices to prove that $P|f^{-1}(Y_0) = 0$ implies $P = 0$. Suppose $P \neq 0$ and put $r = \mathrm{Min}\{q|\alpha_q \neq 0\}$. Consider the degree of P, C_ξ and C_η with respect to the indeterminates. We have

$$\deg P = \begin{cases} m - n - 1 - r & \text{for } G = Z_p , \\ m - n - 2 - 2r & \text{for } G = S^1 , \end{cases}$$

$$\deg C_\xi = m - d, \ \deg C_\eta = n - d .$$

Since $\deg PC_\eta \geq \deg C_\xi$ by Theorem 1, we have $r \leq -1$ which is a contradiction.

§4 Equivariant point theorem

Let X and Y be G-spaces, and let $f : X \longrightarrow Y$ be a continuous map. Then a point $x \in X$ satisfying $f(gx) = g(f(x))$ for any $g \in G$ is called an equivariant point of f. In this section, by applying Theorem 1, we establish theorems concerning the set of equivariant points, which are generalizations of the result in [6] and its improvement in [7].

If $\phi : E \longrightarrow B$ is a (real) G-vector bundle, we denote by $S\phi : SE \longrightarrow B$ the unit sphere bundle (on some metric) of ϕ, equipped with the induced G-action.

Theorem 2. Let $\xi = (X, \pi, B, K)$ be a mod 2 cohomology m-sphere bundle with Z_2-action, and let $\phi : E \longrightarrow B$ be an n-dimensional Z_2-vector bundle such that the fixed point set is a (d+1)-dimensional vector subbundle. Let $f : X \longrightarrow E$ be a fibre preserving map such that f sends $\pi^{-1}(b)^G$ into $\phi^{-1}(b)^G - 0$ and it induces a surjection $H^d(\phi^{-1}(b)^G - 0) \longrightarrow H^d(\pi^{-1}(b)^G)$ where $G = Z_2$, $b \in B$. Put

$$A(f) = \{x \in X | f(\tau x) = -\tau f(x)\} ,$$

where τ is the generator of Z_2. Then, if $P \in H^*(B)[t]$ satisfies $P|A(f) = 0$, we have

$$PC_{S(\phi)} = C_\xi Q$$

for some $Q \in H^*(B)[t]$. In particular, if $m \geqq n$ there is an injection

$$\bigoplus_{q=0}^{m-n} H^{i+q}(B) \longrightarrow H_G^{i+m-n}(A(f)) .$$

<u>Proof</u>. An equivariant fibre preserving map $h : X \longrightarrow E$ is defined by $h(x) = f(x) + \tau f\tau(x)$. Since $h(x) = 2f(x)$ for $x \in X^G$, h satisfies the same properties as f. We have $A(f) = h^{-1}(E_0)$ for the zero section E_0 of $\phi : E \longrightarrow B$. Therefore the result follows from Theorem 1.

<u>Theorem 3</u>. Let (X, π, B, K) be a mod p cohomology m-sphere bundle with Z_p-action (p prime). Let $\phi : E \longrightarrow B$ be a fibre bundle with fibre R^n, and let $\phi|E_0 : E_0 \longrightarrow B$ be a subbundle with fibre R^{n-d-1}. Assume that the local system $\{H^d(\phi'^{-1}(b))\}_{b\in B}$ over B is trivial, where $\phi' = \phi|E - E_0$. Suppose that $f : X \longrightarrow E$ is a fibre preserving map such that $f(X^G) \subset E - E_0$ and $f|\pi^{-1}(b)^G$ $(b \in B)$ induces a surjection $H^d(R^n - R^{n-d-1}) \longrightarrow H^d(K^G)$, where $G = Z_p$. Put

$$A(f) = \{x \in X | f(gx) = f(x) \quad \text{for any} \quad g \in G\} .$$

Then, if $m > n(p - 1) + d$, for any i there is an injection

$$\bigoplus_{0 \leqq q < m-\ell} H^{i+q}(B) \longrightarrow H_G^{i+m-\ell-1}(A(f) \cap f^{-1}(E_0))$$

$(\ell = n(p - 1) + d)$ given by

$$(\alpha_0, \alpha_1, \cdots, \alpha_{m-\ell-1}) \longrightarrow \sum_{0 \leqq q < m-\ell} \alpha_{m-\ell-1-q} u_q (A(f) \cap f^{-1}(E_0)) .$$

<u>Proof</u>. Put

$$Y = \{(e_1, e_2, \cdots, e_p) \in E^p | \phi(e_1) = \phi(e_2) = \cdots = \phi(e_p)\},$$

and regard Y as a G-space by $\tau(e_1, e_2, \cdots, e_p) = (e_2, \cdots, e_p, e_1)$ where τ is a fixed generator of G. Let $d : E \longrightarrow E^p$ denote the diagonal map, and put $Y_0 = dE_0$. Define $\rho : Y \longrightarrow B$ by $\rho(e_1, e_2, \cdots, e_p) = \phi(e_1)$

and put $\rho' = \rho|Y - Y_0$. Then $\eta = \{Y - Y_0, \rho', B, R^{np} - R^{n-d-1}\}$ is a mod p cohomology $(n(p-1)+d)$-sphere bundle with a Z_p-action, and the local system $\{H^d(\rho'^{-1}(b)^G)\}_{b \in B}$ is trivial. Define $\hat{f} : X \longrightarrow Y$ by $\hat{f}(x) = (f(x), f(\tau x), \cdots, f(\tau^{p-1}x))$ $(x \in X)$. Then $\hat{f}$ is an equivariant map such that $\hat{f} : X - f^{-1}(Y_0) \longrightarrow Y - Y_0$ is a fibre preserving map, $\hat{f}(X^G) \subset Y^G - Y_0$, and $\hat{f}|X^G$ induces a surjection $H^d(R^n - R^{n-d-1}) \longrightarrow H^d(K^G)$. Since $A(f) \cap f^{-1}(E_0) = \hat{f}^{-1}(Y_0)$, we have the result by Corollary of Theorem 1.

Next, we shall consider Theorem 3 in the case $\phi : E \longrightarrow B$ is a vector bundle. First we prepare some facts.

The join $X_1 \circ X_2$ of topological spaces X_1 and X_2 is the quotient space of $X_1 \times X_2 \times [0,1]$ in which $(x_1, x_2, 0)$ and $(x_1, x_2, 1)$ are identified with x_1 and x_2 respectively, where $x_1 \in X_1$, $x_2 \in X_2$. The point of $X_1 \circ X_2$ corresponding to $(x_1, x_2, t) \in X_1 \times X_2 \times [0,1]$ is denoted by $(1-t)x_1 + tx_2$. If X_1 and X_2 are G-spaces, then $X_1 \circ X_2$ is regarded as a G-space by $g((1-t)x_1 + tx_2) = (1-t)(gx_1) + t(gx_2)$, $g \in G$.

Let $\xi_i = (X_i, \pi_i, B, K_i)$ $(i = 1,2)$ be fibre bundles with G-action. Then there is the fibre bundle $\xi_1 \circ \xi_2 = (X, \pi, B, K_1 \circ K_2)$ with G-action such that $\pi^{-1}(b)$ $(b \in B)$ is the join of G-spaces $\pi_1^{-1}(b)$ and $\pi_2^{-1}(b)$.

<u>Proposition 5</u>. If ξ_i $(i = 1,2)$ are mod p cohomology m_i-sphere bundles with free Z_p-action, then $\xi_1 \circ \xi_2$ is a mod p cohomology (m_1+m_2+1)-sphere bundle with free Z_p-action, and it holds

$$c_{\xi_1 \circ \xi_2} = c_{\xi_1} c_{\xi_2} .$$

<u>Proof</u>. Let K_1', K_2' denote the part of $K_1 \circ K_2$ with $t \leq 1/2$, $t \geq 1/2$ respectively. Since K_i is a deformation retract of K_i' and $K_1' \cap K_2' = K_1 \times K_2$, the Mayer-Vietoris sequence of $\{K_1', K_2'\}$ yields an exact sequence

$$0 \longrightarrow \tilde{H}^q(K_1) \oplus \tilde{H}^q(K_2) \xrightarrow{p_1^* + p_2^*} \tilde{H}^q(K_1 \times K_2) \longrightarrow \tilde{H}^{q+1}(K_1 \circ K_2) \longrightarrow 0 ,$$

where p_i is the projection. This shows $K_1 \circ K_2 \underset{p}{\sim} S^{m_1+m_2+1}$.

Let X_1', X_2' denote the part of X with $t \leq 1/2$, $t \geq 1/2$ respectively. Then X_i' is a closed invariant subset of X, and X_i is an equivariant deformation retract of X_i'. Consider a commutative diagram

$$\begin{array}{ccc} H_G^*(X, X_1') \otimes H_G^*(X, X_2') & \xrightarrow{\smile} & H_G^*(X, X) \\ \downarrow j^* \otimes j^* & & \downarrow j^* \\ H_G^*(X) \otimes H_G^*(X) & \xrightarrow{\smile} & H_G^*(X) \end{array}$$

and the characteristic polynomials C_{ξ_i}. Then, since $C_{\xi_i}|X_i' = C_{\xi_i}|X_i = 0$, it follows that $C_{\xi_i}|X$ is in $\operatorname{Im} j^*$, and hence $C_{\xi_1}C_{\xi_2}|X = 0$. This proves $C_{\xi_1}C_{\xi_2} = C_{\xi_1 \circ \xi_2}$ by (2.1) and (2.2).

Remark. It follows that

$$c_q(\xi_1 \circ \xi_2) = \sum_{i+j=q} c_i(\xi_1)c_j(\xi_2)$$

if $p = 2$ [2], and

$$c_{2q}(\xi_1 \circ \xi_2) = \sum_{i+j=q} c_{2i}(\xi_1)c_{2j}(\xi_2),$$

$$c_{2q+1}(\xi_1 \circ \xi_2) = \sum_{i+j=2q+1} c_i(\xi_1)c_j(\xi_2)$$

if p is odd.

Let $\phi : E \longrightarrow B$ be a vector bundle, and let V be a real G-module. Then a G-vector bundle $\phi \otimes V$ is defined by $(\phi \otimes V)^{-1}(b) = \phi^{-1}(b) \otimes V$ and $g(e \otimes v) = e \otimes gv$ $(b \in B,\ g \in G,\ v \in V)$ Let L denote a 1-dimensional complex Z_p-module given by $\tau z = z \exp 2\pi\sqrt{-1}/p$, where τ is a fixed generator of Z_p. We write

$$\phi_k = \phi \otimes (L^k)_R,$$

where $(L^k)_R$ denotes the real form of the k-fold tensor product L^k of L.

Theorem 3'. In Theorem 3, assume that $\phi : E \longrightarrow B$ is a vector bundle and $\phi_0 : E_0 \longrightarrow B$ is a vector subbundle.

(i) Case $p = 2$. If $P \in H^*(B)[t]$ satisfies $P|A(f) \cap f^{-1}(E_0) = 0$, it holds that

$$PC_{S(\phi)} = C_\xi Q$$

for some $Q \in H^*(B)[t]$, where the antipodal action is taken in ϕ.

(ii) Case p odd. If $P \in H^*(B)[t,s]/(s^2)$ satisfies $P|A(f) \cap f^{-1}(E_0) = 0$, it holds that

$$p \prod_{k=1}^{(p-1)/2} c_{S(\phi_k)} = c_\xi Q$$

for some $Q \in H^*(B)[t,s]/(s^2)$.

Proof. We use the notations in the proof of Theorem 3. Then $\rho : Y \longrightarrow B$ is a Z_p-vector bundle. Identifying E with the diagonal dE, we regard $\phi : E \longrightarrow B$ and $\phi_0 : E_0 \longrightarrow B$ as subbundles of ρ. Denote by ϕ_1 the normal bundle of ϕ_0 in ρ (on some metric). Since the total space of ϕ_1 is an equivariant deformation retract of $Y - Y_0$, it follows from Proposition 4 such that $c_\eta = c_{S(\phi_1)}$. Put

$$Y_1 = \{(e_1, e_2, \cdots, e_p) \in Y | e_1 + e_2 + \cdots + e_p = 0\} .$$

Then $\rho_1 = \rho | Y_1$ is a Z_p-vector bundle, and it follows from Propositions 3 and 4 that $c_{S(\phi_1)} = c_{S(\rho_1)}$.

(i) Since there is an isomorphism $\rho_1 \cong \phi$ of Z_2-vector bundles, we have the result by Theorem 4.

(ii) Put

$$V = \{(t_1, t_2, \cdots, t_p) \in R^p | t_1 + t_2 + \cdots + t_p = 0\} ,$$

and regard V as a Z_p-module by $\tau(t_1, t_2, \cdots, t_p) = (t_2, \cdots, t_p, t_1)$. Then we have $\rho_1 \cong \phi \otimes V$ as Z_p-vector bundles. Since it is known that

$$V \cong (L \oplus L^2 \oplus \cdots \oplus L^{(p-1)/2})_R$$

there is an isomorphism

$$\rho_1 \cong \phi_1 \oplus \phi_2 \oplus \cdots \oplus \phi_{(p-1)/2}$$

as Z_p-vector bundles. Therefore $S(\rho_1)$ is Z_p-equivalent to the join $S(\phi_1) \circ S(\phi_2) \circ \cdots \circ S(\phi_{(p-1)/2})$, and hence

$$c_{S(\rho_1)} = \prod_{k=1}^{(p-1)/2} c_{S(\phi_k)}$$

by Proposition 5. Thus we have the result by Theorem 1.

Remark. It follows that

$$c_{2i-1}(\phi_k) = k^i c_{2i-1}(\phi_1), \quad c_{2i}(\phi_k) = k^i c_{2i}(\phi_1)$$

for $i = 1, 2, \cdots, n$.

References

[1] G. E. Bredon, Introduction to Compact Transformation Groups, Academic Press, 1972.

[2] P. Conner and E. Floyd, Fixed point free involutions and equivariant maps, Bull. Amer. Math. Soc., 66(1960), 416-441.

[3] A. Dold, Lectures on Algebraic Topology, Springer-Verlag, 1972.

[4] A. Dold, Parametrized Borsuk-Ulam theorems, Comment. Math. Helvetici, 63(1988), 275-285.

[5] E. Fadell, S. Husseini and P. Rabinowitz, Borsuk-Ulam theorems for arbitrary S^1 actions and applications, Trans. Amer. Math. Soc., 274(1982), 345-360.

[6] J. Jaworowski, Fibre-preserving maps of sphere-bundles into vector space bundles, Springer Lecture Notes, 886(1981), 154-162.

[7] M. Nakaoka, Equivariant point theorem for fibre-preserving maps, Osaka J. Math., 21(1984), 809-815.

[8] R. Palais, The classification of G-spaces, Memoir Amer. Math. Soc., 36(1960).

On an example of Weier

DEYU TONG

Peking University

§0. Introduction.

It is now well-known that the Nielsen fixed point theory, generally speaking, cannot give us the best lower bound for the number of fixed points in a homotopy class for surface mappings ([J1],[J2],[Z]). Weier ([W],1956) was the first to propose such a counter-example. His example is as follows. Let M be a two dimensional disc with four holes, ρ_i be the loop at $x_0 \in M$ going around the i-th hole once (see Fig.1) for $i = 1,2,3,4$. His selfmap $f : M, x_0 \to M, x_0$ is such that the induced homomorphism $f_* : \pi_1(M, x_0) \to \pi_1(M, x_0)$ is $f_*(\rho_1) = \rho_4\rho_1\rho_4^{-1}$, $f_*(\rho_2) = \rho_2$, $f_*(\rho_3) = \rho_4\rho_3\rho_4^{-1}$, $f_*(\rho_4) = 1$. He claimed that the Nielsen number $N(f)$ of f is 1, but the minimal number $MF[f]$ of fixed points in the homotopy class of f is 2, therefore $MF[f] > N(f)$. It is easy to deform f into a map $g : M \to M$ with two fixed points, each of index -1 and both in the same fixed point class. So $N(f) = 1$ and $MF[f] \leq 2$. But no proof of $MF[f] = 2$ has been published to date.

Recently, Boju Jiang transformed the fixed point problem of a surface map into a braid equation, and suggested to apply commutator analysis for more information about fixed points ([J3]). The purpose of this paper is to prove

THEOREM 1. *For Weier's map $f : M \to M$, $MF[f] = 2$.*

We shall transform our problem into a braid equation in §1, do commutator analysis to the braid equation in §2, and complete the proof of Weier's claim in §3. The Appendix contains the proof of a generalization, used in §2, of the Magnus representation theorem for free groups (cf.[MKS]) to free group-rings.

This paper is completed under the guidance and help of my advisor, Professor Boju Jiang. I wish to express my sincere thanks to him here.

§1. The algebraic formulation of the problem.

Let Δ be the diagonal in $M\times M$, x_1, x_2 be two distinct preassigned points in $intM = M - \partial M$, and $G = \pi_1(M\times M - \Delta, (x_1, x_2))$ be the pure 2-braid group of M. Let $i_1 : y \mapsto (y, x_2)$ and $i_2 : y \mapsto (x_1, y)$ be the maps from M to $M\times M$.

Let r_{kj} be the loop at x_k going around the j-th hole once (see Fig.2), and $\rho_{kj} \in G$ be the braid represented by the i_k-image of r_{kj}, for $k = 1,2$ and $j = 1,2,3,4$. Let b be the loop at x_2 going around x_1 once (see Fig.2), and $B \in G$ be the braid represented by the i_2-image of b. We use the commutator notation $(u,v) = u^{-1}v^{-1}uv$ for group elements u, v. The presentation of G is well-known (cf.[S]):

Partially supported by a TWAS grant.

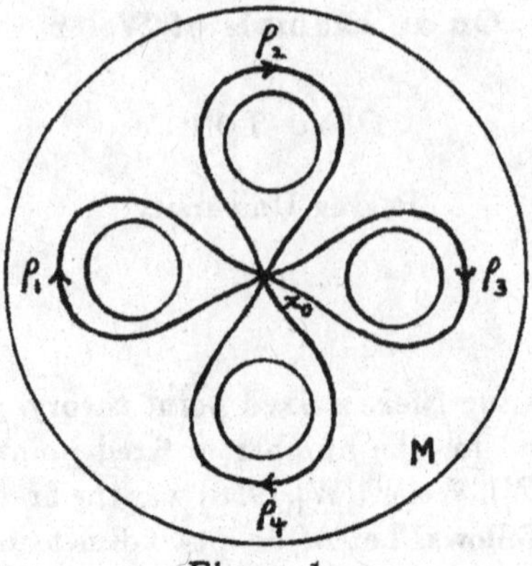

Figure 1

PROPOSITION 1.1. *The group G has a presentation with generators*

$$\rho_{11}, \rho_{12}, \rho_{13}, \rho_{14}, B, \rho_{21}, \rho_{22}, \rho_{23}, \rho_{24},$$

and relations:

$$\begin{aligned} \rho_{1i} B \rho_{1i}^{-1} &= \rho_{2i}^{-1} B \rho_{2i}, \\ \rho_{1i} \rho_{2i} \rho_{1i}^{-1} &= (\rho_{2i}, B^{-1}) \rho_{2i}, \\ \rho_{1i} \rho_{2j} \rho_{1i}^{-1} &= \rho_{2j}, && \text{if } j > i, \\ \rho_{1i} \rho_{2j} \rho_{1j}^{-1} &= (\rho_{2i}, B^{-1}) \rho_{2j} (B^{-1}, \rho_{2j}), && \text{if } j < i, \end{aligned}$$

where $i, j, = 1, 2, 3, 4$. ∎

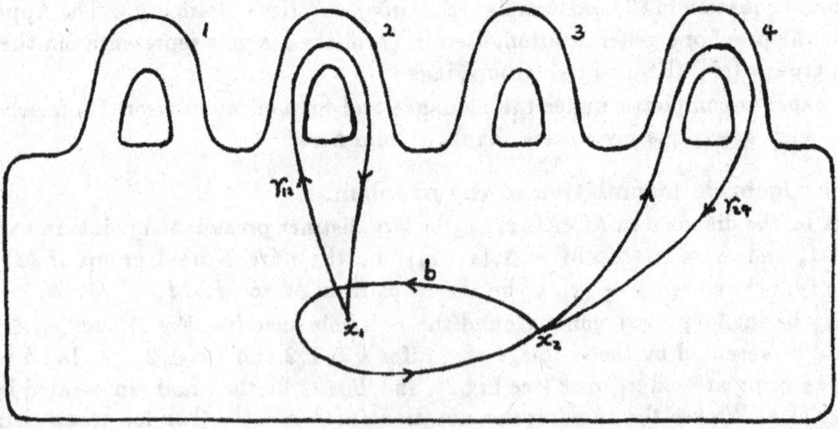

Figure 2

Let H and K be the normal subgroups of G generated by $\{B, \rho_{21}, \rho_{22}, \rho_{23}, \rho_{24}\}$ and $\{B\}$, respectively. Let π be the subgroup of G generated by $\{\rho_{21}, \rho_{22}, \rho_{23}, \rho_{24}\}$. As a corollary of Proposition 1.1, we have (cf.[J3])

PROPOSITION 1.2. *H, π, and K have presentations, respectively:*

$$H = \langle \rho_{21}, \rho_{22}, \rho_{23}, \rho_{24}, B \mid - \rangle,$$
$$\pi = \langle \rho_{21}, \rho_{22}, \rho_{23}, \rho_{24} \mid - \rangle,$$
$$K = \langle \{\alpha^{-1} B \alpha\}_{\alpha \in \pi} \mid - \rangle. \blacksquare$$

Let

$$(1\text{-}1) \quad \begin{aligned} f_1 &= f_1(\rho_{21}, \rho_{22}, \rho_{23}, \rho_{24}) = \rho_{24}\rho_{21}\rho_{24}^{-1}, \\ f_2 &= f_2(\rho_{21}, \rho_{22}, \rho_{22}, \rho_{23}, \rho_{24}) = \rho_{22}, \\ f_3 &= f_3(\rho_{21}, \rho_{22}, \rho_{23}, \rho_{24}) = \rho_{24}\rho_{23}\rho_{24}^{-1}, \\ f_4 &= f_4(\rho_{21}, \rho_{22}, \rho_{23}, \rho_{24}) = 1. \end{aligned}$$

And let

$$(1\text{-}2) \quad \begin{aligned} \sigma_0 &= f_1 f_2 f_3 f_4 \rho_{11}\rho_{12}\rho_{13}\rho_{14} B^{-1}, \\ \sigma_1 &= f_1(B^{-1}\rho_{21}, \rho_{22}, \rho_{23}, \rho_{24})\rho_{11}, \\ \sigma_2 &= f_2(B^{-1}\rho_{21}B, B^{-1}\rho_{22}, \rho_{23}, \rho_{24})\rho_{12}, \\ \sigma_3 &= f_3(B^{-1}\rho_{21}B, B^{-1}\rho_{22}B, B^{-1}\rho_{23}, \rho_{24})\rho_{13}, \\ \sigma_4 &= f_4(B^{-1}\rho_{21}B, B^{-1}\rho_{22}B, B^{-1}\rho_{23}B, B^{-1}\rho_{24})\rho_{14}. \end{aligned}$$

By the Corollary of Theorem 1 in [Z] (see also Proposition 3.3 of [J3]), we have

PROPOSITION 1.3. *If f is homotopic to a map with a single fixed point of index i, then there exist $u_1, u_2, u_3, u_4 \in K$ and $v \in H$ such that the following equation holds:*

$$(1\text{-}3) \quad u_1\sigma_1 u_1^{-1} u_2\sigma_2 u_2^{-1} u_3\sigma_3 u_3^{-1} u_4\sigma_4 u_4^{-1} \sigma_0^{-1} = vB^i v^{-1}. \blacksquare$$

We denote the left hand side of the equation (1-3) by

$$(1\text{-}4) \quad L = u_1\sigma_1 u_1^{-1} u_2\sigma_2 u_2^{-1} u_3\sigma_3 u_3^{-1} u_4\sigma_4 u_4^{-1} \sigma_0^{-1}.$$

By applying the homomorphism $G \to \mathbb{Z}$, $\rho_{ij} \mapsto 0$, $B \mapsto 1$ to both sides of (1-3), we see $-2 = i$. On the other hand, $v \in H$ can be written as $v = u\alpha$ with $u \in K$ and $\alpha \in \pi$, in fact, α is the image of v under the natural projection $H \to \pi$ annihilating B. So, to prove Theorem 1 it suffices to prove the following algebraic proposition:

PROPOSITION 1.4. *There is no $u_1, u_2, u_3, u_4, u \in K$ and $\alpha \in \pi$ such that*

$$L' = 1,$$

where L' is defined by

$$(1\text{-}5) \quad L' = Lu\alpha B^2 \alpha^{-1} u^{-1}.$$

For any group P, let

$$P = P_1 \supset P_2 \supset P_3 \supset \cdots,$$

where $P_m = (P_{m-1}, P)$, $m = 2, 3, \cdots$, be the lower central series of P (cf.[MKS]). The symbol $\equiv_m$ will denote the congruence modulo P_m. The purpose of §2 and §3 is actually to do *mod* K_2 and *mod* K_3 commutator calculus to L respectively, then to draw a contradiction from equation (1-5).

§2. Commutator calculus.

Let $\mathbf{F}[\pi]$ be the free additive group on π, i.e. the additive non-commutative free group with basis π. Left and right multiplication by elements of π are defined naturally and are distributive, and the abelianization of $\mathbf{F}[\pi]$ is the integral group-ring $\mathbb{Z}[\pi]$. This notation is adopted from [J3]. Note that $(x, y) = -x - y + x + y$ and $-(x+y) = -y - x$ for $x, y \in \mathbf{F}[\pi]$. As in [J3], we have a group isomorphism $\lambda^* : K \to \mathbf{F}[\pi]$, $gBg^{-1} \mapsto g$ for $g \in \pi$. It is an isomorphism from a multiplicative group into an additive group, thus $\lambda^*(k^{-1}) = -\lambda^*(k)$ and $\lambda^*(kh) = \lambda^*(k) + \lambda^*(h)$. It induces a group isomorphism $\lambda^*_m : K/K_m \to \mathbf{F}[\pi]/(\mathbf{F}[\pi])_m$ for $m = 2, 3, \cdots$.

We denote $y \bullet g = f_*(g^{-1})yg$ for $g \in \pi$ and $y \in \mathbf{F}[\pi]$. This defines a right action of π on $\mathbf{F}[\pi]$. Let $u_i \in K$ be as in (1-3), and let $\lambda^*(u_i) = \xi_i$. For convenience of calculation, we introduce the following changes of variables in $\mathbf{F}[\pi]$:

$$\text{(2-1)} \qquad \begin{cases} \varsigma_1 = -\xi_1 + \xi_2 - 1, \\ \varsigma_2 = -\xi_2 + \xi_3 - \rho_{24}, \\ \varsigma_3 = -\xi_3 + \xi_4, \\ \varsigma_4 = -\xi_4 + 1, \end{cases}$$

and

$$\text{(2-2)} \qquad \begin{cases} \eta_1 = \varsigma \bullet \rho_{21}^{-1} + \varsigma_2 \bullet \rho_{22}^{-1}\rho_{21}^{-1} + \varsigma_1 \bullet \rho_{23}^{-1}\rho_{22}^{-1}\rho_{21}^{-1} + \varsigma_4 \bullet \rho_{24}^{-1}\rho_{23}^{-1}\rho_{22}^{-1}\rho_{21}^{-1}, \\ \eta_2 = \varsigma_2 \bullet \rho_{22}^{-1} + \varsigma_3 \bullet \rho_{23}^{-1}\rho_{22}^{-1} + \varsigma_4 \bullet \rho_{24}^{-1}\rho_{23}^{-1}\rho_{22}^{-1}, \\ \eta_3 = \varsigma_3 \bullet \rho_{23}^{-1} + \varsigma_4 \bullet \rho_{24}^{-1}\rho_{23}^{-1}, \\ \eta_4 = \varsigma_4 \bullet \rho_{24}^{-1}. \end{cases}$$

Doing *mod* K_2 commutator calculus to L, we have

LEMMA 2.1.

$$\lambda^*(L) \equiv_2 \sum_{i=1}^{4} (\eta_i - \eta_i \bullet \rho_{2i}) - 2\rho_{24}.$$

PROOF: Let

$$\text{(2-3)} \qquad \begin{aligned} w_1 &= f_1(B^{-1}\rho_{21}, \rho_{22}, \rho_{23}, \rho_{24})f_1^{-1}, \\ w_2 &= f_2(B^{-1}\rho_{21}B, B^{-1}\rho_{22}, \rho_{23}, \rho_{24})f_2^{-1}, \\ w_3 &= f_3(B^{-1}\rho_{21}B, B^{-1}\rho_{22}B, B^{-1}\rho_{23}, \rho_{24})f_3^{-1}, \\ w_4 &= f_4(B^{-1}\rho_{21}B, B^{-1}\rho_{22}B, B^{-1}\rho_{23}B, B^{-1}\rho_{24})f_4^{-1}. \end{aligned}$$

Then $w_i \in K$ and $\sigma_i = w_i f_i \rho_{1i}$ by (1-2) for $i = 1, 2, 3, 4$. One calculates $\lambda^*(w_1) = -\rho_{24}$, $\lambda^*(w_2) = -1$, $\lambda^*(w_3) = -\rho_{24}$, $\lambda^*(w_4) = 0$.

We can write L as:

$$\text{(2-4)} \qquad L = L_1 L_2 L_3 L_4 L_5 L_0,$$

where

$$
\begin{aligned}
L_1 &= u_1 w_1,\\
L_2 &= f_1\rho_{11}(u_1^{-1}u_2w_3)\rho_{11}^{-1}f_1^{-1},\\
L_3 &= f_1\rho_{11}f_2\rho_{12}(u_2^{-1}u_3w_3)\rho_{12}^{-1}f_2^{-1}\rho_{11}^{-1}f_1^{-1},\\
L_4 &= f_1\rho_{11}f_2\rho_{12}f_3\rho_{13}(u_3^{-1}u_4w_4)\rho_{13}^{-1}f_3^{-1}\rho_{12}^{-1}f_2^{-1}\rho_{11}^{-1}f_1^{-1},\\
L_5 &= f_1\rho_{11}f_2\rho_{12}f_3\rho_{13}f_4\rho_{14}(u_4^{-1}B)\rho_{14}^{-1}f_4^{-1}\rho_{13}^{-1}\rho_{12}^{-1}f_2^{-1}\rho_{11}^{-1}f_1^{-1},\\
L_0 &= f_1\rho_{11}f_2\rho_{12}f_3\rho_{13}f_4\rho_{13}^{-1}\rho_{12}^{-1}\rho_{11}^{-1}f_4^{-1}f_3^{-1}f_2^{-1}f_1^{-1}.
\end{aligned} \tag{2-5}
$$

By the definition of λ^* and the relations in Proposition 1.1, we have

$$\lambda^*(\rho_{2i}^{\pm}u\rho_{2i}^{\mp}) = \rho_{2i}^{\pm}\lambda^*(u), \quad \lambda^*(\rho_{1i}u\rho_{1i}^{-1}) \equiv_2 \lambda^*(u)\rho_{2i}^{-1}. \tag{2-6}$$

(Cf. Lemma 4.2 in [J3]). So, directly computing $\lambda^*(L_i)$ with (2-6), by (1-1) and (2-1), we obtain

$$
\begin{aligned}
\lambda^*(L_1) &\equiv_2 -\varsigma_1-\varsigma_2-\varsigma_3-\varsigma_4-2\rho_{24},\\
\lambda^*(L_{j+1}) &\equiv_2 \varsigma_j\bullet\rho_{2j}^{-1}\cdots\rho_{21}^{-1}, \quad \text{for } j=1,2,3,4,\\
\lambda^*(L_0) &= 0.
\end{aligned}
$$

Therefore, by (2-2), we have

$$
\begin{aligned}
\lambda^*(L) &= \sum_{i=1}^{5}\lambda^*(L_i)+\lambda^*(L_0)\\
&\equiv_2 \sum_{j=1}^{4}(\varsigma_j\bullet\rho_{2j}^{-1}\cdots\rho_{21}^{-1}-\varsigma_j)-2\rho_{24}\\
&\equiv_2 \sum_{j=1}^{4}(\eta_j-\eta_j\bullet\rho_{2j})-2\rho_{24}. \blacksquare
\end{aligned}
$$

Now we introduce a quotient group $\overline{G}$ of G. $\overline{G}$ is defined by the presentation:

$$\overline{G} = \langle\, B, \rho_{14}, \rho_{24} \mid \rho_{14}B\rho_{14}^{-1} = \rho_{24}^{-1}B\rho_{24}, \rho_{14}\rho_{24}\rho_{14}^{-1} = (\rho_{24}, B^{-1})\rho_{24}\,\rangle.$$

There is a natural epimorphism $\phi: G\to\overline{G}$ with $\rho_{1i}, \rho_{2i}\mapsto 1$ for $i\neq 4$ and $\rho_{24}\mapsto\rho_{24}$, $\rho_{14}\mapsto\rho_{14}$, $B\mapsto B$. We denote $\overline{L}=\phi(L)$, $\overline{\pi}=\phi(\pi)$, and $\overline{K}=\phi(K)$. It is easy to see that

LEMMA 2.2. *(1)*

$$
\begin{aligned}
\overline{\pi} &= \langle\, \rho_{24} \mid - \,\rangle,\\
\overline{K} &= \langle\, \{\alpha^{-1}B\alpha\}_{\alpha\in\overline{\pi}} \mid - \,\rangle.
\end{aligned}
$$

(2) Let $\mathbf{F}[\overline{\pi}]$ be the free additive group on $\overline{\pi}$. Then $\lambda^* : K \to \mathbf{F}[\pi]$ induces an isomorphism $\lambda^* : \overline{K} \to \mathbf{F}[\overline{\pi}]$, $\overline{g}^{-1}B\overline{g} \mapsto \overline{g}$ for $\overline{g} \in \overline{\pi}$, such that the following diagram commutes:

$$\begin{array}{ccc} K & \xrightarrow{\phi} & \overline{K} \\ {\scriptstyle \lambda^*}\downarrow & & \downarrow{\scriptstyle \lambda^*} \\ \mathbf{F}[\pi] & \xrightarrow{\phi} & \mathbf{F}[\overline{\pi}], \end{array}$$

where $\phi : \mathbf{F}[\pi] \to \mathbf{F}[\overline{\pi}]$ is the homomorphism induced by $\phi : \pi \to \overline{\pi}$. ∎

Notation. We shall use a bar to denote the image under ϕ, e.g. $\overline{u}_i = \phi(u_i)$ and $\overline{\xi}_i = \phi(\xi_i)$.

LEMMA 2.3. *Suppose $X_1, X_2, X_3, X_4 \in \mathbb{Z}[\pi]$ such that*

$$\sum_{i=1}^{4} X_i \bullet (1 - \rho_{2i}) = 0. \tag{2-7}$$

Then there exist $a_1^0, a_2^0, a_3^0 \in \mathbb{Z}$ such that

$$\overline{X}_1 = a_1^0 \rho_{24}, \quad \overline{X}_2 = a_2^0, \quad \overline{X}_3 = a_3^0 \rho_{24}, \quad \overline{X}_4 = 0 \tag{2-8}$$

in $\mathbb{Z}[\overline{\pi}]$.

PROOF: Let

$$w = \sum_{i=1}^{4} X_i \bullet (1 - \rho_{2i}),$$

and

$$X_1' = \rho_{21}^{-1}\rho_{24}^{-1}X_1, \quad X_2' = \rho_{22}^{-1}X_2, \quad X_3' = \rho_{23}^{-1}\rho_{24}^{-1}X_3, \quad X_4' = X_4. \tag{2-9}$$

Then

$$\begin{aligned} w =& \rho_{24}(\rho_{21}X_1' - X_1'\rho_{21}) + (\rho_{22}X_2' - X_2'\rho_{22}) \\ &+ \rho_{24}(\rho_{23}X_3' - X_3'\rho_{23}) + (X_4' - X_4'\rho_{24}). \end{aligned}$$

Let $A(\mathbb{Z}, 4)$ be the ring of integral formal power series in non-commuting variables x_1, x_2, x_3, x_4 (cf. [MKS]) and $\psi : \mathbb{Z}[\pi] \to A(\mathbb{Z}, 4)$, $\rho_{2i} \mapsto 1 + x_i$ for $i = 1, 2, 3, 4$ be the extension of the Magnus representation as in Theorem A (see Appendix). Similarly, let $A(\mathbb{Z}, 1)$ be the ring of integral formal power series in one variable x_4. Then, it is easy to see that there is a ring epimorphism $\phi' : A(\mathbb{Z}, 4) \to A(\mathbb{Z}, 1)$ with $x_i \mapsto 0$ for $i \neq 4$ and $x_4 \mapsto x_4$ such that the following diagram commutes:

$$\begin{array}{ccc} \mathbb{Z}[\pi] & \xrightarrow{\psi} & A(\mathbb{Z}, 4) \\ {\scriptstyle \phi}\downarrow & & \downarrow{\scriptstyle \phi'} \\ \mathbb{Z}[\overline{\pi}] & \xrightarrow{\overline{\psi}} & A(\mathbb{Z}, 1), \end{array}$$

where $\phi : \mathbb{Z}[\pi] \to \mathbb{Z}[\overline{\pi}]$ is the homomorphism induced by $\phi : \pi \to \overline{\pi}$.

We use the Einstein summation convention in $A(\mathbb{Z}, 4)$. Let

$$Y_j = \psi(X_j') = a_j^0 + \sum_{s=1}^{+\infty} a_j^{i_1 \cdots i_s} x_{i_1} \cdots x_{i_s},$$

where $a_j^0, a_j^{i_1 \cdots i_s} \in \mathbb{Z}$. The assumption (2-7) implies $\psi(w) = 0$, i.e.

(2-10) $\quad (1 + x_4)(x_1 Y_1 - Y_1 x_1) + (x_2 Y_2 - Y_2 x_2) + (1 + x_4)(x_3 Y_3 - Y_3 x_3) - Y_4 x_4 = 0.$

Consider the coefficients of $x_4^s x_j$ and x_4 in (2-10), we have

$$\begin{aligned} a_j^{44\cdots 4} &= -a_j^{4\cdots 4} = \pm a_j^4 = 0, && \text{if } j = 1, 3, \\ a_j^{44\cdots 4} &= 0, && \text{if } j = 2, 4, \\ a_4^0 &= 0. \end{aligned}$$

So $\phi'(Y_j) = a_j^0$. Then

$$\begin{aligned} \overline{\psi}(\overline{X}_1) &= \overline{\psi}\phi(\rho_{24}\rho_{21} X_1') = \overline{\psi}(\rho_{24})\phi'(Y_1) = \overline{\psi}(a_1^0 \rho_{24}), \\ \overline{\psi}(\overline{X}_2) &= \overline{\psi}\phi(\rho_{22} X_2') = \phi'(Y_2) = a_2^0 = \overline{\psi}(a_2^0), \\ \overline{\psi}(\overline{X}_3) &= \overline{\psi}\phi(\rho_{24}\rho_{23} X_3') = \overline{\psi}(\rho_{24})\phi'(Y_3) = \overline{\psi}(a_3^0 \rho_{24}), \\ \overline{\psi}(\overline{X}_4) &= \overline{\psi}\phi(X_4') = \phi'(Y_4) = a_4^0 = 0. \end{aligned}$$

Since $\overline{\psi}$ is injective by Theorem A of the Appendix, (2-8) holds. ∎

PROPOSITION 2.4. *Suppose (1-5) holds for some $u_1, u_2, u_3, u_4, u \in K$ and $\alpha \in \pi$. Then there exist $\overline{Y}_i \in \mathbb{Z}[\overline{\pi}]$ and $a_i^0 \in \mathbb{Z}$ such that*

(2-11)
$$\begin{aligned} \overline{\eta}_1 &\equiv_2 a_1^0 \rho_{24} + 2\overline{Y}_1, \\ \overline{\eta}_2 &\equiv_2 a_2^0 + 2\overline{Y}_2, \\ \overline{\eta}_3 &\equiv_2 a_3^0 \rho_{24} + 2\overline{Y}_3, \\ \overline{\eta}_4 &\equiv_2 2\overline{Y}_4. \end{aligned}$$

PROOF: By (1-5) and Lemma 2.1,

(2-12)
$$\sum_{i=1}^{4} (\eta_i - \eta_i \bullet \rho_{2i}) - 2\rho_{24} + 2\alpha \equiv_2 0.$$

The right π-action on $\mathbf{F}[\pi]$ defined at the beginning of this section induces a right π-action on $\mathbb{Z}[\pi]$. Looking at the π-orbits of both sides of (2-12), we see α is in the same orbit as ρ_{24}. So there is $y \in \pi$ such that $\alpha = \rho_{24} \bullet y$.

Let $\frac{\partial y}{\partial \rho_{2i}}$ denote the Fox derivative of y with respect to the basis $\{\rho_{2i}\}$ (cf. [LS]). We have

$$\begin{aligned}-2\rho_{24}+2\alpha &= -2\rho_{24}\bullet(1-y)\\ &= -2\rho_{24}\bullet\Big(\sum_{i=1}^{4}\frac{\partial y}{\partial\rho_{2i}}(1-\rho_{2i})\Big)\\ &= -\sum_{i=1}^{4}\Big(2\rho_{24}\bullet\frac{\partial y}{\partial\rho_{2i}}\Big)\bullet(1-\rho_{2i})\end{aligned}$$

Therefore, by (2-12), we obtain

$$\sum_{i=1}^{4}\Big(\eta_i-2\rho_{24}\bullet\frac{\partial y}{\partial\rho_{2i}}\Big)\bullet(1-\rho_{2i})\equiv_2 0.$$

Then (2-11) follows from Lemma 2.3. ∎

§3. The proof of Weier's example.

LEMMA 3.1. *For any $\overline{u}\in\overline{K}$, we have $\lambda_*(\rho_{14}\overline{u}\rho_{14}^{-1})=\rho_{24}^{-1}+(\lambda^*\overline{u})\rho_{24}^{-1}-\rho_{24}^{-1}$ in $\mathbf{F}[\overline{\pi}]$.*

PROOF: It suffices to prove the case $\overline{u}=\rho_{24}^{k}B\rho_{24}^{-k}$ because $\{\rho_{24}^{k}B\rho_{24}^{-k}\}_{k\in\mathbb{Z}}$ is a free basis of $\overline{K}$. Noting that $(\rho_{14}B^{-1})\rho_{24}=\rho_{24}(\rho_{14}B^{-1})$ by Proposition 1.1, we have

$$\begin{aligned}&\lambda^*(\rho_{14}(\rho_{24}^{k}B\rho_{24}^{-k})\rho_{14}^{-1})\\ &=\lambda^*(\rho_{14}B\rho_{14}^{-1})+\lambda^*(\rho_{14}B^{-1}\rho_{24}^{k}B\rho_{24}^{-k}B\rho_{14}^{-1})+\lambda^*(\rho_{14}B^{-1}\rho_{14}^{-1})\\ &=\rho_{24}^{-1}+\overline{\lambda}^*(\rho_{24}^{k}\rho_{14}B\rho_{14}^{-1}\rho_{24}^{-k})-\rho_{24}^{-1}\\ &=\rho_{24}^{-1}+\rho_{24}^{k-1}-\rho_{24}^{-1}\\ &=\rho_{24}^{-1}+\lambda^*(\rho_{24}^{k}B\rho_{24}^{-k})\rho_{24}^{-1}-\rho_{24}^{-1}.\ \blacksquare\end{aligned}$$

PROPOSITION 3.2. *In $\mathbf{F}[\overline{\pi}]$ we have*

$$\begin{aligned}\lambda^*(\overline{L}')=1-\overline{\eta}_4\rho_{24}+\overline{\eta}_4+(\overline{\eta}_3,\rho_{24})-\rho_{24}+(\overline{\eta}_2,1)-1&\\ +(\overline{\eta}_1,\rho_{24})-\rho_{24}+(\overline{\eta}_4,-\rho_{24}^{-1})+2\overline{\alpha}+(2\overline{\alpha},-\overline{\xi}),&\qquad(3\text{-}1)\end{aligned}$$

where $\xi=\lambda^*(u)$.

PROOF: By (1-5),

$$\lambda^*(\overline{L}')=\lambda^*(\overline{L})+\overline{\xi}+2\overline{\alpha}-\overline{\xi}=\lambda^*(\overline{L})+2\overline{\alpha}+(2\overline{\alpha},-\overline{\xi}).\qquad(3\text{-}2)$$

By (1-1), (2-3) and (2-5), it is easy to compute that

$$\begin{aligned}&\overline{f}_1=\overline{f}_2=\overline{f}_3=\overline{f}_4=1,\\ &\overline{w}_1=\rho_{24}B^{-1}\rho_{24}^{-1},\quad \overline{w}_2=B^{-1},\quad \overline{w}_3=\rho_{24}B^{-1}\rho_{24}^{-1},\quad \overline{w}_4=1,\\ &\overline{L}_1=\overline{u}_1\overline{w}_1,\quad \overline{L}_2=\overline{u}_1^{-1}\overline{u}_2\overline{w}_2,\quad \overline{L}_3=\overline{u}_2^{-1}\overline{u}_3\overline{w}_3,\\ &\overline{L}_4=\overline{u}_3^{-1}\overline{u}_4\overline{w}_4,\quad \overline{L}_5=\rho_{14}\overline{u}_4^{-1}B\rho_{14}^{-1},\quad \overline{L}_0=1.\end{aligned}$$

So, by (2-1), (2-2) and Lemma 3.1,

$$\begin{aligned}
\lambda^*(\overline{L}_1) &= \overline{\xi}_1 - \rho_{24} = 1 - \overline{\zeta}_4 - \overline{\zeta}_3 - \rho_{24} - \overline{\zeta}_2 - 1 - \overline{\zeta}_1 - \rho_{24} \\
&= 1 - \overline{\eta}_4\rho_{24} + \overline{\eta}_4 - \overline{\eta}_3 - \rho_{24} + \overline{\eta}_3 - \overline{\eta}_2 - 1 + \overline{\eta}_2 - \overline{\eta}_1 - \rho_{24}, \\
\lambda^*(\overline{L}_2) &= -\overline{\xi}_1 + \overline{\xi}_2 - 1 = \overline{\zeta}_1 = \overline{\eta}_1 - \overline{\eta}_2, \\
\lambda^*(\overline{L}_3) &= -\overline{\xi}_2 + \overline{\xi}_3 - \rho_{24} = \overline{\zeta}_2 = \overline{\eta}_2 - \overline{\eta}_3, \\
\lambda^*(\overline{L}_4) &= -\overline{\xi}_3 + \overline{\xi}_4 = \overline{\zeta}_3 = \overline{\eta}_3 - \overline{\eta}_4, \\
\lambda^*(\overline{L}_5) &= \rho_{24}^{-1} + \overline{\zeta}_4\rho_{24}^{-1} - \rho_{24}^{-1} = \rho_{24}^{-1} + \overline{\eta}_4 - \rho_{24}^{-1}, \\
\lambda^*(\overline{L}_0) &= 0.
\end{aligned}$$

Our conclusion now follows from (2-4) and (3-2). ∎

Let A be the additive group obtained from $\mathbf{F}[\overline{\pi}]$ by adding the relations $2\rho_{24}^k = 0$ for all $k \in \mathbb{Z}$, and let $\theta : \mathbf{F}[\overline{\pi}] \to A$ be the projection. We shall use a tilde to denote the image under θ, e.g. $\widetilde{\eta} = \theta(\eta)$ and $\widetilde{\lambda}^* = \theta \circ \lambda^*$.

It is a standard fact in commutator calculus that the group $(\mathbf{F}[\overline{\pi}])_2/(\mathbf{F}[\overline{\pi}])_3$ is a free abelian group with basis $\{(\rho_{24}^i, \rho_{24}^j)\}_{i,j \in \mathbb{Z} \text{ and } i>j}$. Define a homomorphism

$$\delta : (\mathbf{F}[\overline{\pi}])_2/(\mathbf{F}[\overline{\pi}])_3 \to \mathbb{Z}/2\mathbb{Z}$$

by $\delta : (\rho_{24}^i, \rho_{24}^j) \mapsto i - j \ (mod 2)$. It is not difficult to check that δ induces a homomorphism $\widetilde{\delta} : A_2/A_3 \to \mathbb{Z}/2\mathbb{Z}$.

Proposition 3.3. $\widetilde{\delta}\widetilde{\lambda}^*(\overline{L}') = 1 \in \mathbb{Z}/2\mathbb{Z}$.

Proof: We first show that $(\widetilde{\eta}_1, \rho_{24}) \equiv_3 0$ in A_2. In fact, by Proposition 2.4, $\overline{\eta}_1$ can be written in the form

$$\overline{\eta}_1 = a_1^0\rho_{24} + \sum_{i=1}^{s} 2m_i\rho_{24}^{k_i} + \sum_{j=1}^{t} n_j(\overline{x}_i, \overline{y}_j)$$

with all $\overline{x}_i, \overline{y}_j \in \mathbf{F}[\overline{\pi}]$. Hence

$$\widetilde{\eta}_1 = a_1^0\rho_{24} + \sum_{j=1}^{t} n_j(\widetilde{x}_j, \widetilde{y}_j),$$

so that

$$(\widetilde{\eta}_1, \rho_{24}) \equiv_3 (a_1^0\rho_{24}, \rho_{24}) \equiv_3 0.$$

Similarly we can show that $(\widetilde{\eta}_2, 1) \equiv_3 (\widetilde{\eta}_3, \rho_{24}) \equiv_3 0$ in A_2, and $\widetilde{\eta}_4 \in A_2$. The latter implies $(\widetilde{\eta}_4, \rho_{24}^{-1}) \equiv_3 0$, and $\widetilde{\delta}(\widetilde{\eta}_4\rho_{24}) = \widetilde{\delta}(\widetilde{\eta}_4)$. Substituting these into (3-1) we get

$$\widetilde{\lambda}^*(\overline{L}') \equiv_3 1 - \widetilde{\eta}_4\rho_{24} + \widetilde{\eta}_4 - \rho_{24} - 1 - \rho_{24} \equiv_3 -\widetilde{\eta}_4\rho_{24} + \widetilde{\eta}_4 + (\rho_{24}, 1)$$

and

$$\widetilde{\delta}\widetilde{\lambda}^*(\overline{L}') = \widetilde{\delta}((\rho_{24}, 1)) = 1 \in \mathbb{Z}/2\mathbb{Z}. \quad ∎$$

The Proposition 3.3 certainly implies Proposition 1.4. This completes the proof of our Theorem 1.

Appendix: A representation theorem.

Let $F(r) = \langle e_1, \cdots, e_r \mid - \rangle$ be a free group of rank r, $\mathbb{Z}[F(r)]$ be the integral group-ring of $F(r)$, and $A(\mathbb{Z}, r)$ be the $\mathbb{Z}$-algebra of formal power series in non-commuting variables $x_1, \cdots, x_r$ as in Chapter 5 of [MKS]. Let $\psi_F : F(r) \to A(\mathbb{Z}, r)$, $e_i \mapsto 1 + x_i$ be the Magnus representation which is a monomorphism into the multiplicative group of the units of $A(\mathbb{Z}, r)$. A special case ($r = 1$) of the following theorem is used in the proof of Lemma 2.3.

THEOREM A. *The group representation* $\psi_F : F(r) \to A(\mathbb{Z}, r)$ *extends to a ring monomorphism* $\psi : \mathbb{Z}[F(r)] \to A(\mathbb{Z}, r)$.

PROOF: It suffices to prove that ψ is injective.

Suppose $Y \in \mathbb{Z}[F(r)]$, and $\psi(Y) = 0$. If $Y \neq 0$, we can write it uniquely as

$$Y = \sum_{j=1}^{t} l_j e_{i_{j1}}^{n_{j1}} \cdots e_{i_{js(j)}}^{n_{js(j)}},$$

where $t > 0$, $s(j) \geq 0$, $n_{jk} \in \mathbb{Z} - \{0\}$, $i_{jk} \in \{1, \cdots, r\}$, $i_{jk} \neq i_{j,k+1}$, and

$$(i_{j1}, \cdots, i_{js(j)}, n_{j1}, \cdots, n_{js(j)}) \neq (i_{h1}, \cdots, i_{hs(h)}, n_{h1}, \cdots, n_{hs(h)})$$

whenever $j \neq h$. Then

$$\psi(Y) = \sum_{j=1}^{t} l_j (1 + x_{i_{j1}})^{n_{j1}} \cdots (1 + x_{i_{js(j)}})^{n_{js(j)}}.$$

For convenience we need a special ordering of the t terms. We introduce an ordering $\prec$ in $\mathbb{Z}$ by setting $0 \prec 1 \prec 2 \prec \cdots \prec -1 \prec -2 \prec \cdots$. Then we define an ordering in the set $\mathbb{Z}^\infty$ of infinite sequences of integers, still denoted by $\prec$, to be the lexicographic ordering induced by $(\mathbb{Z}, \prec)$. Finite sequences will be regarded as infinite by attaching a tail of 0's. Now rearrange the t terms in the expression of Y according to the ordering of sequences $(s(j), i_{j1}, \cdots, i_{js(j)}, n_{j1}, \cdots, n_{js(j)})$. Let τ be the largest integer such that $(s(\tau), i_{\tau 1}, \cdots, i_{\tau s(\tau)}) = (s(1), i_{11}, \cdots, i_{1s(1)})$, and let $s = s(1)$.

Recall the "binomial expansion"

$$(1 + x)^m = 1 + \sum_{m=1}^{+\infty} C_n^m x^m,$$

where C_n^m is the binomial coefficient $\frac{n(n-1)\cdots(n-m+1)}{1.2\cdots m}$. Note that $C_n^m = 0$ iff $m > n \geq 0$. Consider the coefficients of $x_{i_{11}}^{m_1} \cdots x_{i_{1s}}^{m_s}$ in $\psi(Y)$=0, we have

$$l_1 C_{n_{11}}^{m_1} \cdots C_{n_{1s}}^{m_s} = -\sum_{j=2}^{\tau} l_j C_{n_{j1}}^{m_1} \cdots C_{n_{js}}^{m_s}.$$

Set m_k to be n_{ik} if $n_{1k} > 0$. Then $C_{n_{11}}^{m_1} \cdots C_{n1_s}^{m_s} \neq 0$, so that

$$l_1 = -\sum_{j=2}^{\tau} l_j \frac{C_{n_{j1}}^{m_1} \cdots C_{n_{js}}^{m_s}}{C_{n_{11}}^{m_1} \cdots C_{n_{1s}}^{m_s}}.$$

Let $1 \le k_1 < k_2 < \cdots < k_p \le s$ be all the k's with $n_{1k} < 0$. The special ordering $\prec$ guarantees

$$\lim_{m_{k_p}\to+\infty} \cdots \lim_{m_{k_1}\to+\infty} \frac{C_{n_{j1}}^{m_1}\cdots C_{n_{js}}^{m_s}}{C_{n_{11}}^{m_1}\cdots C_{n_{1s}}^{m_s}} = 0, j = 2, 3, \cdots, r.$$

So $l_1 = 0$. This is a contradiction! ∎

REFERENCES

[J1] Jiang, B., *Fixed points and braids*, Invent. Math. **75** (1984), 69–74.

[J2] Jiang, B., *Fixed points and braids, II*, Math. Ann. **272** (1985), 249–256.

[J3] Jiang, B., *Surface maps and braid equations, I*, in "Contributions to Geometry and Topology," Springer Lecture Notes in Math. series, to appear.

[LS] Lyndon, R.C., Schupp, P.E., "Combinatorial Group Theory," Springer, Berlin, Heidelberg, New York, 1977.

[MKS] Magnus, W., Karras, A., Solita, D., "Combinatorial Group Theory," Dover, Now York, 1976.

[S] Scott, G.P., *Braid groups and the group of homeomorphisms of surfaces*, Proc. Camb. Phil. Soc. **68** (1970), 605–617.

[W] Weier, J., *Über probleme aus der topologie der ebene und der fläche*, Math. Japon. **4** (1956), 101–105.

[Z] Zhang, X.-G., *The least number of fixed points can be arbitrarily larger than the Nielsen number*, Acta Sci. Natur. Univ. Pekin. 1986:3, 15–25.

1980 *Mathematics subject classifications*: 55M20, 57M99

Department of Mathematics, Peking University, Beijing 100871, China

The Positive Solutions of Quasilinear Elliptic Boundary Value Problem

QINGYU YU AND TIAN MA

Lanzhou University

The existence of positive solutions of the following equations is an open problem

(1) $$-\Delta u = \lambda u + u e^{u^2}, \quad x \in \Omega \subset R^2,$$

(2) $$u|_{\partial\Omega} = 0.$$

In this paper, we obtain some results for the problem (1) and (2) as a special example of a more general problem. We consider the existence of positive solutions of the following quasilinear elliptic equations

(3) $$Au = g(x, u, \nabla u) + \lambda(u + f(x, u, \nabla u)), \quad x \in \Omega,$$

(4) $$u|_{\partial\Omega} = 0.$$

where $\Omega \subset R^m$ is a bounded $C^{2,\gamma}$ domain, and

$$Au = -a_{ij}(x, u, \nabla u) D_{ij} u + b_i(x, u, \nabla u) D_i u + c(x, u, \nabla u) u.$$

In [1], we also discussed the problem (3) and (4), in which we assumed that $g(x, z, \xi) = o(|z|, |\xi|)$. In this paper we shall remove this assumption and sharpen the conclusions of [1].

The existence of positive solutions of (3) and (4) is an interesting problem. It arises in a variety of situations, for example, in differential geometry, in the theory of nonlinear diffusion generated by nonlinear sources and in quantum field theory. For semilinear elliptic boundary value problems, we refer to [4] and [5].

Our results for the equations (3), (4) have no restrictions on the growth of z and ξ in the coefficients $a_{ij}(x, z, \xi), b_i(x, z, \xi), c(x, z, \xi), g(x, z, \xi), f(x, z, \xi)$. In addition, the form of (3) is rather general.

Let X be a Banach space, we consider the operator equation

(5) $$x = p(x) + h(\lambda, x).$$

Suppose that

(H_1) $p: X \to X$ and $h: X \times R \to X$ are completely continuous, $p(0) = 0$, $h(0, x) = 0$, $h(\lambda, 0) = 0$ for $x \in X, \lambda \in R$.

(H_2) There exist a real number λ_1 and a nondecreasing function $r_\lambda > 0$ of $\lambda > \lambda_1$ with the following property: for any $\epsilon > 0$, there is $q \in X, \|q\| = \epsilon$, such that the equation

$$x - p(x) - h(\lambda, x) = q$$

Research partially supported by National Natural Science Fund (PRC).

has no solution in $\overline{B}_{r_\lambda}$, where $B_{r_\lambda} = \{x \in X \mid \|x\| < r_\lambda\}$. Moreover, the equation

$$x - p(x) - h(\lambda, x) = 0$$

has no nontrivial solution in $\overline{B}_{r_\lambda}$.

(H_3) There exist real numbers $\lambda_0 < \lambda_1$ and $\varepsilon > 0$ such that for any $0 \le t \le 1$ the equation

$$x - t(p(x) + h(\lambda_0, x)) = 0$$

has no nontrivial solution in B_ε.

Denote

$$\Sigma = \{(\lambda, x) \in R \times X \mid x = p(x) + h(\lambda, x), x \neq 0\},$$

and let $\overline{\Sigma}$ be the closure of Σ in $R \times X$.

Theorem 1. Under the assumptions (H_1) – (H_3), the equation (5) has a bifurcation point λ^* in $[\lambda_0, \lambda_1]$ at least. Moreover, if $\lambda^* > \lambda_0$ is the unique bifurcation point of (5) in $[\lambda_0, \lambda_1]$, then the connected component C of $\overline{\Sigma}$ which contains $(\lambda^*, 0)$ either is unbounded in $R \times X$, or $(\{\lambda\} \times X) \cap C \neq \emptyset$ for any $\lambda \in (\lambda_0, \lambda^*)$.

Proof. We assume C is bounded. If there exists $\lambda \in (\lambda_0, \lambda^*)$ such that $(\{\lambda\} \times X) \cap C = \emptyset$, then there exists a real number $k > \max\{\lambda_1, 0\}$ such that $\partial Q_k \cap C = \emptyset$, where $Q_k = [\lambda_0, k] \times \overline{B}_k$. We take $\varepsilon > 0$ sufficiently small so that $\varepsilon < r_k$ and satisfies (H_3), where r_k is the same as in (H_2).

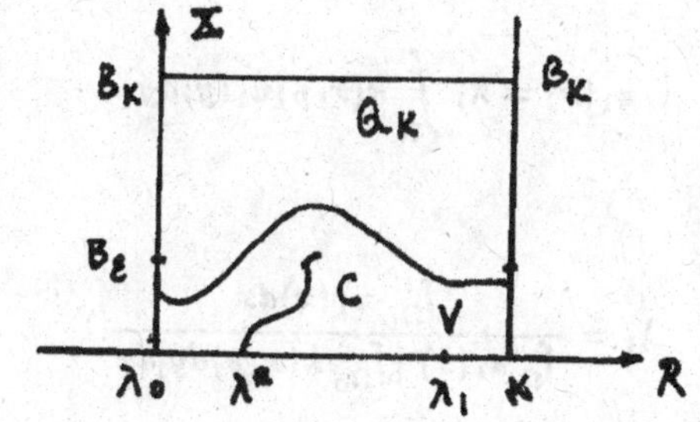

Denote $C_1 = C \cup ([\lambda_0, k] \times \{0\})$ and $D = (\{\lambda_0\} \times (\overline{B}_k \setminus B_\varepsilon)) \cup ([\lambda_0, k] \times \partial B_k) \cup (\{k\} \times \overline{B}_k \setminus B_\varepsilon))$. By ($H_2$), if λ^* is the unique bifurcation point of (5) in $[\lambda_0, \lambda_1]$, then λ^* is also the unique bifurcation point in $[\lambda_0, \infty]$. Hence, C_1 and $D \cup (\overline{\Sigma} \setminus C)$ are two disjoint closed subsets of $R \times X$. By the theory of general topology, we know that there exists a relatively open set V, such that $C_1 \subset V$ and $\overline{V} \cap (D \cup (\overline{\Sigma} \setminus C)) = \emptyset$. This means $\partial V \cap (\Sigma \cup ([\lambda_0, k] \times \{0\})) = \emptyset$ and $V_0, V_k \subset \overline{B}_\varepsilon$, where V_λ denotes the section of V at $\lambda \in [\lambda_0, k]$.

By the homotopy invariance and excision property of the topological degree, we have

$$\begin{aligned} \deg(id - p - h(\lambda_0, \cdot), B_\varepsilon, 0) &= \deg(id - p - h(\lambda_0, \cdot), V_0, 0) \\ &= \deg(id - p - h(k, \cdot), V_k, 0) \\ &= \deg(id - p - h(k, \cdot), B_{r_k}, 0). \end{aligned}$$

By (H_3),

$$\deg(id - p - h(\lambda_0, \cdot), B_\varepsilon, 0) = \deg(id, B_\varepsilon, 0) = 1.$$

By (H_2),

$$\deg(id - p - h(k, \cdot), B_{r_k}, 0) = 0.$$

It is a contradiction. The theorem is proved.

Now we use Theorem 1 to discuss the problem (3) and (4). Let λ_1 be the first eigenvalue of the following problem

$$-a_{ij}(x)D_{ij}u + b_i(x)D_iu + c(x)u = \lambda u, \quad x \in \Omega, \tag{6}$$

$$u|_{\partial\Omega} = 0, \tag{7}$$

where $a_{ij}(x) = a_{ji}(x)$, $c(x) \geq 0$, $a_{ij}, b_i, c \in C^\gamma(\overline{\Omega})$, $0 < \gamma < 1$, and there is a constant $\beta > 0$ such that

$$a_{ij}(x)\xi_i\xi_j \geq \beta|\xi|^2.$$

Lemma 2. λ_1 satisfies the following inequality

$$0 < \lambda_1 \leq \frac{1}{\sup_{\Omega_0 \subset \Omega} \inf_{x \in \Omega_0} \int_{\Omega_0} k(x,y)dy}, \tag{8}$$

where $k(x,y)$ is the Green function.

Proof. It is well known that $k(x,y)$ is a symmetric positive function and the eigenvector $u_1(x)$ corresponding to λ_1 is positive. $u_1(x)$ satisfies

$$u_1(x) = \lambda_1 \int_\Omega k(x,y)u_1(y)dy.$$

For any $\Omega_0 \subset \Omega$, we have

$$\lambda_1 = \frac{\int_{\Omega_0} u_1(x)dx}{\int_\Omega u_1(x)\,[\int_{\Omega_0} k(x,y)dy]dx}.$$

Since $u_1(x) > 0$ for $x \in \Omega$, we deduce that

$$\begin{aligned}\lambda_1 &\leq \frac{\int_{\Omega_0} u_1(x)dx}{\int_{\Omega_0} u_1(x)[\int_{\Omega_0} k(x,y)dy]dx}\\ &\leq \frac{1}{\inf_{x\in\Omega_0}[\int_{\Omega_0} k(x,y)dy]}.\end{aligned}$$

Hence the inequality (8) holds.

For the problem (3) and (4), we assume that:

(A_1) $a_{ij}(x,z,\xi) = a_{ji}(x,z,\xi)$,
$a_{ij}(x,z,\xi)\eta_i\eta_j \geq \beta|\eta|^2$.

(A_2) $a_{ij}, b_i, c, g, f \in C^\gamma(\overline{\Omega} \times I \times I^m)$, where $0 < \gamma < 1$, $I = [a,b] \subset R$ is an arbitrary bounded interval.

(A_3) For any $z \geq 0, \xi \in R^m$, we have $f(x,z,\xi), g(x,z,\xi), c(x,z,\xi) \geq 0$ and $f(x,0,0) = 0$, $g(x,0,0) = 0$.

(A_4) There exists a constant $\alpha \geq 0$ such that

$$\lim_{\substack{z\to 0\\ \xi\to 0}} \frac{g(x,z,\xi)}{z} = \alpha.$$

Denote

$$\Sigma_1 = \{(\lambda, u_\lambda) \in R \times C^{2,\gamma}(\overline{\Omega}) \mid (\lambda, u_\lambda) \text{ satisfy (3), (4) and } u_\lambda > 0 \text{ in } \Omega\}.$$

Let $\lambda_1(v)$ be the first eigenvalue of the following problem

$$\text{(9)} \qquad \begin{aligned} &-a_{ij}(x,v,\nabla v)D_{ij}u + b_i(x,v,\nabla v)D_i u + c^+(x,v,\nabla v)u = \lambda u, \quad x \in \Omega, \\ &u|_{\partial\Omega} = 0, \end{aligned}$$

where the function $c^+(x)$ is defined by

$$c^+(x) = \begin{cases} c(x), & \text{for } c(x) \geq 0, \\ 0, & \text{for } c(x) < 0, \end{cases}$$

$\lambda_1(0)$ is the first eigenvalue of (9) with $v = 0$.

Now we are in a position to state and prove our main theorems in this paper.

Theorem 3. Under the assumptions $(A_1) - (A_4)$, if $\lambda_1(0) > \alpha$, then there exists a real number $\lambda_0 > 0$ such that the problem (3) and (4) has at least a bifurcation point in $[0, \lambda_0]$ which corresponds to a positive solution. Moreover, if $\lambda^* > 0$ is the unique bifurcation point of (3) and (4) in $[0, \lambda_0]$, then the connected component C of Σ_1 which contains $(\lambda^*, 0)$ either is unbounded in $R \times C^{2,\gamma}(\overline{\Omega})$ or for any $\lambda \in (0, \lambda^*)$ the problem (3), (4) has a positive solution in $C^{2,\gamma}(\overline{\Omega})$.

Theorem 4. Under the assumptions $(A_1)-(A_4)$, if $f(x,z,\xi) \equiv 0$, then the problem (3) and (4) has a unique bifurcation point $\lambda^* = \lambda_1(0) - \alpha$, and the connected component C of Σ_1 which contains $(\lambda^*, 0)$ either is unbounded in $R \times C^{2,\gamma}(\overline{\Omega})$ or for any $\lambda \in (-\alpha, \lambda^*)$, $(\{\lambda\} \times C^{2,\gamma}(\overline{\Omega})) \cap C \neq \emptyset$.

Remark 5. If $f(x,u,\nabla u) = o(\|u\|_{C^{1,\gamma}})$ in Theorem 3, then the problem (3), (4) has a unique bifurcation point $\lambda^* = \lambda_1(0) - \alpha$ which corresponds to positive solution.

Proof of Theorem 3. By the theory of linear elliptic equations, we know that for $p \in C^\gamma(\overline{\Omega} \times I \times I^m)$ and $v \in C^{1,\gamma}(\overline{\Omega})$, the following problem has a unique solution $w \in C^{2,\gamma}(\overline{\Omega})$.

$$\text{(10)} \qquad -a_{ij}(x,v,\nabla v)D_{ij}w + b_i(x,v,\nabla v)D_i w + c^+(x,v,\nabla v)w = p(x,v,\nabla v), \quad x \in \Omega.$$

$$\text{(11)} \qquad w|_{\partial\Omega} = 0.$$

We define Tp: $C^{1,\gamma}(\overline{\Omega}) \to C^{1,\gamma}(\overline{\Omega})$ by $Tp(v) = w$.

By Theorem 10.4 in [2], Tp is a completely continuous mapping. It is easy to see that for $p_1, p_2 \in C^\gamma(\overline{\Omega} \times I \times I^m)$ and any real numbers λ_1, λ_2 the following assertion holds

$$T(\lambda_1 p_1 + \lambda_2 p_2)(v) = \lambda_1 T p_1(v) + \lambda_2 T p_2(v).$$

On the other hand, for each given $v \in C^{1,\gamma}(\overline{\Omega})$, under the boundary value condition (11), there is a Green function $k_v(x,y)$ such that the solution w of (10), (11) can be expressed as

$$w(x) = \int_\Omega k_v(x,y)p(y,v(y),\nabla v(y))dy.$$

By the uniqueness of solution, we have

$$Tp(v) = \int_\Omega k_v(x,y)p(y,v,\nabla v)dy.$$

By the strong maximum principle, the Schauder estimate and the conditions (A_1)–(A_3), the existence of positive solutions $u_\lambda \in C^{2,\gamma}(\overline{\Omega})$ of (3), (4) is equivalent to the existence of nontrivial solutions of the following equation in $C^{1,\gamma}(\overline{\Omega})$,

$$u = Tg^+(u) + \lambda Tu^+ + \lambda Tf^+(u). \tag{12}$$

where $Tu^+ = \int_\Omega k_u(x,y)u^+(y)dy$.

By (A_4), $g^+(x,u,\nabla u) - \alpha u^+ = o(\|u\|_{C^{1,\gamma}})$, and $\alpha < \lambda_1(0)$, hence there exists an $\varepsilon > 0$ such that for any $0 \le t \le 1$, the equation $u = tTg^+(u)$ has no nontrivial solution in $B_\varepsilon = \{v \in C^{1,\gamma}(\overline{\Omega}),\ \|v\|_{C^{1,\gamma}} < \varepsilon\}$.

Now we need to check the condition (H_2) of Theorem 1. It is known that for any $\lambda > \lambda_1(v)$ and $Q(x) \ge 0, x \in \Omega$, the following problem has no positive solution in $C^{2,\gamma}(\overline{\Omega})$ (see [3])

$$a_{ij}(x,v,\nabla v)D_{ij}u + b_i(x,v,\nabla v)D_iu + c^+(x,v,\nabla v)u - \lambda u = Q(x), \quad x \in \Omega,$$
$$u|_{\partial\Omega} = 0.$$

Let $\lambda_0 = [\inf_{x\in\tilde\Omega}\{\int_{\tilde\Omega} k_0(x,y)dy\}]^{-1}$, where $\tilde\Omega \subset\subset \Omega$ is an open subset. Below we prove that $\lambda_1(0) \le \lambda_0 < +\infty$. By means of Lemma 2 we can obtain that $\lambda_1(0) \le \lambda_0$.

Taking $e \in C_0^\infty(\Omega)$ with $0 \le e(x) \le 1$, $e \not\equiv 0$ and supp $e \subset \tilde\Omega$, we can get

$$u_e(x) = \int_\Omega k_0(x,y)e(y)dy = \int_{\tilde\Omega} k_0(x,y)e(y)dy$$

which is a solution of the problem

$$-a_{ij}(x,0,0)D_{ij}u + b_i(x,0,0)D_iu + c(x,0,0)u = e, \quad x \in \Omega,$$
$$u|_{\partial\Omega} = 0.$$

Clearly, we have

$$\int_{\tilde\Omega} k_0(x,y)dy \ge \int_{\tilde\Omega} k_0(x,y)e(y)dy = u_e(x).$$

By means of the strong maximum principle and $\tilde\Omega \subset\subset \Omega$, there is $\beta_0 > 0$ such that

$$\inf_{x\in\tilde\Omega} u_e(x) \ge \beta_0.$$

Hence, $\lambda_0 < +\infty$.

We define a mapping $\lambda : C^{1,\gamma}(\overline{\Omega}) \to C(\overline{\Omega})$ as follows

$$\lambda(v) = \int_{\tilde{\Omega}} k_v(x, y)dy.$$

If we can prove that $\lambda(v)$ is continuous, it is easy to get that for any $\lambda > \lambda_0$, there exists $r_\lambda > 0$ such that when $v \in C^{1,\gamma}(\overline{\Omega}), \|v\|_{C^{1,\gamma}} \le r_\lambda$, we have $\lambda_1(v) < \lambda$. Obviously, $r_\lambda > 0$ is an increasing function of $\lambda > \lambda_0$.

Using the method in the proof of Theorem 10.4 in [2], we can prove that when $v_n(x) \to v_0(x)$ in $C^{1,\gamma}(\overline{\Omega})$,

$$\int_\Omega k_{v_n}(x, y)dy \longrightarrow \int_\Omega k_{v_0}(x, y)dy$$

uniformly for $x \in \overline{\Omega}$. Hence $\lambda : C^{1,\gamma}(\overline{\Omega}) \to C(\overline{\Omega})$ is continuous.

By the proof above, we see for any $\lambda > \lambda_0$, the following equation has no solution in $\overline{B}_{r_\lambda}$,

$$u = Tg^+(u) + \lambda Tu^+ + \lambda Tf^+(u) + q(x)$$

where $q \in C^\gamma(\overline{\Omega}), q(x) > 0$ is arbitrary. Moreover, the equation (12) has the unique solution $u = 0$ in $\overline{B}_{r_\lambda}$. By Theorem 1, the theorem is proved.

Proof of Theorem 4. In fact, we consider the problem

$$(13)\quad Au = t[g(x, u, \nabla u) + \lambda u], \quad x \in \Omega,$$
$$(14)\quad u|_{\partial\Omega} = 0.$$

When $\lambda = -\alpha$, it is clear that there is a constant $\varepsilon > 0$, such that the problem (13), (14) has no positive solution in B_ε for any $0 \le t \le 1$. It means that the equation below has no nontrivial solution in B_ε for any $0 \le t \le 1$.

$$u = t(Tg^+(u) + \lambda Tu^+).$$

On the other hand, $\lambda^* = \lambda_1(0) - \alpha$ is obviously the unique bifurcation point corresponding to positive solution of the problem

$$Au = \lambda u + g(x, u, \nabla u), \quad x \in \Omega,$$
$$u|_{\partial\Omega} = 0.$$

We can complete the proof of Theorem 4 in the same way as for Theorem 3.

Example 6. We consider the following problem

$$(15)\quad \begin{aligned} &-\Delta u = u^p + \lambda u + \lambda|\nabla u|^\alpha, \quad 0 < \alpha, \quad 1 < p, \quad x \in \Omega, \\ &u|_{\partial\Omega} = 0. \end{aligned}$$

By Theorem 3, we can obtain that the problem (15) has a bifurcation point $\lambda_0 \ge 0$ which corresponds to positive solution.

Let λ_1 be the first eigenvalue of the Laplace operator, we have

<u>Corollary 7.</u> For the problem (1), (2), $\lambda^* = \lambda_1 - 1$ is a bifurcation point corresponding to positive solution, and the connected component C from $(\lambda^*, 0)$ either is unbounded or for any $\lambda \in (-1, \lambda^*)$ the problem (1), (2) has a positive solution.

<u>Corollary 8.</u> λ_1 is a bifurcation point corresponding to positive solution of the problem

$$(16) \qquad \begin{aligned} &-\Delta u = u^p + \lambda u, \quad x \in \Omega, \\ &u|_{\partial\Omega} = 0, \end{aligned}$$

where $p > 1, \Omega \subset R^m$ is a bounded $C^{2,\gamma}$ domain. Moreover the connected component C from $(\lambda_1, 0)$ either is unbounded, or for any $\lambda \in (0, \lambda_1)$, the problem (16) has a positive solution.

<u>Remark 9.</u> In Theorem 3, the coefficient $f(x, z, \xi)$ needs not to be differentiable on z and ξ. Hence Theorem 3 cannot be obtained from Rabinowitz's global bifurcation theorem.

REFERENCES

1. Q.Yu and T.Ma, The alternative principle of nonlinear operators and its applications, in: Nonlinear Analysis, ed. by W.Chen, Lanzhou University Press, 1988.

2. D.Gilbarg and N.S.Trudinger, Elliptic partial differential equation of second order, Springer-Verlag, 1977.

3. H.Amann, Fixed point equations and nonlinear eigenvalue problems in ordered Banach space, SIAM Review, **18**, 1976, 620-710.

4. H.Brezis and L.Nirenberg, Positive solutions of nonlinear elliptic equations, Comm. Pure Appl. Math. **36**, 1983, 437-477.

5. P.L.Lions, On the existence of positive solutions of semilinear elliptic equations, SIAM Review, **24**, 1982, 441-467.

Department of Mathematics, Lanzhou University, Lanzhou, P.R. of China

A relative Nielsen number for the complement

XUEZHI ZHAO

Peking University

§1. Introduction.

Classical Nielsen fixed point theory (see [1], [2]) is concerned with the determination of the minimal number $MF[f]$ of fixed points for all maps in the homotopy class of a given map $f: X \to X$. The Nielsen number $N(f)$ is always a lower bound for $MF[f]$. It is the best lower bound when X is a polyhedron which has no local cut point and is not a surface of negative Euler characteristic ([3]). However, when we consider selfmaps of a pair of spaces, i.e. $f: (X, A) \to (X, A)$ and homotopies of the form $H: (X \times I, A \times I) \to (X, A)$, $N(f)$ would be a poor lower bound for $MF[f]$. The relative Nielsen number $N(f; X, A)$ is introduced in [8], which is a lower bound for the number of fixed points for all maps in the relative homotopy class of f, and $N(f; X, A) \geq N(f)$. It is known [7] that the position of minimal fixed points on X is usually arbitrary on X for a map $f: X \to X$. But this is no longer true if $A \neq \emptyset$ and $\tilde{N}(f; X, A)$ is defined in [9], which is a lower bound for the number of fixed points on the closure of $X - A$.

It is the purpose of this paper to determine the minimal number $MF[f; X - A]$ of fixed points on the complement $X - A$. In section 2, the Nielsen number on the complementary space, $N(f; X-A)$, is defined, which is a lower bound for $MF[f; X-A]$, and has the same basic properties as $N(f; X, A)$. In section 3, we shall prove that $N(f; X, A) = MF[f; X - A]$ for a pair of compact polyhedra (X, A) if A can be "by-passed" in X and if $X - A$ has no local cut point and is not a 2-manifold. A counter-example is given to show the necessity of the "by-passing" condition. Finally, in section 4, a method to compute $N(f; X - A)$ is given.

To determine $MF[f; X-A]$ for pairs without the by-passing condition, we introduced in [13] a stronger relative Nielsen number $SN(f; X-A)$. It will be published elsewhere.

Throughout this paper, $f: (X, A) \to (X, A)$ will be a selfmap of a pair of compact polyhedra with X connected. The definition of $N(f; X - A)$ can easily be generalized to compact maps or admissible maps of pairs of ANR's in the same way as in [9] or [10].

We follow the notations and terminology of [2].

I would like to thank Professor Boju Jiang for his guidance and Professor C. Y. You for some important suggestions about this paper.

§2. Weakly common fixed point class.

Let $f: (X, A) \to (X, A)$ be a selfmap of a pair of compact polyhedra. We shall write $\bar{f}: A \to A$ for the restriction of f to A and write $f: X \to X$ if the condition that $f(A) \subset A$ is immaterial. The homotopies of f are maps of the form $H: (X \times I, A \times I) \to (X, A)$. For this map f, let $\hat{A} = \cup_1^n A_k$ be the disjoint union of all components of A which are

Partially supported by a TWAS grant.

mapped by f into themselves, and we shall write $f_k : A_k \to A_k$ for the restriction of f to A_k. Then we have a morphism of selfmaps

$$\begin{array}{ccc} A_k & \xrightarrow{f_k} & A_k \\ {\scriptstyle i_k}\downarrow & & \downarrow{\scriptstyle i_k} \\ X & \xrightarrow{f} & X \end{array}$$

where i_k is the inclusion. X and the components of $\widehat{A}$ have universal coverings

$$p : \tilde{X} \to X$$
$$p_k : \tilde{A}_k \to A_k \qquad k = 1, 2, \cdots, n.$$

For each k, we pick a lifting $\tilde{i}_k$ of i_k such that the diagram

$$\begin{array}{ccc} \tilde{A}_k & \xrightarrow{\tilde{i}_k} & \tilde{X} \\ {\scriptstyle p_k}\downarrow & & \downarrow{\scriptstyle p} \\ A_k & \xrightarrow{i_k} & X \end{array}$$

commutes. This $\tilde{i}_k$ determines a correspondence $\tilde{i}_{k,\text{lift}}$ from liftings of f_k to liftings of $f, \tilde{i}_{k,\text{lift}}(\tilde{f}_k) = \tilde{f}$ if $\tilde{i}_k \circ \tilde{f}_k = \tilde{f} \circ \tilde{i}_k$. And $\tilde{i}_{k,\text{lift}}$ induces a correspondence from lifting classes of f_k to lifting

classes of f, which is independent of the choice of the lifting $\tilde{i}_k$ of i_k and is determined by i_k itself. It is denoted

$$i_{k,\text{FPC}} : \text{FPC}(f_k) \to \text{FPC}(f)$$

where FPC(f) is the fixed point class data of f, the weighted set of lifting classes of f, the weight of a class $[\tilde{f}]$ being $index(f, p\text{Fix}\tilde{f})$ [2; Ch.III, Sec.1].

Recall that a fixed point class of a selfmap is always labelled by a lifting class, and we have

Proposition 2.1 Every fixed point class of $f_k : A_k \to A_k$ belongs to some fixed point class of $f : X \to X$. When $p_k\text{Fix}\tilde{f}_k$ is non-empty, $p_k\text{Fix}\tilde{f}_k$ belongs to $p\text{Fix}\tilde{f}$ if and only if $i_{k,\text{FPC}}[\tilde{f}_k] = [\tilde{f}]$. ∎

Thus, we shall say $p\text{Fix}\tilde{f}$ contains $p_k\text{Fix}\tilde{f}_k$ if and only if $i_{k,\text{FPC}}[\tilde{f}_k] = [\tilde{f}]$, even when $p_k\text{Fix}\tilde{f}_k$ is empty.

Definition 2.2 A fixed point class $p\text{Fix}\tilde{f}$ of $f : X \to X$ is a weakly common fixed point class of f and $\overline{f}$ if it contains a fixed point class of $f_k : A_k \to A_k$ for some k. It is an essential weakly common fixed point class of f and $\overline{f}$ if it is an essential fixed point class of f as well as a weakly common fixed point class of f and $\overline{f}$. We write $E(f, \overline{f})$ for the number of essential weakly common fixed point classes of f and $\overline{f}$.

Theorem 2.3 A fixed point x_0 of f belongs to a weakly common fixed point class of f and $\overline{f}$ if and only if there is a path α from x_0 to A such that $\alpha \simeq f \circ \alpha : I, 0, 1 \to X, x_0, A$.

Proof: "Only if". Let x_0 belong to a weakly common fixed point class $p\mathrm{Fix}\tilde{f}$ of f and $\overline{f}$. Suppose $\tilde{x}_0 \in p^{-1}(x_0)$ and $\tilde{f}(\tilde{x}_0) = \tilde{x}_0$. By assumption there exists a lifting $\tilde{f}_k$ of $f_k : A_k \to A_k$ so that $i_{k,\mathrm{lift}}(\tilde{f}_k) = \tilde{f}$. Pick a point $\tilde{a} \in \tilde{i}_k(\tilde{A}_k)$, then $\tilde{f}(\tilde{a}) \in \tilde{i}_k(\tilde{A}_k)$. Take a path $\tilde{\alpha}$ in $\tilde{X}$ from $\tilde{x}_0$ to $\tilde{a}$. Since $\tilde{X}$ is 1-connected, there is a homotopy of the form

$$\tilde{\alpha} \simeq \tilde{f} \circ \tilde{\alpha} : I, 0, 1 \to \tilde{X}, \tilde{x}_0, \tilde{i}_k(\tilde{A}_k).$$

Projecting down to X, we have

$$\alpha \simeq f \circ \alpha : I, 0, 1 \to X, x_0, A$$

where $\alpha = p \circ \tilde{\alpha}$.

"If". Suppose $x_0 \in p\mathrm{Fix}\tilde{f}, \tilde{x}_0 \in p^{-1}(x_0)$ and $\tilde{f}(\tilde{x}_0) = \tilde{x}_0$. Lift the path α from $\tilde{x}_0$ to get a path $\tilde{\alpha}$ in $\tilde{X}$. Let $a = \alpha(1) \in A_k$, and pick $\tilde{a} \in p_k^{-1}(a)$, then there is a lifting $\tilde{i}_k$ of i_k such that

$$\begin{array}{ccc} (\tilde{A}_k, \tilde{a}) & \xrightarrow{\tilde{i}_k} & (\tilde{X}, \tilde{\alpha}(1)) \\ {\scriptstyle p_k}\downarrow & & \downarrow{\scriptstyle p} \\ (A_k, a) & \xrightarrow{i_k} & (X, a) \end{array}$$

commutes [2; p.42, Proposition 1.2(i)]. Let $H : I \times I \to X$ be the homotopy from α to $f \circ \alpha$, i.e. $H(t,0) = \alpha(t), H(t,1) = f \circ \alpha(t)$. Then $\tilde{\alpha}$ determines a lifting $\tilde{H} : I \times I \to \tilde{X}$ of H. Denote β the path $\{H(1,s)\}_{0 \le s \le 1}$ in A_k. Lift the path $\beta : I \to A_k$ from $\tilde{a}$ to get a path $\tilde{\beta}$ in $\tilde{A}_k$, then $\tilde{i}_k \circ \tilde{\beta} : I \to \tilde{X}$ is a lifting from $\tilde{\alpha}(1)$ in $\tilde{X}$ of the path $i_k \circ \beta$. By the unique lifting property of covering spaces, we have $\tilde{H}(1,s) = \tilde{i}_k \circ \tilde{\beta}(s)$. Then $\tilde{i}_k \circ \tilde{\beta}(1) = \tilde{H}(1,1) = \tilde{f} \circ \tilde{\alpha}(1)$, there exists a unique lifting $\tilde{f}_k$ of $f_k : A_k \to A_k$ such that $\tilde{f}_k(\tilde{a}) = \tilde{\beta}(1)$. Thus $\tilde{i}_k \circ \tilde{f}_k(\tilde{a}) = \tilde{f} \circ \tilde{i}_k(\tilde{a})$. By the unique lifting property of covering spaces, we have $\tilde{i}_k \circ \tilde{f}_k = \tilde{f} \circ \tilde{i}_k$, i.e. $\tilde{f} = i_{k,\mathrm{lift}}(\tilde{f}_k)$. This implies $[\tilde{f}] = i_{k,\mathrm{FPC}}[\tilde{f}_k]$, i.e. $p\mathrm{Fix}\tilde{f}$ is a weakly common fixed point class of f and $\overline{f}$. ∎

Corollary 2.4 A fixed point class of $f : X \to X$ containing a fixed point on A is a weakly common fixed point class of f and $\overline{f}$. ∎

In [8], the number $N(f, \overline{f})$ of essential common fixed point classes of f and $\overline{f}$ is introduced, and we have

Proposition 2.5 $N(f, \overline{f}) \le E(f, \overline{f}) \le N(f)$.

Proof: By Corollary 2.4 and [8, Definition 2.1], we know that a common fixed point class is always a weakly common fixed point class. This implies the left inequality. The right one is obvious. ∎

In general, $E(f, \overline{f})$ is different from $N(f, \overline{f})$. A simple example is the identity map $f : (D^2, S^1) \to (D^2, S^1)$ of the pair of a 2-disc and its boundary, it is easy to see $N(f) = 1$ and $N(\overline{f}) = 0$, then $N(f, \overline{f}) = 0$, but $E(f, \overline{f}) = 1$.

Theorem 2.6 (Homotopy invariance) If two maps $f \simeq g : (X, A) \to (X, A)$ are homotopic, then $E(f, \overline{f}) = E(g, \overline{g})$.

Proof: Let $H = \{h_t, \overline{h}_t\} : (X, A) \to (X, A)$ be a homotopy between f and g. There exists an index-preserving bijection $\{h_t\} : \mathrm{FPC}(f) \to \mathrm{FPC}(g)$. It suffices to show $\{h_t\}$ sends weakly common fixed point classes to weakly common fixed point classes.

Let $p\mathrm{Fix}\tilde{f}$ be a weakly common fixed point class of f and $\overline{f}$, then there exist a component A_k of $\widehat{A}$ and a lifting class $[\tilde{f}_k]$ of $f_k : A_k \to A_k$ such that $i_{k,\mathrm{FPC}}[\tilde{f}_k] = [\tilde{f}]$. Let $\{h_{k,t}\}$, which is the restriction of h_t to A_k, send $[\tilde{f}_k]$ to $[\tilde{g}_k]$, then we have a commutative diagram

$$\begin{array}{ccc} [\tilde{f}_k] & \xrightarrow{\{h_{k,t}\}} & [\tilde{g}_k] \\ {\scriptstyle i_{k,\mathrm{FPC}}}\downarrow & & \downarrow{\scriptstyle i_{k,\mathrm{FPC}}} \\ [\tilde{f}] & \xrightarrow{\{h_t\}} & [\tilde{g}] \end{array}$$

[2; p.47, Theorem 1.16]. Thus, $\{h_t\}$ sends $[\tilde{f}]$ to $[\tilde{g}] = i_{k,\mathrm{FPC}}[\tilde{g}_k]$, we get the conclusion. ∎

We also have the commutativity and homotopy type invariance of $E(f,\overline{f})$, similar to [8, Theorem 3.4 and Theorem 3.5].

Definition 2.7 The number of essential fixed point classes of $f : X \to X$ which are not weakly common fixed point classes is called the Nielsen number of f on the complementary space $X - A$, denoted $N(f; X - A)$.

By definition, $N(f; X - A)$ is a non-negative integer, and $N(f; X - A) + E(f, \overline{f}) = N(f)$. Hence, it shares the basic properties of $E(f,\overline{f})$: homotopy invariance, commutativity and homotopy type invariance. Its importance lies in:

Theorem 2.8 Any map $f : (X, A) \to (X, A)$ has at least $N(f; X - A)$ fixed points on $X - A$. Thus $N(f; X - A) \leq MF[f; X - A]$.

Proof: Recall that each essential fixed point class contains at least one fixed point. By Corollary 2.4, we shall get the conclusion. ∎

Theorem 2.9 $N(f; X - A) \leq \tilde{N}(f; X, A)$. ∎

§3. Minimum theorem.

In this section we consider the question whether $N(f; X - A) = MF[f; X - A]$, i.e. whether there exists a map $g : (X, A) \to (X, A)$ homotopic to a given map $f : (X, A) \to (X, A)$ which has precisely $N(f; X - A)$ fixed points on $X - A$. The proof of the minimum theorem is based on the proof of [8, Theorem 6.2] or, for the case $A = \emptyset$, of [3, Theorem 5.2].

Lemma 3.1 Let (X, A) be a pair of compact polyhedra, then any map $f : (X, A) \to (X, A)$ is homotopic to a map $g : (X, A) \to (X, A)$ such that g is fix-finite and all fixed points on $X - A$ lie in maximal simplexes.

Proof: Compare [8, Theorem 4.1], in which the additional assumption on A is responsible for the additional conclusion (i) that we do not need. ∎

Definition 3.2 ([8, Definition 5.1]) A subspace A of a space X can be by-passed if every path in X with end points in $X - A$ is homotopic to a path in $X - A$ keeping end points fixed.

Theorem 3.3 ([8, Theorem 5.2]) Let (X, A) be a pair of spaces and X be arcwise connected. Then A can be by-passed in X if and only if $X - A$ is arcwise connected and $i_\pi : \pi_1(X - A) \to \pi_1(X)$ is onto. ∎

The easy proof of the next Lemma is left to the reader.

Lemma 3.4 Let (X, A) be a pair of compact polyhedra such that A can be by-passed in X, then the boundary $\mathrm{Bd}A_k$ of every component of A is a connected complex, and

$\mathrm{Bd}A_k$ is 0-dimensional only if A_k is 1-connected. ∎

Lemma 3.5 Let (X,A) be a pair of compact polyhedra, where A can be by-passed in X. A fixed point $x_0 \in X - A$ of a selfmap $f:(X,A) \to (X,A)$ belongs to a weakly common fixed point class of f and $\overline{f}$ if and only if there exists a path

$$\alpha : I, 0, I - \{1\}, 1 \to X, x_0, X - A, A$$

such that

$$\alpha \simeq f \circ \alpha : I, 0, 1 \to X, x_0, A.$$

Moreover, when f is fix-finite, we can get either $\alpha(1) \notin \mathrm{Fix}f$ or the homotopy with the form

$$\alpha \simeq f \circ \alpha : I, 0, 1 \to X, x_0, \alpha(1).$$

Proof: "If". It is obvious by Theorem 2.3.

"Only if". From Theorem 2.3 there a path $\beta : I, 0, 1 \to X, x_0, A_k$, with $\beta \simeq f \circ \beta : I, 0, 1 \to X, x_0, A_k$, for some k. Let a be an arbitrary point of $\mathrm{Bd}A_k$. Since A is a subpolyhedron of X, there is a small arc $\gamma : I \to X$ touching A only in its starting point a. Let β' be a path in A_k from $\beta(1)$ to a. Then the path $\beta\beta'\gamma$ has both ends in $X - A$, so by Definition 3.2 there is a path δ in $X - A$ such that δ is homotopic to $\beta\beta'\gamma$ in X rel end points. Define $\alpha = \delta\gamma^{-1}$. Then it is evident that $\alpha([0,1)) \in X - A$ and $\alpha \simeq \beta\beta' \simeq \beta \simeq f \circ \beta \simeq f \circ (\beta\beta') \simeq f \circ \alpha : I, 0, 1 \to X, x_0, A$.

We have full freedom in choosing $a \in \mathrm{Bd}A_k$, preferrably $a \notin \mathrm{Fix}f$. When $\mathrm{Bd}A_k \subset \mathrm{Fix}f$ and $\mathrm{Bd}A_k$ is 0-dimensional, by Lemma 3.4 A_k is 1-connected. In this case $\alpha \simeq f \circ \alpha : I, 0, 1 \to X, x_0, a$. ∎

Lemma 3.6 Let (X,A) be a pair of compact polyhedra, where every component of $X - A$ has no local cut point and is not a 2-manifold. Let x_0 be an isolated fixed point of $f:(X,A) \to (X,A)$ lying in a maximal simplex of $X - A$. Suppose $\delta : I, 0, I - \{1\}, 1 \to X, x_0, X - A, A$ is a path from x_0 to $a \in A$ with $\mathrm{Fix}f \cap \delta(I) = \{x_0\}$ and

$$\delta \simeq f \circ \delta : I, 0, 1 \to X, x_0, A.$$

If f is fix-finite, then f is homotopic to a map $f' : (X,A) \to (X,A)$ with $\mathrm{Fix}f' = (\mathrm{Fix}f - \{x_0\}) \cup \{a\}$.

Proof: As in [8, Lemma 5.3], we can assume that δ is a normal PL arc in (X,A) (for definition see [8, p.469]). Apply the method of [3, Lemma 3.4] to ensure that δ satisfies the conditions (α) and (β) of [3, Lemma 3.4], where τ and σ_1 are chosen in $X - A$. Let $H : (I \times I, \{1\} \times I) \to (X,A)$ be the homotopy from δ to $f \circ \delta$, i.e. $H(t,0) = \delta(t)$ and $H(t,1) = f \circ \delta(t)$. Since (X,A) is a simplicial pair, we may assume that a is a vertex of X and $\beta = \{H(1,s)\}_{0 \le s \le 1}$ is a PL arc in A from a to $f(a)$. Choose a conic neighborhood $U(a,\varepsilon) = \{x \in X \mid d(x,a) < \varepsilon\}$ of a such that $f(\mathrm{Cl}U) \cap \mathrm{Cl}U = \emptyset$ and $U \cap \beta(I)$ is a line segment. We define a homotopy $G : (X \times I, A \times I) \to (X,A)$ by

$$G(x,t) = \begin{cases} f(x) & \text{if } x \notin U(a,\varepsilon t) \\ f((\frac{2}{\varepsilon t}d(x,a) - 1)x + (2 - \frac{2}{\varepsilon t}d(x,a))a) & \\ & \text{if } 0 < \frac{\varepsilon t}{2} < d(x,a) \le \varepsilon t \\ H(1, 1 - t + \frac{2}{\varepsilon}d(x,a)) & \text{if } 0 \le d(x,a) \le \frac{\varepsilon t}{2} \end{cases}$$

(cf. [5; p.75, Lemma 1.1]). We define $g : (X, A) \to (X, A)$ by $g(x) = G(x, 1)$, then g is homotopic to f with $\text{Fix}g = \text{Fix}f \cup \{a\}$ and $g \circ \delta = (f \circ \delta)\beta^{-1} \simeq \delta$ rel$\{0, 1\}$.

Choose $\epsilon' > 0$ so that [8, Lemma 6.1] applies. Define a path $\delta' : I \to X$ homotopic to δ by

$$\delta'(t) = \delta(t + \lambda \sin t\pi)$$

where $\lambda > 0$ is so small that $d(\delta(t), \delta'(t)) < \epsilon'$ for all $t \in I$. Then δ' and $g \circ \delta$ are special paths with respect to δ and are homotopic. As in the proof of [3, Lemma 5.1], it follows that δ' and $g \circ \delta$ are specially homotopic. Thus $g|_{\delta(I)} : \delta(I) \to X$ and $\delta' \cdot \delta^{-1} : \delta(I) \to X$ are specially homotopic [3, p.755]. Apply [8, Lemma 6.1] to get a map $f' : (X, A) \to (X, A)$ homotopic to g with $\text{Fix}f' = \text{Fix}g - \{x_0\}$. Thus, $\text{Fix}f' = (\text{Fix}f - \{x_0\}) \cup \{a\}$. ∎

Implicit in the above proof is also the following

Lemma 3.7 Let (X, A) be as in Lemma 3.6. Let x_0 and x_1 be two isolated fixed points of $f : (X, A) \to (X, A)$ with x_0 lying in a maximal simplex of $X - A$ and x_1 lying on BdA. Suppose $\delta : I, 0, I - \{1\}, 1 \to X, x_0, X - A, x_1$ is a path with $\text{Fix}f \cap \delta(I) = \{x_0, x_1\}$ and $\delta \simeq f \circ \delta$ rel$\{0, 1\}$. Then f is homotopic to a map $f' : (X, A) \to (X, A)$ with $\text{Fix}f' = \text{Fix}f - \{x_0\}$. ∎

We now prove the main theorem of this section.

Theorem 3.8 Let (X, A) be a pair of compact polyhedra, where A can be by-passed in $X, X - A$ has no local cut point and is not a 2-manifold. Then every map $f : (X, A) \to (X, A)$ is homotopic to a map $g : (X, A) \to (X, A)$ with $N(f; X - A)$ fixed points on $X - A$.

Proof: By Lemma 3.1, we can assume that f is fix-finite and that all fixed points of f on $X - A$ lie in maximal simplexes. We can unite fixed points on $X - A$ belonging to the same fixed point class of f as in the proof of [8, Theorem 6.2]. Move all fixed points of f on $X - A$ belonging to weakly common fixed point classes to BdA by using Lemma 3.5 and Lemma 3.6, and delete fixed point classes which consist of a single fixed point on $X - A$ of index zero by the usual method [1; p.123, Theorem 4]. Then we get a map $g : (X, A) \to (X, A)$ with $N(f; X - A)$ fixed points on $X - A$. ∎

A relative Nielsen number $N(f; X, A)$ is defined in [8], which is a lower bound for the number of fixed points for all maps in the homotopy class of $f : (X, A) \to (X, A)$ and

$$N(f; X, A) = N(f) + N(\overline{f}) - N(f, \overline{f}).$$

We can combine Theorem 3.8 with the Minimum Theorem [8, Theorem 6.2] and [9, Theorem 5.1]. Recall from [8] that a space Y is called a Nielsen space if every map $f : Y \to Y$ is homotopic to a map $g : Y \to Y$ which has $N(f)$ fixed points, and if these fixed points can lie anywhere in Y. We have

Theorem 3.9. Let (X, A) be a pair of compact polyhedra satisfying the conditions of Theorem 3.8. Suppose every component of A is a Nielsen space with non-empty interior. Then every map $f : (X, A) \to (X, A)$ is homotopic to a map $g : (X, A) \to (X, A)$ with $N(f; X, A)$ fixed points on X, $\tilde{N}(f; X, A)$ fixed points on Cl$(X - A)$ and $N(f; X - A)$ fixed points on $X - A$.

Proof: By [9, Theorem 5.1], we can homotope f to a map $g : (X, A) \to (X, A)$ with $N(f; X, A)$ fixed points on X and $\tilde{N}(f; X, A)$ fixed points on Cl$(X - A)$. Move all fixed points of g on $X - A$ belonging to weakly common fixed point classes to BdA, then we get the conclusion. ∎

In Theorem 3.8, it is necessary to assume $X - A$ has no local cut point and is not a 2-manifold, since it is needed even in the case $A = \emptyset$, see [4] and [11]. An example which shows that Theorem 3.9 is false without by-passing follows.

Example 3.10 Let $X = \{e^{\theta i}\}_{0 \le \theta \le 2\pi}$ be the unit circle in the complex plane, and let A be two points $\{1, e^{\pi i}\}$. The selfmap $f : (X, A) \to (X, A)$ is define by

$$f(e^{\theta i}) = e^{-2|\theta - \pi| i}$$

It is easy to see $\deg(f) = 0$, then $N(f) = R(f) = 1$, i.e. $f : X \to X$ has only one fixed point class that is essential. This class contains the fixed point $\{1\}$ of $\overline{f} : A \to A$. By Corollary 2.4, it is a weakly common fixed point class of f and $\overline{f}$. Thus, we get $N(f; X - A) = 0$.

On the other hand, if a map $g : (X, A) \to (X, A)$ is homotopic to f, then there exists a point $e^{\xi i}$ $(\pi < \xi < 2\pi)$ such that $g(e^{\xi i}) = e^{\xi i}$. As f has two fixed points, we have $MF[f; X - A] = 1$.

Higher-dimensional examples can easily be obtained by using a solid torus $S^1 \times D^n$, where D^n is a n-ball.

From this we know $N(f; X - A)$ is not equal to $MF[f; X - A]$ in general.

§4. Computation of $N(f; X - A)$.

Theorem 4.1 Let $f : (X, A) \to (X, A)$ be a selfmap of a pair of compact polyhedra. If there is a component A_k of $\widehat{A}$ such that $i_{k,\pi} : \pi_1(A_k) \to \pi_1(X)$ is onto, then $N(f; X - A) = 0$.

Proof: By [2; p.46, Theorem 1.14(i)], $i_{k,\mathrm{FPC}}$ is surjective. Then every fixed point class of $f : X \to X$ is a weakly common fixed point class of f and $\overline{f}$. ∎

The computation of $N(f; X - A)$ is similar to the corresponding results for the Nielsen number $N(f)$ [2, Ch.II]. Pick a base point $a_k \in A_k$ for each $A_k \subset \widehat{A}$ and a base point $x_0 \in X$. Recall that points of universal covering spaces $\tilde{A}_k$ and $\tilde{X}$ of A_k and X are respectively in one-to-one correspondence with the path classes in A_k and X starting from a_k and x_0. Under this identification, let $\tilde{a_k} = \langle e_k \rangle \in \tilde{A}_k$ and $\tilde{x_0} = \langle e \rangle \in \tilde{X}$ be the constant paths. Pick a path w_k in A_k from a_k to $f_k(a_k)$ for each k and a path w_0 in X from x_0 to $f(x_0)$. Then there are unique liftings $\tilde{f}_k$ and $\tilde{f}$ of maps $f_k : A_k \to A_k$ and $f : X \to X$ such that $\tilde{f}_k(\langle e_k \rangle) = \langle w_k \rangle \in \tilde{A}_k$ and $\tilde{f}(\langle e \rangle) = \langle w_0 \rangle \in \tilde{X}$. If the liftings $\tilde{f}_k$ and $\tilde{f}$ are chosen as references, then we have

Lemma 4.2 There exist one-to-one correspondences

$$\rho_k : \nabla(f_k; a_k, w_k) \longrightarrow \mathrm{FPC}(f_k)$$
$$\rho : \nabla(f; x_0, w_0) \longrightarrow \mathrm{FPC}(f)$$

defined by

$$\rho_k([\alpha]) = [\alpha \circ \tilde{f}_k]$$
$$\rho([\beta]) = [\beta \circ \tilde{f}]$$

where $\nabla(f_k; a_k, w_k)$ is the set of $f_{k,\pi}$-conjugacy classes in $\pi_1(A_k, a_k)$ and $\nabla(f; x_0, w_0)$ is the set of f_π-conjugacy classes in $\pi_1(X, x_0)$ [12; p.218 and p.220].

Proof: See [2; p.26, Theorem 1.7]. ▮

Pick a path u_k from x_0 to a_k and take a lifting $\tilde{i}_k$ of $i_k : A_k \to X$ such that $\tilde{i}_k(\langle e_k\rangle) = \langle u_k\rangle$. Define a function $\nu_{k,\pi} : \pi_1(A_k, a_k) \to \pi_1(X, x_0)$ by

$$\begin{aligned}\nu_{k,\pi}\langle\alpha\rangle &= \langle u_k(i_k \circ \alpha)w_k(f \circ u_k)^{-1}w_0^{-1}\rangle \\ &= (\tilde{i}_{k,\pi}\langle\alpha\rangle)\langle\gamma_k\rangle\end{aligned}$$

where $\langle\gamma_k\rangle = \langle u_k w_k (f \circ u_k)^{-1} w_0^{-1}\rangle$. Note that $\nu_{k,\pi}$ is the composition of the homomorphism $\tilde{i}_{k,\pi}$ followed by a right translation.

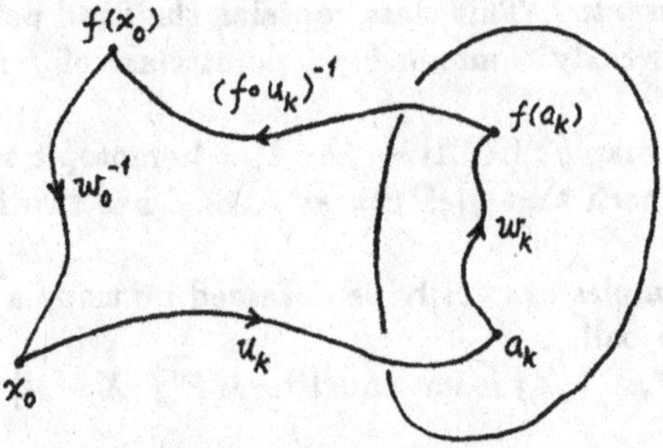

Lemma 4.3 The function $\nu_{k,\pi}$ induces a transformation

$$\nu_k : \nabla(f_k; a_k, w_k) \longrightarrow \nabla(f; x_0, w_0)$$

and ν_k is independent of the choice of the path u_k.

Proof: See [12; Lemma 1.2]. ▮

Theorem 4.4 Let $f : (X, A) \to (X, A)$. An element in $\nabla(f; x_0, w_0)$ corresponds to a weakly common fixed point class if and only if it is in the image of some ν_k.

Proof: By Lemma 4.2, it suffices to check that the diagram

$$\begin{array}{ccc}\nabla(f_k; a_k, w_k) & \xrightarrow{\rho_k} & \mathrm{FPC}(f_k) \\ \nu_k\big\downarrow & & \big\downarrow i_{k,\mathrm{FPC}} \\ \nabla(f; x_0, w_0) & \xrightarrow{\rho} & \mathrm{FPC}(f)\end{array}$$

commutes.

Let $[\langle\alpha\rangle] \in \nabla(f_k; a_k, w_k)$, then

$$\begin{aligned}\rho \circ \nu_k[\langle\alpha\rangle] &= \rho[\langle u_k(i_k \circ \alpha)w_k(f \circ u_k)^{-1}w_0^{-1}\rangle] \\ &= [\langle u_k(i_k \circ \alpha)w_k(f \circ u_k)^{-1}w_0^{-1}\rangle\tilde{f}]\end{aligned}$$

$$\begin{aligned}&\langle u_k \circ \alpha)w_k(f \circ u_k)^{-1}w_0^{-1}\rangle\tilde{f} \circ \tilde{i}_k\langle e_k\rangle \\ &= \langle u_k(i_k \circ \alpha)w_k(f \circ u_k)^{-1}w_0^{-1}\rangle\langle w_0(f \circ u_k)\rangle \\ &= \langle u_k(i_k \circ \alpha)w_k\rangle\end{aligned}$$

and

$$\tilde{i}_k \circ (\langle\alpha\rangle \tilde{f}_k)\langle e_k\rangle = \tilde{i}_k \circ \langle \alpha w_k\rangle = \langle u_k(i_k \circ \alpha) w_k\rangle.$$

By the unique lifting property of covering space, we have

$$\langle u_k(i_k \circ \alpha) w_k (f \circ u_k)^{-1} w_0^{-1}\rangle \tilde{f} \circ \tilde{i}_k = \tilde{i}_k \circ (\langle\alpha\rangle \tilde{f}_k),$$

i.e.

$$\tilde{i}_{k,\mathrm{lift}}(\langle\alpha\rangle \tilde{f}_k) = \langle u_k(i_k \circ \alpha) w_k (f \circ u_k)^{-1} w_0^{-1}\rangle \tilde{f}.$$

Hence

$$i_{k,\mathrm{FPC}}[\langle\alpha\rangle \tilde{f}_k] = [\langle u_k(i_k \circ \alpha) w_k (f \circ u_k)^{-1} w_0^{-1}\rangle \tilde{f}],$$

and we get

$$i_{k,\mathrm{FPC}} \circ \rho_k[\langle\alpha\rangle] = \rho \circ \nu_k[\langle\alpha\rangle]. \quad \blacksquare$$

Let us consider the commutative diagram

$$\begin{array}{ccccc}
\pi_1(A_k, a_k) & \xrightarrow{\theta_k} & H_1(A_k) & \xrightarrow{\eta_k} & \mathrm{coker}(1 - f_{k*} : H_1(A_k) \to H_1(A_k)) \\
\downarrow{\scriptstyle i_{k,\pi}} & & \downarrow{\scriptstyle i_{k*}} & & \downarrow{\scriptstyle i_{k*}} \\
\pi_1(X, a_k) & \xrightarrow{\theta} & H_1(X) & \xrightarrow{\eta} & \mathrm{coker}(1 - f_* : H_1(X) \to H_1(X))
\end{array}$$

where θ_k, θ are abelianization and η_k, η are the natural projection.

<u>Lemma 4.5</u> The compositions $\eta_k \circ \theta_k$ and $\eta \circ \theta$ induce correspondences

$$\tau_k : \nabla(f_k; a_k, w_k) \longrightarrow \mathrm{coker}(1 - f_{k*})$$
$$\tau : \nabla(f; x_0, w_0) \longrightarrow \mathrm{coker}(1 - f_*)$$

and the diagram

$$\begin{array}{ccc}
\nabla(f_k; a_k, w_k) & \xrightarrow{\tau_k} & \mathrm{coker}(1 - f_{k*}) \\
\downarrow{\scriptstyle \nu_k} & & \downarrow{\scriptstyle \mu_k} \\
\nabla(f; x_0, w_0) & \xrightarrow{\tau} & \mathrm{coker}(1 - f_*)
\end{array}$$

commutes, where $\mu_k(c) = i_{k*}(c) + \eta \circ \theta\langle\gamma_k\rangle$.

Proof: By [2; p.27, Theorem 2.1], we can get the correspondences τ_k and τ. Let $\langle\alpha\rangle \in \pi_1(A_k, a_k)$. Since $\nu_{k,\pi}\langle\alpha\rangle = (\tilde{i}_{k,\pi}\langle\alpha\rangle)\langle\gamma_k\rangle$, we have

$$\tau \circ \nu_k[\langle\alpha\rangle] = \eta \circ \theta\langle i_k \circ \alpha\rangle + \eta \circ \theta\langle\gamma_k\rangle.$$

On the other hand,

$$\begin{aligned}
\tau_k[\langle\alpha\rangle] &= \eta_k \circ \theta_k\langle\alpha\rangle \\
\mu_k \circ \tau_k[\langle\alpha\rangle] &= i_{k*} \circ \eta_k \circ \theta_k\langle\alpha\rangle + \eta \circ \theta\langle\gamma_k\rangle \\
&= \eta \circ \theta\langle i_k \circ \alpha\rangle + \eta \circ \theta\langle\gamma_k\rangle
\end{aligned}$$

Hence,

$$\mu_k r \circ \tau_k = \tau \circ \nu_k. \quad \blacksquare$$

Theorem 4.6 Let $f : (X, A) \to (X, A)$. Suppose $f_\pi(\pi_1(X)) \subset J(f)$ (for definition see [2; Ch.II, Sec. 3]). If $L(f) = 0$ then $N(f; X - A) = 0$; if $L(f) \neq 0$ then

$$N(f; X - A) = \sharp\{\text{coker}(1 - f_*)\} - \sharp\{\cup_{k=1}^{n} \mu_k \text{coker}(1 - f_{k*})\}.$$

Proof: By [2; p.33, Theorem 4.2], the correspondence τ is bijective when $f_\pi(\pi_1(X)) \subset J(f)$. Apply Theorem 4.4 and Lemma 4.5 to get the conclusion. ∎

Theorem 4.7 Let $f : (X, A) \to (X, A)$ be a selfmap of a pair of compact polyhedra with $\widehat{A} = \cup_{k=1}^{n} A_k$. Suppose $f_\pi(\pi_1(X)) \subset J(f)$ and $f_{k,\pi}(\pi_1(A_k)) \subset J(f_k)$. If $L(f) \cdot \prod_{k=1}^{m} L(f_k) \neq 0$ and $L(f_q) = 0$ for $m < q \leq n$, then

$$\begin{aligned} N(f; X, A) &= N(f) + N(\overline{f}) - N(f, \overline{f}) \\ &= \sharp\{\text{coker}(1 - f_*)\} + \sum_{k=1}^{m} \sharp\{\text{coker}(1 - f_{k*})\} \\ &\quad - \sharp\{\cup_{k=1}^{m} \mu_k \text{coker}(1 - f_{k*})\}. \quad \blacksquare \end{aligned}$$

If $n = 1$, i.e. $\widehat{A}$ is connected, we can take $w_0 = w_1$ and $x_0 = a_1$. Then $\nu_1[\langle\alpha\rangle] = [\langle i_{1,\pi}\alpha\rangle]$ and $\mu_1 = i_* : \text{coker}(1 - f_{1*}) \to \text{coker}(1 - f_*)$. We shall get

Corollary 4.8 Let $f : (X, A) \to (X, A)$. Suppose $\widehat{A}$ is connected and $f_\pi(\pi_1(X)) \subset J(f)$. If $L(f) = 0$ then $N(f; X - A) = 0$; if $L(f) \neq 0$ then

$$N(f; X - A) = \sharp\{\text{coker}(1 - f_*)\} - \sharp\{i_* \text{coker}(1 - f_{1*})\}. \quad \blacksquare$$

Corollary 4.9 Let $f : (X, A) \to (X, A)$ be a selfmap of a pair of compact polyhedra with $\widehat{A}$ connected. Suppose $f_\pi(\pi_1(X)) \subset J(f)$ and $f_{1,\pi}(\pi_1(A_1)) \subset J(f_1)$. If $L(f) \cdot L(f_1) \neq 0$, then

$$\begin{aligned} N(f; X, A) &= \sharp\{\text{coker}(1 - f_*)\} + \sharp\{\text{coker}(1 - f_{1*})\} \\ &\quad - \sharp\{i_* \text{coker}(1 - f_{1*})\}. \quad \blacksquare \end{aligned}$$

In general, $\mu_k \text{coker}(1 - f_{k*})$ is different from $i_{k*} \text{coker}(1 - f_{k*})$. An example is given as follows

Example 4.10 Let $X = \{e^{\theta i}\}$ be the unit circle in the complex plane, and let A be two points $\{1, -1\}$. The selfmap $f : (X, A) \to (X, A)$ is the reflection on the real axis.

Take $a_1 = 1, a_2 = -1$ and $x_0 = 1$. Let w_1, w_2 and w_0 be the constant paths. Since $H_1(A_1) = H_1(A_2) = 0$, then

$$\begin{aligned} & i_* \text{coker}(1 - f_{1*}) \cup i_{2*} \text{coker}(1 - f_{2*}) \\ &= \eta(i_* H_1(A_1)) \cup \eta(i_{2*} H_1(A_2)) \\ &= \{0\} \end{aligned}$$

is a single element in $\text{coker}(1 - f_*) = Z_2$.

On the other hand, let u_1 be constant path at $a_1 = x_0$ and let $u_2 = \{e^{\pi t i}\}_{0\leq t\leq 1}$. We have

$$\begin{aligned}
&\mu_1 \mathrm{coker}(1-f_{1*}) \\
&\quad = i_*\mathrm{coker}(1-f_{1*}) + \eta\circ\theta\langle u_1 w_1 (f\circ u_1)^{-1} w_0^{-1}\rangle \\
&\quad = i_*(0) + 0 \\
&\quad = \{0\}
\end{aligned}$$

and

$$\begin{aligned}
&\mu_2 \mathrm{coker}(1-f_{2*}) \\
&\quad = i_{2*}\mathrm{coker}(1-f_{2*}) + \eta\circ\theta\langle u_2 w_2 (f\circ u_2)^{-1} w_0^{-1}\rangle \\
&\quad = i_{2*}(0) + 1 \\
&\quad = \{1\}.
\end{aligned}$$

Thus,

$$\begin{aligned}
&\mu_1 \mathrm{coker}(1-f_{1*}) \cup \mu_2 \mathrm{coker}(1-f_{2*}) \\
&\quad = \mathrm{coker}(1-f_*).
\end{aligned}$$

References

[1] R.F.Brown, The Lefschetz Fixed Point Theorem, Scott, Foresman and Co., Glenview, Ill., 1971.

[2] Boju Jiang, Lectures on Nielsen Fixed Point Theory, Contemporary Mathematics Vol.14, Amer. Math. Soc., Providence, Rhode Island, 1983.

[3] Boju Jiang, On the least number of fixed points, Amer. J. Math., 102 (1980), 749-763.

[4] Boju Jiang, Fixed points and braids II, Math. Ann., 272 (1985), 249-256.

[5] T.H.Kiang, The Theory of Fixed Point Classes, Scientific Press, Peking, 1979.

[6] W.S.Massey, Algebraic Topology: An Introduction, Harcout-Brace-World, New York, 1967.

[7] H.Schirmer, Mappings of polyhedron with prescribed fixed points and fixed point indices, Pacific J. Math., 63 (1976), 521-530.

[8] H.Schirmer, A relative Nielsen number, Pacific J. Math., 122 (1986), 459- 473.

[9] H.Schirmer, On the location of fixed points on pairs of spaces, Topology and its application, to appear.

[10] U.K.Scholz, The Nielsen fixed point theory for non-compact spaces, Rocky Mountain J. Math., 4 (1974), 81-87.

[11] G.H.Shi, Least number of fixed points of the identity class, Acta Math. Sinica, 18 (1975), 192-202.

[12] C.Y.You, Fixed point classes of a fiber map, Picific J. Math., 100 (1982), 217-241.

[13] X.Z.Zhao, Estimation of the number of fixed points on the complementary space for selfmaps of a space pair, Thesis, Peking University, 1988.

1980 *Mathematics subject classifications*: 55M20, 57M99

Current address:
Department of Mathematics, Liaoning University, Shengyang 110036, P. R. China

LIST OF PARTICIPANTS

BARTSCH, Thomas J.	Mathematical Institute, University of Heidelberg, 6900 Heidelberg, FED REP GERMANY
BORISOVICH, Yurii G.	Department of Mathematics, Voronezh University, Voronezh 394693, USSR
BROWDER, Felix E.	Department of Mathematics, Rutgers University, New Brunswick, NJ 08903, USA
BROWN, Robert F.	Department of Mathematics, UCLA, Los Angeles, CA 90024, USA
CARBONE, Antonio	Department of Mathematics, University of Calabria, 87036 Arcavacata di Rende (Cosenza), ITALY
CHANG, Shih-sen	Department of Mathematics, Sichuan University, Chengdu, Sichuan, CHINA
CHEN, Wei	Department of Mathematics, Beijing Normal College, Beijing, CHINA
CHEN, Yansong	Department of Mathematics, Fujian Normal University, Fuzhou, Fujian, CHINA
CHEN, Yuqing	Department of Mathematics, Sichuan University, Chengdu, Sichuan, CHINA
CHEN, Zhonghe	Department of Mathematics, Peking University, Beijing, CHINA
DING, Xieping	Department of Mathematics, Sichuan Normal University, Chengdu, Sichuan, CHINA
DOLD, Albrecht E.	Mathematical Institute, University of Heidelberg, 6900 Heidelberg, FED REP GERMANY
DU, Xuguang	Department of Mathematics, Sichuan Normal University, Chengdu, Sichuan, CHINA
DUAN, Haibao	Institute of System Science, Academy of Sciences, Beijing, CHINA
FADELL, Edward R.	Department of Mathematics, University of Wisconsin, Madison, WI 53706, USA
FENSKE, Christian C.	Mathematical Institute, University of Giessen, 6300 Giessen, FED REP GERMANY
FOURNIER, Gilles	Department of Mathematics and Statistics, University of Sherbrooke, Sherbrooke, Québec J1K 2R1, CANADA
FRANKS, John M.	Department of Mathematics, Northwestern University, Evanston, IL 60201, USA

FRIED, David	Department of Mathematics, Boston University, Boston, MA 02215, USA
GAO, Wenfeng	Nankai Institute of Mathematics, Nankai University, Tianjin, CHINA
GÓRNIEWICZ, Lech	Institute of Mathematics, Nicholas Copernicus University, 87-100 Toruń, POLAND
GRANAS, Andrzej	Department of Mathematics and Statistics, University of Montréal, Montréal, Québec H3C 3J7, CANADA
GUZZARDI, Renato	Department of Mathematics, University of Calabria, 87036 Arcavacata di Rende (Cosenza), ITALY
HEATH, Philip R.	Department of Mathematics and Statistics, Memorial University of Newfoundland, St. John's, Newfoundland A1C 5S7, CANADA
HOU, Zixin	Department of Mathematics, Nankai University, Tianjin, CHINA
HUANG, Haihua	Nankai Institute of Mathematics, Nankai University, Tianjin, CHINA
HUANG, Houben	Institute of System Science, Academy of Sciences, Beijing, CHINA
HUSSEINI, Sufian Y.	Department of Mathematics, University of Wisconsin, Madison, WI 53706, USA
ILLMAN, Soren A.	Department of Mathematics, University of Helsinki, 00100 Helsinki, FINLAND
JIANG, Boju	Department of Mathematics, Peking University, Beijing 100871, CHINA
KOMIYA, Katsuhiro	Department of Mathematics, Yamaguchi University, Yamaguchi 753, JAPAN
LEI, Chengfeng	Jiangxi College of Transportation, Nanchang, Jiangxi, CHINA
LI, Bingyou	Department of Mathematics, Hebei Teachers University, Shijiazhuang, Hebei, CHINA
LI, Chaoqun	Department of Mathematics, Peking University, Beijing, CHINA
LI, Shujie	Institute of Mathematics, Academy of Sciences, Beijing, CHINA
LI, Weigang	Department of Mathematics, Sichuan Normal University, Chengdu, Sichuan, CHINA
LIN, Youhao	Department of Mathematics, Beijing Normal College, Beijing, CHINA

LIU, Xiyu	Department of Mathematics, Shandong University, Jinan, Shandong, CHINA
LUO, Qun	Department of Mathematics, Sichuan University, Chengdu, Sichuan, CHINA
MASSABO', Ivar	Department of Mathematics, University of Calabria, 87036 Arcavacata di Rende (Cosenza), ITALY
MATSUOKA, Takashi	Department of Mathematics, Naruto University of Teacher Education, Takashima, Naruto-shi 772, JAPAN
NAKAOKA, Minoru	Department of Mathematics, Okayama University of Science, Okayama 700, JAPAN
PAN, Xingbin	Department of Applied Mathematics, Zhejiang University, Hangzhou, Zhejiang, CHINA
PRIETO, Carlos	Institute of Mathematics, U.N.A.M., 04510 Mexico, D.F., MEXICO
QI, Guijie	Department of Mathematics, Shandong University, Jinan, Shandong, CHINA
SCHIRMER, Helga H.	Department of Mathematics and Statistics, Carleton Universit Ottawa, Ontario K1S 5B6, CANADA
SHI, Chuan	Department of Applied Mathematics, East China Institute of Technology, Nanjing, Jiangsu, CHINA
SHI, Shuzhong	Nankai Institute of Mathematics, Nankai University, Tianjin, CHINA
TONG, Deyu	Department of Mathematics, Peking University, Beijing, CHINA
VIGNOLI, Alfonso	Department of Mathematics, University of Roma II, 00173 Roma, ITALY
WANG, Zhiqiang	Department of Mathematics, Peking University, Beijing, CHINA
WILLE, Friedrich	Mathematical Institute, University of Kassel, 3500 Kassel, FED REP GERMANY
WU, Shaoping	Department of Applied Mathematics, Zhejiang University, Hanghou, Zhejiang, CHINA
WU, Yingqing	Department of Mathematics, Nanjing Normal University, Nanjing, Jiangsu, CHINA
XU, Hongkun	Department of Mathematics, Xi'an Jiaotong University, Xi'an, Shaanxi, CHINA
XUE, Tong	Nanjing Communication Engineering Institute, Nanjing, Jiangsu, CHINA

YU, Baozhen	Department of Mathematics, Peking University, Beijing, CHINA
YU, Bo	Department of Mathematics, Jilin University, Changchun, Jilin, CHINA
YU, Qingyu	Department of Mathematics, Lanzhou University, Lanzhou, Gansu, CHINA
YU, Yanlin	Nankai Institute of Mathematics, Nankai University, Tianjin, CHINA
YU, Zhishou	Department of Mathematics, Jilin University, Changchun, Jilin, CHINA
ZHANG, Qingyong	Department of Mathematics, Sichuan Normal University, Chengdu, Sichuan, CHINA
ZHAO, Xuezhi	Department of Mathematics, Peking University, Beijing, CHINA
ZHAO, Yichun	Department of Applied Mathematics, Northeastern Technology University, Shenyang, Liaoning, CHINA
ZHONG, Chengkui	Department of Mathematics, Lanzhou University, Lanzhou, Gansu, CHINA
ZHU, Jun	Department of Mathematics, Suzhou University, Suzhou, Jiangsu, CHINA
ZHU, Yuanguo	Department of Mathematics, Sichuan University, Chengdu, Sichuan, CHINA

Vol. 1232: P.C. Schuur, Asymptotic Analysis of Soliton Problems. VIII, 180 pages. 1986.

Vol. 1233: Stability Problems for Stochastic Models. Proceedings, 1985. Edited by V.V. Kalashnikov, B. Penkov and V.M. Zolotarev. VI, 223 pages. 1986.

Vol. 1234: Combinatoire énumérative. Proceedings, 1985. Edité par G. Labelle et P. Leroux. XIV, 387 pages. 1986.

Vol. 1235: Séminaire de Théorie du Potentiel, Paris, No. 8. Directeurs: M. Brelot, G. Choquet et J. Deny. Rédacteurs: F. Hirsch et G. Mokobodzki. III, 209 pages. 1987.

Vol. 1236: Stochastic Partial Differential Equations and Applications. Proceedings, 1985. Edited by G. Da Prato and L. Tubaro. V, 257 pages. 1987.

Vol. 1237: Rational Approximation and its Applications in Mathematics and Physics. Proceedings, 1985. Edited by J. Gilewicz, M. Pindor and W. Siemaszko. XII, 350 pages. 1987.

Vol. 1238: M. Holz, K.-P. Podewski and K. Steffens, Injective Choice Functions. VI, 183 pages. 1987.

Vol. 1239: P. Vojta, Diophantine Approximations and Value Distribution Theory. X, 132 pages. 1987.

Vol. 1240: Number Theory, New York 1984–85. Seminar. Edited by D.V. Chudnovsky, G.V. Chudnovsky, H. Cohn and M.B. Nathanson. V, 324 pages. 1987.

Vol. 1241: L. Gårding, Singularities in Linear Wave Propagation. III, 125 pages. 1987.

Vol. 1242: Functional Analysis II, with Contributions by J. Hoffmann-Jørgensen et al. Edited by S. Kurepa, H. Kraljević and D. Butković. VII, 432 pages. 1987.

Vol. 1243: Non Commutative Harmonic Analysis and Lie Groups. Proceedings, 1985. Edited by J. Carmona, P. Delorme and M. Vergne. V, 309 pages. 1987.

Vol. 1244: W. Müller, Manifolds with Cusps of Rank One. XI, 158 pages. 1987.

Vol. 1245: S. Rallis, L-Functions and the Oscillator Representation. XVI, 239 pages. 1987.

Vol. 1246: Hodge Theory. Proceedings, 1985. Edited by E. Cattani, F. Guillén, A. Kaplan and F. Puerta. VII, 175 pages. 1987.

Vol. 1247: Séminaire de Probabilités XXI. Proceedings. Edité par J. Azéma, P.A. Meyer et M. Yor. IV, 579 pages. 1987.

Vol. 1248: Nonlinear Semigroups, Partial Differential Equations and Attractors. Proceedings, 1985. Edited by T.L. Gill and W.W. Zachary. IX, 185 pages. 1987.

Vol. 1249: I. van den Berg, Nonstandard Asymptotic Analysis. IX, 187 pages. 1987.

Vol. 1250: Stochastic Processes – Mathematics and Physics II. Proceedings 1985. Edited by S. Albeverio, Ph. Blanchard and L. Streit. VI, 359 pages. 1987.

Vol. 1251: Differential Geometric Methods in Mathematical Physics. Proceedings, 1985. Edited by P.L. García and A. Pérez-Rendón. VII, 300 pages. 1987.

Vol. 1252: T. Kaise, Représentations de Weil et GL_2 Algèbres de division et GL_n. VII, 203 pages. 1987.

Vol. 1253: J. Fischer, An Approach to the Selberg Trace Formula via the Selberg Zeta-Function. III, 184 pages. 1987.

Vol. 1254: S. Gelbart, I. Piatetski-Shapiro, S. Rallis. Explicit Constructions of Automorphic L-Functions. VI, 152 pages. 1987.

Vol. 1255: Differential Geometry and Differential Equations. Proceedings, 1985. Edited by C. Gu, M. Berger and R.L. Bryant. XII, 243 pages. 1987.

Vol. 1256: Pseudo-Differential Operators. Proceedings, 1986. Edited by H.O. Cordes, B. Gramsch and H. Widom. X, 479 pages. 1987.

Vol. 1257: X. Wang, On the C*-Algebras of Foliations in the Plane. V, 165 pages. 1987.

Vol. 1258: J. Weidmann, Spectral Theory of Ordinary Differential Operators. VI, 303 pages. 1987.

Vol. 1259: F. Cano Torres, Desingularization Strategies for Three-Dimensional Vector Fields. IX, 189 pages. 1987.

Vol. 1260: N.H. Pavel, Nonlinear Evolution Operators and Semigroups. VI, 285 pages. 1987.

Vol. 1261: H. Abels, Finite Presentability of S-Arithmetic Groups. Compact Presentability of Solvable Groups. VI, 178 pages. 1987.

Vol. 1262: E. Hlawka (Hrsg.), Zahlentheoretische Analysis II. Seminar, 1984–86. V, 158 Seiten. 1987.

Vol. 1263: V.L. Hansen (Ed.), Differential Geometry. Proceedings, 1985. XI, 288 pages. 1987.

Vol. 1264: Wu Wen-tsün, Rational Homotopy Type. VIII, 219 pages. 1987.

Vol. 1265: W. Van Assche, Asymptotics for Orthogonal Polynomials. VI, 201 pages. 1987.

Vol. 1266: F. Ghione, C. Peskine, E. Sernesi (Eds.), Space Curves. Proceedings, 1985. VI, 272 pages. 1987.

Vol. 1267: J. Lindenstrauss, V.D. Milman (Eds.), Geometrical Aspects of Functional Analysis. Seminar. VII, 212 pages. 1987.

Vol. 1268: S.G. Krantz (Ed.), Complex Analysis. Seminar, 1986. VII, 195 pages. 1987.

Vol. 1269: M. Shiota, Nash Manifolds. VI, 223 pages. 1987.

Vol. 1270: C. Carasso, P.-A. Raviart, D. Serre (Eds.), Nonlinear Hyperbolic Problems. Proceedings, 1986. XV, 341 pages. 1987.

Vol. 1271: A.M. Cohen, W.H. Hesselink, W.L.J. van der Kallen, J.R. Strooker (Eds.), Algebraic Groups Utrecht 1986. Proceedings. XII, 284 pages. 1987.

Vol. 1272: M.S. Livšic, L.L. Waksman, Commuting Nonselfadjoint Operators in Hilbert Space. III, 115 pages. 1987.

Vol. 1273: G.-M. Greuel, G. Trautmann (Eds.), Singularities, Representation of Algebras, and Vector Bundles. Proceedings, 1985. XIV, 383 pages. 1987.

Vol. 1274: N. C. Phillips, Equivariant K-Theory and Freeness of Group Actions on C*-Algebras. VIII, 371 pages. 1987.

Vol. 1275: C.A. Berenstein (Ed.), Complex Analysis I. Proceedings, 1985–86. XV, 331 pages. 1987.

Vol. 1276: C.A. Berenstein (Ed.), Complex Analysis II. Proceedings, 1985–86. IX, 320 pages. 1987.

Vol. 1277: C.A. Berenstein (Ed.), Complex Analysis III. Proceedings, 1985–86. X, 350 pages. 1987.

Vol. 1278: S.S. Koh (Ed.), Invariant Theory. Proceedings, 1985. V, 102 pages. 1987.

Vol. 1279: D. Ieşan, Saint-Venant's Problem. VIII, 162 Seiten. 1987.

Vol. 1280: E. Neher, Jordan Triple Systems by the Grid Approach. XII, 193 pages. 1987.

Vol. 1281: O.H. Kegel, F. Menegazzo, G. Zacher (Eds.), Group Theory. Proceedings, 1986. VII, 179 pages. 1987.

Vol. 1282: D.E. Handelman, Positive Polynomials, Convex Integral Polytopes, and a Random Walk Problem. XI, 136 pages. 1987.

Vol. 1283: S. Mardešić, J. Segal (Eds.), Geometric Topology and Shape Theory. Proceedings, 1986. V, 261 pages. 1987.

Vol. 1284: B.H. Matzat, Konstruktive Galoistheorie. X, 286 pages. 1987.

Vol. 1285: I.W. Knowles, Y. Saitō (Eds.), Differential Equations and Mathematical Physics. Proceedings, 1986. XVI, 499 pages. 1987.

Vol. 1286: H.R. Miller, D.C. Ravenel (Eds.), Algebraic Topology. Proceedings, 1986. VII, 341 pages. 1987.

Vol. 1287: E.B. Saff (Ed.), Approximation Theory, Tampa. Proceedings, 1985–1986. V, 228 pages. 1987.

Vol. 1288: Yu. L. Rodin, Generalized Analytic Functions on Riemann Surfaces. V, 128 pages, 1987.

Vol. 1289: Yu. I. Manin (Ed.), K-Theory, Arithmetic and Geometry. Seminar, 1984–1986. V, 399 pages. 1987.

Vol. 1290: G. Wüstholz (Ed.), Diophantine Approximation and Transcendence Theory. Seminar, 1985. V, 243 pages. 1987.

Vol. 1291: C. Mœglin, M.-F. Vignéras, J.-L. Waldspurger, Correspondances de Howe sur un Corps p-adique. VII, 163 pages. 1987

Vol. 1292: J.T. Baldwin (Ed.), Classification Theory. Proceedings, 1985. VI, 500 pages. 1987.

Vol. 1293: W. Ebeling, The Monodromy Groups of Isolated Singularities of Complete Intersections. XIV, 153 pages. 1987.

Vol. 1294: M. Queffélec, Substitution Dynamical Systems – Spectral Analysis. XIII, 240 pages. 1987.

Vol. 1295: P. Lelong, P. Dolbeault, H. Skoda (Réd.), Séminaire d'Analyse P. Lelong – P. Dolbeault – H. Skoda. Seminar, 1985/1986. VII, 283 pages. 1987.

Vol. 1296: M.-P. Malliavin (Ed.), Séminaire d'Algèbre Paul Dubreil et Marie-Paule Malliavin. Proceedings, 1986. IV, 324 pages. 1987.

Vol. 1297: Zhu Y.-l., Guo B.-y. (Eds.), Numerical Methods for Partial Differential Equations. Proceedings. XI, 244 pages. 1987.

Vol. 1298: J. Aguadé, R. Kane (Eds.), Algebraic Topology, Barcelona 1986. Proceedings. X, 255 pages. 1987.

Vol. 1299: S. Watanabe, Yu.V. Prokhorov (Eds.), Probability Theory and Mathematical Statistics. Proceedings, 1986. VIII, 589 pages. 1988.

Vol. 1300: G.B. Seligman, Constructions of Lie Algebras and their Modules. VI, 190 pages. 1988.

Vol. 1301: N. Schappacher, Periods of Hecke Characters. XV, 160 pages. 1988.

Vol. 1302: M. Cwikel, J. Peetre, Y. Sagher, H. Wallin (Eds.), Function Spaces and Applications. Proceedings, 1986. VI, 445 pages. 1988.

Vol. 1303: L. Accardi, W. von Waldenfels (Eds.), Quantum Probability and Applications III. Proceedings, 1987. VI, 373 pages. 1988.

Vol. 1304: F.Q. Gouvêa, Arithmetic of p-adic Modular Forms. VIII, 121 pages. 1988.

Vol. 1305: D.S. Lubinsky, E.B. Saff, Strong Asymptotics for Extremal Polynomials Associated with Weights on ℝ. VII, 153 pages. 1988.

Vol. 1306: S.S. Chern (Ed.), Partial Differential Equations. Proceedings, 1986. VI, 294 pages. 1988.

Vol. 1307: T. Murai, A Real Variable Method for the Cauchy Transform, and Analytic Capacity. VIII, 133 pages. 1988.

Vol. 1308: P. Imkeller, Two-Parameter Martingales and Their Quadratic Variation. IV, 177 pages. 1988.

Vol. 1309: B. Fiedler, Global Bifurcation of Periodic Solutions with Symmetry. VIII, 144 pages. 1988.

Vol. 1310: O.A. Laudal, G. Pfister, Local Moduli and Singularities. V, 117 pages. 1988.

Vol. 1311: A. Holme, R. Speiser (Eds.), Algebraic Geometry, Sundance 1986. Proceedings. VI, 320 pages. 1988.

Vol. 1312: N.A. Shirokov, Analytic Functions Smooth up to the Boundary. III, 213 pages. 1988.

Vol. 1313: F. Colonius, Optimal Periodic Control. VI, 177 pages. 1988.

Vol. 1314: A. Futaki, Kähler-Einstein Metrics and Integral Invariants. IV, 140 pages. 1988.

Vol. 1315: R.A. McCoy, I. Ntantu, Topological Properties of Spaces of Continuous Functions. IV, 124 pages. 1988.

Vol. 1316: H. Korezlioglu, A.S. Ustunel (Eds.), Stochastic Analysis and Related Topics. Proceedings, 1986. V, 371 pages. 1988.

Vol. 1317: J. Lindenstrauss, V.D. Milman (Eds.), Geometric Aspects of Functional Analysis. Seminar, 1986–87. VII, 289 pages. 1988.

Vol. 1318: Y. Felix (Ed.), Algebraic Topology – Rational Homotopy. Proceedings, 1986. VIII, 245 pages. 1988

Vol. 1319: M. Vuorinen, Conformal Geometry and Quasiregular Mappings. XIX, 209 pages. 1988.

Vol. 1320: H. Jürgensen, G. Lallement, H.J. Weinert (Eds.), Semigroups, Theory and Applications. Proceedings, 1986. X, 416 pages. 1988.

Vol. 1321: J. Azéma, P.A. Meyer, M. Yor (Eds.), Séminaire de Probabilités XXII. Proceedings. IV, 600 pages. 1988.

Vol. 1322: M. Métivier, S. Watanabe (Eds.), Stochastic Analysis. Proceedings, 1987. VII, 197 pages. 1988.

Vol. 1323: D.R. Anderson, H.J. Munkholm, Boundedly Controlle Topology. XII, 309 pages. 1988.

Vol. 1324: F. Cardoso, D.G. de Figueiredo, R. Iório, O. Lopes (Eds.), Partial Differential Equations. Proceedings, 1986. VIII, 433 pages. 1988.

Vol. 1325: A. Truman, I.M. Davies (Eds.), Stochastic Mechanics an Stochastic Processes. Proceedings, 1986. V, 220 pages. 1988.

Vol. 1326: P.S. Landweber (Ed.), Elliptic Curves and Modular Forms i Algebraic Topology. Proceedings, 1986. V, 224 pages. 1988.

Vol. 1327: W. Bruns, U. Vetter, Determinantal Rings. VII,236 pages 1988.

Vol. 1328: J.L. Bueso, P. Jara, B. Torrecillas (Eds.), Ring Theor Proceedings, 1986. IX, 331 pages. 1988.

Vol. 1329: M. Alfaro, J.S. Dehesa, F.J. Marcellan, J.L. Rubio d Francia, J. Vinuesa (Eds.): Orthogonal Polynomials and their Applic tions. Proceedings, 1986. XV, 334 pages. 1988.

Vol. 1330: A. Ambrosetti, F. Gori, R. Lucchetti (Eds.), Mathematic Economics. Montecatini Terme 1986. Seminar. VII, 137 pages. 1988

Vol. 1331: R. Bamón, R. Labarca, J. Palis Jr. (Eds.), Dynamic Systems, Valparaiso 1986. Proceedings. VI, 250 pages. 1988.

Vol. 1332: E. Odell, H. Rosenthal (Eds.), Functional Analysis. Pr ceedings, 1986–87. V, 202 pages. 1988.

Vol. 1333: A.S. Kechris, D.A. Martin, J.R. Steel (Eds.), Cabal Semin 81–85. Proceedings, 1981–85. V, 224 pages. 1988.

Vol. 1334: Yu.G. Borisovich, Yu. E. Gliklikh (Eds.), Global Analys – Studies and Applications III. V, 331 pages. 1988.

Vol. 1335: F. Guillén, V. Navarro Aznar, P. Pascual-Gainza, F. Puert Hyperrésolutions cubiques et descente cohomologique. XII, 19 pages. 1988.

Vol. 1336: B. Helffer, Semi-Classical Analysis for the Schröding Operator and Applications. V, 107 pages. 1988.

Vol. 1337: E. Sernesi (Ed.), Theory of Moduli. Seminar, 1985. VIII, 2 pages. 1988.

Vol. 1338: A.B. Mingarelli, S.G. Halvorsen, Non-Oscillation Domai of Differential Equations with Two Parameters. XI, 109 pages. 1988.

Vol. 1339: T. Sunada (Ed.), Geometry and Analysis of Manifol Procedings, 1987. IX, 277 pages. 1988.

Vol. 1340: S. Hildebrandt, D.S. Kinderlehrer, M. Miranda (Eds Calculus of Variations and Partial Differential Equations. Proceeding 1986. IX, 301 pages. 1988.

Vol. 1341: M. Dauge, Elliptic Boundary Value Problems on Corn Domains. VIII, 259 pages. 1988.

Vol. 1342: J.C. Alexander (Ed.), Dynamical Systems. Proceeding 1986–87. VIII, 726 pages. 1988.

Vol. 1343: H. Ulrich, Fixed Point Theory of Parametrized Equivari Maps. VII, 147 pages. 1988.

Vol. 1344: J. Král, J. Lukeš, J. Netuka, J. Veselý (Eds.), Poten Theory – Surveys and Problems. Proceedings, 1987. VIII, 271 pag 1988.

Vol. 1345: X. Gomez-Mont, J. Seade, A. Verjovski (Eds.), Holomorp Dynamics. Proceedings, 1986. VII, 321 pages. 1988.

Vol. 1346: O. Ya. Viro (Ed.), Topology and Geometry – Ro Seminar. XI, 581 pages. 1988.

Vol. 1347: C. Preston, Iterates of Piecewise Monotone Mappings an Interval. V, 166 pages. 1988.

Vol. 1348: F. Borceux (Ed.), Categorical Algebra and its Applicatio Proceedings, 1987. VIII, 375 pages. 1988.

Vol. 1349: E. Novak, Deterministic and Stochastic Error Bounds Numerical Analysis. V, 113 pages. 1988.